1st International Symposium on Green Technology for Value Chains 2016

IOP Conference Series: Earth and Environmental Science
Volume 60

Tangerang, Indonesia
3 – 5 October 2016

ISBN: 978-1-5108-4026-3
ISSN: 1755-1307

Printed from e-media with permission by:

Curran Associates, Inc.
57 Morehouse Lane
Red Hook, NY 12571

Some format issues inherent in the e-media version may also appear in this print version.

Printed by Curran Associates, Inc. (2017)

For permission requests, please contact the Institute of Physics
at the address below.

Institute of Physics
Dirac House, Temple Back
Bristol BS1 6BE UK

Phone: 44 1 17 929 7481
Fax: 44 1 17 920 0979

techtracking@iop.org

Additional copies of this publication are available from:

Curran Associates, Inc.
57 Morehouse Lane
Red Hook, NY 12571 USA
Phone: 845-758-0400
Fax: 845-758-2633
Email: curran@proceedings.com
Web: www.proceedings.com

Table of contents

Volume 60

1st International Symposium on Green Technology for Value Chains 2016
3–5 October 2016, Tangerang, Indonesia

Accepted papers received: 24 March 2017
Published online: 2 May 2017

Preface

Plenary

Keynote

Green Materials and Process

Waste Treatment and Management

Green Energy

Socio-Technological Assessment on Green Production

Life Cycle and Environmental Monitoring

IOP Conf. Series: Earth and Environmental Science **60** (2017) 011001 doi:10.1088/1755-1315/60/1/011001

Proceedings of 1st International Symposium on Green Technology for Value Chains (Green VC) 2016

Preface

The First International Symposium on Green Technology for Value Chains 2016 was held on October 2-3, 2016 in Indonesia Convention Exhibition (ICE), BSD City, Indonesia. It was organized by the Research Unit for Clean Technology, Indonesian Institute of Sciences (LIPI) and mainly supported by Deputy for Engineering and Sciences, LIPI. Studies on (1) Waste Treatment and Management, (2) Green Materials and Process, (3) Green Energy, (4) Life Cycle and Environmental Monitoring, and (5) Socio-technological Assessment on Green Production were all presented orally.

Keynote presentations were provided to show the contribution of scientist in to sustainable subject of research. Value Chains for Sustainable Development was presented by Prof. Naoyuki Funamizu from Hokkaido University, Japan whereas presentation about Towards Sustainable Pollution Management was provided by Prof. Ng Wun Jern from Nanyang Technology University, Singapore. Recent issues on Green Materials for Sustainable Development was delivered by Prof. Bambang Sunendar Purwasasmita from Bandung Institute of Technology, Indonesia. In addition, presentation on Indonesia Municiple Solid Waste Life Cycle and Environmental Monitoring was provided by Ir. Susmono from Kemitraan Habitat, Indonesia.

Participants from Japan, Zambia, Burkina Faso, and Indonesia had many fruitful discussions and exchanges that contributed to the success of the symposium.

This Proceedings, cover 33 papers selected from 46 scientific presentations presented in the symposium.

We would like to thank our sincere gratitude to our Advisory Board, Scientific and Editorial Board for their positive contribution to keep the quality of the articles presented in this proceedings. We also would like to appreciate to all participants attending this symposium as well as to Organizing Committee whom made the symposium successful.

In the continuation of the success of the first event, the Second International Symposium on Green Technology for Value Chains will take place in Jakarta, Indonesia, 23-24 October 2017.

March 15, 2017
Editor in Chiefs
Prof. Myrtha Karina
Dr. Muhamad Nasir

IOP Conf. Series: Earth and Environmental Science **60** (2017) 011001 doi:10.1088/1755-1315/60/1/011001

International Symposium on Green Technology for Value Chains 2016 (GreenVC 2016)
Value Chains for Sustainable Development
Tangerang Banten, Indonesia
October 3-5, 2016

ORGANIZING COMMITTEE

Advisory Board

- Prof. Iskandar Zulkarnaen (Chairman of Indonesian Institute of Sciences, LIPI, Indonesia)
- Dr. Laksana Tri Handoko (Deputy Chairman for Engineering Sciences, LIPI, Indonesia)
- Dr. Rike Yudianti (Head for Research Unit for Clean Technology, LIPI, Indonesia)
- Prof. Naoyuki Funamizu (University of Hokkaido, Japan)
- Prof. Ng Wun Jern (Nanyang Technology University, Singapore)
- Prof. Bambang Sunendar (Bandung Institute of Technology, ITB, Indonesia)
- Ir. Susmono (Kemitraan Habitat, Indonesia)

Scientific and Editorial (S&E) Board

- Prof. Myrtha Karina (Research Unit for Clean Technology, LIPI, Indonesia)
- Prof. Tarzan Sembiring (Research Unit for Clean Technology, LIPI, Indonesia)
- Prof. M. Hanafi (Research Center for Chemistry, LIPI, Indonesia)
- Prof. Rahmiana Zein (Andalas University, Indonesia)
- Prof. Safni (Andalas University, Indonesia)
- Prof. Syukri Arif (Andalas University, Indonesia)
- Prof. Naoyuki Funamizu (Hokkaido University, Japan)
- Prof. Imasiku Nyambe (Zambia University, Zambia)
- Dr. Sri Pudjiraharti (Research Unit for Clean Technology, LIPI, Indonesia)
- Dr. Neni Sintawardani (Research Unit for Clean Technology, LIPI, Indonesia)
- Dr. Muhammad Nasir (Research Unit for Clean Technology, LIPI, Indonesia)
- Dr. Hari R. Hariyadi (Research Unit for Clean Technology, LIPI, Indonesia)
- Dr. Tiny A. Koesmawati (Research Unit for Clean Technology, LIPI, Indonesia)
- Dr. Nanik Indayaningsih (Research Center for Physics, LIPI, Indonesia)
- Dr. Haznan Abimanyu (Research Center for Chemistry, LIPI, Indonesia)
- Dr. Edi Iswanto Wiloso (Research Center for Chemistry, LIPI, Indonesia)
- Dr. Rizna Triana Dewi (Research Center for Chemistry, LIPI, Indonesia)
- Dr. Dieni Mansur (Research Center for Chemistry, LIPI, Indonesia)
- Dr. Wahyu Dwianto (Research Center for Biomaterial, LIPI, Indonesia)
- Dr. Indriana Kartini (Gadjah Mada University, Indonesia)
- Ir. Imam Djunaedi (Research Center for Electrical Power and Mechatronics, LIPI, Indonesia)
- Ir. I Made Astina M.Eng., Ph.D. (Institute Technology of Bandung, Indonesia)

IOP Conf. Series: Earth and Environmental Science **60** (2017) 011001 doi:10.1088/1755-1315/60/1/011001

Organizing Committee

Chair person	: Dr. Akbar Hanif Dawam Abdullah
Co-chair	: Dewi Nilawati, S.Si., M.T.
Secretary	: - Dr. Widyarani, M. Sc.
	- Diana Rahayuningwulan, M.T.
	- Elsy Rahimi Chaldun, S.Si, M.T.

Technical Program Chair : Indriyati, M.Eng.

Members :

- Dr. Desak Gede Sri Andayani
- Dr. Indah Primadona
- Ir. J. Tri Astuti
- Dra. Lucia Indrarti
- Muchlis, MT
- Anung Syampurwadi, S. Mat.

IOP Conf. Series: Earth and Environmental Science **60** (2017) 011002 doi:10.1088/1755-1315/60/1/011002

Peer review statement

All papers published in this volume of *IOP Conference Series: Earth and Environmental Science* have been peer reviewed through processes administered by the proceedings Editors. Reviews were conducted by expert referees to the professional and scientific standards expected of a proceedings journal published by IOP Publishing.

IOP Conf. Series: Earth and Environmental Science **60** (2017) 012001 doi:10.1088/1755-1315/60/1/012001

Towards sustainable pollution management

N G W Jern
Executive Director & Professor, Nanyang Environment & Water Research Institute (NEWRI)
Nanyang Technological University, Singapore
Email : wjng@ntu.edu.sg

Abstract. It is often overlooked pollution control itself may not be entirely free from adverse impact on the environment if considered from a more holistic perspective. For example mechanised wastewater treatment is energy intensive and so has a carbon footprint because of the need to move air to supply oxygen to the aerobic treatment process. The aerobic treatment process then results in excess bio-sludge which requires disposal and if such is not appropriately performed, then there is risk of surface and groundwater contamination. This presentation explores the changes which have been investigated and are beginning to be implemented in wastewater, sludge, and agro-industrial wastes management which are more environmentally benign. Three examples shall be used to illustrate the discussion. The first example uses the conventional sewage treatment system with a unit process arrangement which converts carbonaceous pollutants from soluble and colloidal forms to particulate forms with an aerobic process before attempting energy recovery with an anaerobic process. Such an arrangement does, however, result in a negative energy balance. This is not withstanding the fact there is potentially more energy in sewage than is required to treat it if that energy can be effectively harvested. The latter can be achieved by removing the carbonaceous pollutants before the aerobic process and thereby using the aerobic process for polishing instead of treating. The carbonaceous pollutants so recovered then becomes the feed for the anaerobic process. Unfortunately conventional anaerobic sludge digestion only removes 35-45% of the organic material fed. Since biogas production (and hence energy recovery) is linked to the amount of organic material which can be degraded anaerobically, the effectiveness of the anaerobic digestion process needs to be improved. Contrary to a commonly held belief wherein methanogenesis is the “bottleneck” in anaerobic processes, hydrolysis is in sludge digestion. Hydrolysis can be enhanced thermally, chemically and biologically. With better anaerobic digestion, the digestate would have more N and P released. The digestate can be blended with organic fertilizers which have been augmented with microbes capable of producing phytohormones. The latter enable crop plants to use inorganic fertilizers more effectively and hence reducing the quantity of inorganic fertilizers required significantly. Use of such organic fertilizers rejuvenates soils which have been subjected to prolonged application of inorganic fertilisers since the latter can result in stagnating or even declining crop yields.

IOP Conf. Series: Earth and Environmental Science **60** (2017) 012002 doi:10.1088/1755-1315/60/1/012002

The sanitation value chain: its concept and new research collaboration project

N Funamizu

Graduated School of Engineering, Hokkaido University, Sapporo, Japan

Email : funamizu@eng.hokudai.ac.jp

Abstract. Sanitation is essential for promoting health, preventing environment pollution, conserving ecosystem, and recovering and recycling resources. Therefore, it can be said that sanitation is closely related to such current global issues as poverty, urban slum, conservation of ecosystem, and resources management. Namely, the question, "How can we handle the waste from 10 billion people in future?" is a global environmental problem to be solved. In developing world, population is growing rapidly especially in urban slums and they have still high under 5 mortality and poverty issues. It also reported that 2.4 billion people are still using unimproved sanitation facilities, including 946 million people who are still practicing open defecation in 2015 (UN, 2015). On the other hand, depopulation and aging are progressing especially in rural area of developed world. Based on the above mentioned background, new research project on sanitation value chain has started. This is a collaboration project with LIPI, RIHN (Research Institute of Humanity and Nature, Kyoto) and HU (Hokkaido University). The concept of the sanitation value chain and the brief summary of the project are discussed in the keynote presentation. The concept of sanitation value chain proposed in the project : The project is proposing new concept, Sanitation Value Chain, which has the following basic policies: 1) Put values of people/ and community in the centre of discussion, and prepare sanitation system to drive this value chain; 2) Design the sanitation system by focusing on incentive for individual users and community; 3) Recognize a sanitation system as an integrated system with social and technical systems; 4) Design the sanitation system by making a good matching between social characteristics and prerequisites of the technologies. The goals of the research are 1) To propose the Sanitation Value Chain as a common solution for both developing and developed countries, 2) To show the validity of the Sanitation Value Chain by pilot study on co-creation of the value chain at developing countries and Japan, 3) To contribute to making interdisciplinary academic foundation on sanitation.

IOP Conf. Series: Earth and Environmental Science **60** (2017) 012003 doi:10.1088/1755-1315/60/1/012003

Indonesia municiple solid waste life cycle and environmental monitoring: current situation, before and future challenges

Susmono
Habitat Partnership Organization, Jakarta, Indonesia
Email : susmono_2@yahoo.com

Abstract. Indonesia is a big country with circa 250 million population, with more than 500 Local Governments and they are going to improve their municiple solid waste dumping method from Open Dumping to Sanitary Landfill (SLF) and to promote Reduce-Reuse-Recycling (3R) since many years ago, and it is strengthened by issuing of Solid Waste Management Act No.18/2008, MSW Government Regulation No.12/2012 and other regulations which are issued by Central Government and Local Governments. During "Water and Sanitation Decade 1980-1990" through "Integrated Urban Infrastructures Development Program" some pilot project such as 30 units of 3R station were developed in the urban areas, and modified or simplification of SLF call Controlled Landfill (CLF) were implemented. In the year of 2002 about 45 units of composting pilot projects were developed under "Western Java Environmental Management Project", and the result was notified that some of them are not sustain because many aspects. At the beginning of 2007 until now, some pilot projects of 3R were continued in some cities and since 2011 some Waste Banks are growing fast. In the year of 2014 was recorded that of 70 % of 3Rs in Java Island well developed (2014, Directorate of Environment Sanitation Report), and in the year of 2012 was recorded that development of Communal Waste Banks were growing fast during two months from 400 units to 800 units (2012, Ministry of Environment report), now more Communal Waste Banks all ready exist. After the last overview monitoring activity by Ministry of Environment and JICA (2008), because of lack of data is very difficult to give current accurate information of Municiple Solid Waste Handling in Indonesia. Nevertheless some innovation are developed because of impact of many pilot projects, Adipura City Cleanest Competition among Local Governments and growing of the spirit of autonomous policy of Local Governments, but some Local Governments still dependence on Central Government support, both technically and non technically aspects such as new appropriate technology development, new integration management especially between formal and informal organizations, acceleration of community education/empowerment, new required regulations development and law enforcement support.
Political will of government. In the beginning, government and people of Indonesia follow the paradigm that municipal solid waste management could be managed by Collecting-Transferring-Dumping system only. This paradigm is appropriate if no problem increase of land providing for solid waste dumping site. Most of local governments are not able to decide it because so many aspects and complexity of problems such as choosing an appropriate technology, finding location for solid waste transfer stations and dumping site, developing of waste management, limitation of affordability, improving people behaviour to increase their low health environment consciousness, as well as lack of professional staffs. Indonesia Ministry of Environment who is responsible for solid waste handling regulations and Ministry of Public Works who is responsible for urban infrastructures development have changed their paradigm that in municipal solid waste handling it is better to reduce as soon as possible. The new approach is to introduce 3R methods from the sources to the solid waste dumping site for minimizing cost of transportation and dumping site area. The Municipal Solid Waste Management Law no 18/2008 stated that municipal solid waste handling consists of Reduction-Reuse-Recycling of waste and running waste management services such as collection of the rest to transport, treat and dumping in the end of the system. Based on the

Autonomous Law, the local governments are still the main responsible governments to handle municipal solid waste management in their administrative area.

Community participation. During the last few years many solid waste communal and non-governmental organizations were grown and developed, some solid waste communal leaders were born, and solid waste handling motivation and participation of community are grown. To accelerate this situation, the government introduces many training and education to produce more municipal solid waste handling facilitators. Since 2007, environment sanitation motivation activities runs through the yearly Sanitation Jamboree that educate, short train, motivate junior school children and competition among other.

Technology innovation. Local governments, with or without central government support, are being to make some improvement how to handle municipal solid waste and through Sister City Program, many innovations were developed such as in Surabaya City (home Takakura composter), Depok (waste separation and composting), Bogor City (management), Malang City, Makasar City and others. The new Closing the Loops of solid waste handling approaches should be introduced in the future to break the bottle neck that always happened in the past. Integration between solid waste management and the farming activities, land plantation rehabilitations, city landscaping and gardening is very urgent to develop, including integration of 3R stakeholders in the region.

The challenges. The municipal solid waste problem in urban areas is relative more complicated compared with the same problem in the rural areas. Accurate data collection and analyzing periodically is very important. Road map development and mobilizing of all stake holders both in central government and in local government such as NGOs, private sectors, education and research institutions, civil societies and the community are very urgent. New research action is required to find our new urban municipal solid waste characteristic and our appropriate technology and management to give some input to the central government, local governments and the community or others who involve in the municipal solid waste handling due to the recent fast growing of urban people income and changing of their life style.

Conclusion. For the future, the strengthening of central and local governments' political will is still required including financial mobilization, community education and/or empowerment, law enforcement, technical innovations, management development, providing required urban and regional solid waste management infrastructures, and Public Private Partnership promotion.

IOP Conf. Series: Earth and Environmental Science **60** (2017) 012004 doi:10.1088/1755-1315/60/1/012004

Green materials for sustainable development

B S Purwasasmita
Bandung Institute of Technology, Indonesia
Email : purwa@tf.itb.ac.id

Abstract. Sustainable development is an integrity of multidiscipline concept combining ecological, social and economic aspects to construct a liveable human living system. The sustainable development can be support through the development of green materials. Green materials offers a unique characteristic and properties including abundant in nature, less toxic, economically affordable and versatility in term of physical and chemical properties. Green materials can be applied for a numerous field in science and technology applications including for energy, building, construction and infrastructures, materials science and engineering applications and pollution management and technology. For instance, green materials can be developed as a source for energy production. Green materials including biomass-based source can be developed as a source for biodiesel and bioethanol production. Biomass-based materials also can be transformed into advanced functionalized materials for advanced bio-applications such as the transformation of chitin into chitosan which further used for biomedicine, biomaterials and tissue engineering applications. Recently, cellulose-based material and lignocellulose-based materials as a source for the developing functional materials attracted the potential prospect for biomaterials, reinforcing materials and nanotechnology. Furthermore, the development of pigment materials has gaining interest by using the green materials as a source due to their unique properties. Eventually, Indonesia as a large country with a large biodiversity can enhance the development of green material to strengthen our nation competitiveness and develop the materials technology for the future.

IOP Conf. Series: Earth and Environmental Science **60** (2017) 012005 doi:10.1088/1755-1315/60/1/012005

Kerong fish (*Terapon jarbua*) peptone production using papain enzyme as nitrogen source in bacterial media

Y Srikandace*, S Priatni, S Pudjiraharti, W Kosasih and L Indrarti
Research Unit for Clean Technology, Indonesian Institute of Sciences (LIPI) Jalan Cisitu 21/154 D, Bandung 40135, Indonesia
*Email : yoice.srikandace.s@gmail.com

Abstract. *Terapon jarbua* fish is a widespread species distributed mainly in the Indo-West Pacific region and occurs in shallow coastal waters, mangroves and freshwater. In West-Java of Indonesia, the consumption of *T. jarbua* was not bound to arouse interest, but otherwise the fish contain the important protein, such as peptone. Peptone is used to support the growth of bacteria as nitrogen source. The present study was aimed to optimize the fish peptone of *T. jarbua* using papain enzyme concentration and to characterize the fish peptone product. The parameter of fish peptone development was enzyme concentration (0.1-0.5%), temperature (50-65°C), hydrolysis time (4-7 h) and pH (5-8). The fish peptone product was characterized on the content of soluble protein, N-amino acid, AN/TN, and growth curve of *Escherichia coli* and *Staphylococcus aureus*. The optimum fish peptone production was obtained with 0.2% of papain enzyme pH 7 at 65°C for 4 hr. The soluble protein content was of 3,63 mg/ml, N-amino content 0.29%, and AN/TN 0.54%. Fish peptone of *T. jarbua* was more selective for *E. coli* than *S. aureus*.

Introduction

Fish as one of marine resources occupies the certain and significant role in the aquatic contaminant studies and the monitoring water quality. Nowadays, fish production is still as one of producers of insignificant garbage due to waste mainly composed of fish body's parts [1]. Currently, fish and by products discarded by fisheries are becoming large in quantity and creating the serious pollution [2]. On the other hand, the by products contain protein-rich material and normally processes into low market such as animal feed, fish meal even laboratories support such as growth media for microbes. Fish waste can also be used for production of various value added product such as protein, oil, amino acids, mineral, enzymes, bioactive peptides, collagen and gelatine [3].

Terapon jarbua as one of fish in West Java Indonesia is over production, but the consumption of the fish is low due to the high salt level and commonly it is used as fishing bait [4]. In food biotechnology, material with the high protein content, good amino acid, and bioactive peptides have attracted much attention biotechnologist to explore the compounds for antioxidant, antihypertensive, immune-modulatory, and antimicrobial peptides application [1].

Peptone is a protein hydrolysate soluble in mixture of polypeptides and amino acids which is widely used in many biological and biotechnological applications such as microbial biomass [3]. Meanwhile, the literatures of fish peptone using papain enzyme was limited and it was needed to increase the economic value of the fish protein called peptone because it can be as nitrogen source in grow media of bacteria [5]. It has been showed that the trash fish peptone and marine fish peptone from ponyfish (*Eubleekeria splenders*), Yellow stripe travally (*Selaroides leptolipis*) and Mackarel

(*Decapterus maruadsi*) could be hydrolized using papain enzyme with various concentration [6][7]. The aim of the present study was to optimize of the production of fish peptone of *T. jarbua* using papain enzyme and to characterize the fish peptone product.

Experimental

2.1. Fish peptone preparation

T. jarbua fish was purchased from Indramayu, West-Java and transported to the laboratory in sealed foamed polystyrene boxes containing flaked ice. A total of 100 g of fish was mixed with water in a equal percentage of weight with water (ratio 1:4), and blended. The hydrolysis was carried out at various pH of 6-8 and at varied papain enzyme concentration (0.2-1.0% v/w). Papain enzyme was added into the mixture, heated at 50-65°C for 4-8 h and inactivated at 85°C for 15 min. The hydrolyzed mixtures were centrifuged (7000g, 20 min) at 4°C. Finally, the supernatant was collected and dried to obtain soluble peptone.

2.2.Proximate composition

The proximate composition of fish samples was determined based on the Association of Official Analytical Chemist (AOAC) method. The moisture content was determined by drying samples overnight at 105°C for 2 hr 30 min. The crude protein content was determined by the Kjeldahl method. The fat content was determined using soxhlet method and the ash content was obtained in furnace at 500°C until fully becoming ash. The carbohydrate content was achieved by calculating the difference of total moisture, protein, fat, ash content.

2.3.Assay for total protein

The total protein of *T. jarbua* fish peptone was carried out based on Lowry's method. A mount of 500 μl samples were added with 2.5 mL solution C (Lowry A 50 mL + 1 mL Lowry B) and dissolved for 10 min. Samples were then added with 0.5 mL solution D (Folin : dH_20 with ratio 1:1), dissolved for 30 min, and measured at 500 nm.

2.4.Assay for total N-amino

The total N-amino of *T. jarbua* fish peptone was estimated by formol titration method. A total of 10 mL sample was added with 20 mL dH_20, 40 mL potassium oxalate (1:3) and 3-4 drops of phenol phtalein indicator (PP). Sample was then titrated using 0.1N NaOH, added with 2 mL of 40% formalin and 3-4 drops of PP indicator. Sample was titrated using 0.1N NaOH and measured its protein content.

2.5.Bacterial growth

Bacteria of *Escherichia coli* and *Staphylococcus aureus* were used for the bacterial growth performance test. The sterilized test medium consisted 1 g of fish peptone extract, 100 mL dH_2O, 0.5% yeast extract, and 1% NaCl with pH 7. The sterilized commercial medium contained 1g of commercial peptone, 100 mL dH_2O, 0.5% yeast extract, 1% NaCl with pH 7. A mount of 1mL pure culture was added with 9 mL medium, incubated at 37°C for 24 hr. The optical density of culture was measured at 600 nm every 2 hr for 24 hr.

3. Result and discussion

3.1. Chemical compounds

The proximate analysis showed that the *T. jarbua* fish contained total protein and fat were of 58.8 and 9.6% respectively. The fish *T. puta* contains 20-30% protein and 2-12% lipid [4]. The chemical compounds of fish can be varied between species. The proximate analysis showed the variation of results according to environment, age, sex species, migration, nutrient, and preservation method [8]. The value of total protein and N-amino content from various temperatures and time incubations for screening the optimal production of fish peptone was described in Table 1.

Table 1. Total protein and N-amino content from fish peptone on pH 7.

	Incubation time (h)	Protein(mg/mL)	N-amino (%)
50°C	4	2.8	0.07
	5	2.5	0.06
	6	2.6	0.07
	7	2.4	0.05
55°C	4	2.3	0.05
	5	2.5	0.06
	6	2.0	0.06
	7	2.4	0.06
60°C	4	2.8	0.07
	5	2.8	0.07
	6	2.9	0.07
	7	2.9	0.08
65°C*	4	3.6	0.07
	5	3.1	0.07
	6	2.9	0.07
	7	2.9	0.07

*Optimal condition for the highest yield of total protein and N-amino of fish peptone

Table 1 shows that the optimal condition was achieved at pH 7, incubation at 65°C time for 4 hr with total protein and N-amino were of 3.6 mg/mL and 0.07% respectively. Various protein and N-amino obtained might be due to the variety of achievement from protein whereas N-amino from the activity of papain enzyme. The papain enzyme activity was affected by the concentration and purity of enzyme, temperature, and time of hydrolysis. The recent study showed that 0.2% enzyme papain produced high peptone. In addition, fish carp contained high peptone using 0.26% papain enzyme, at 60°C for 3 hr hydrolysis [1]. Preparation of fish peptone as growth media for bacteria with the determination of enzyme concentration is one factor degradation speed rate of proteolytic enzyme [8]. Soluble protein content and N-amino of fish peptone *T. jarbua* product was of 3,63 mg/ml, 0.29%, respectively, and AN/TN 0.54%. Almost complete protein solubilisation could be achieved within an hour at 40°C, 1 % enzyme/substrate ratio (w/w) with papain and bromelain [9].

3.2. Bacterial growth analysis

The growth curve of *E. coli* and *S. aureus* on two kinds of peptone, peptone commercial and fish peptone of *T. jarbua.* were shown in Figure 1 and Figure 2.

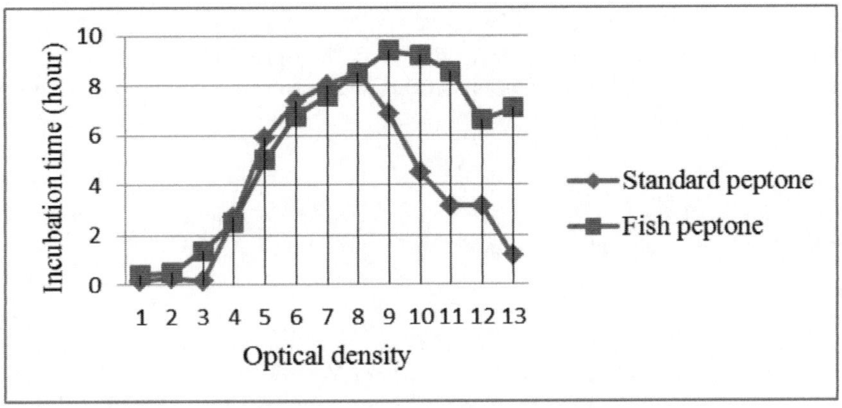

Figure 1. The growth curve of *E. coli* cultured in standard peptone and fish peptone.

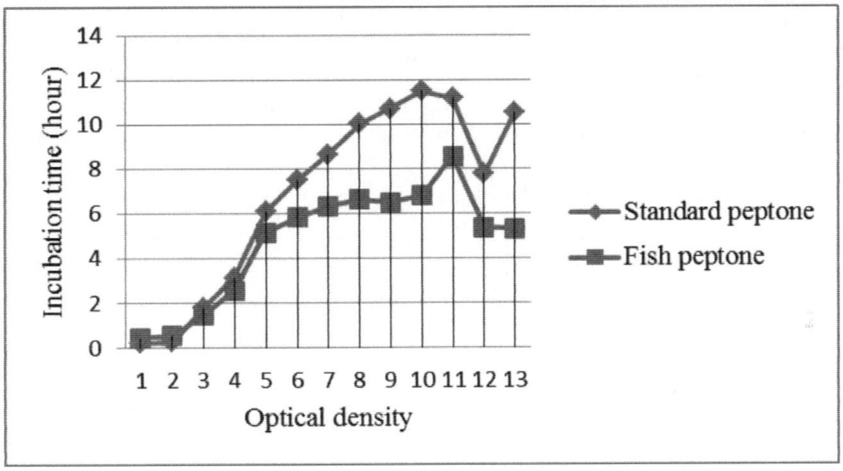

Figure 2. The growth curve of *S. aureus* cultured in standard peptone and fish peptone.

Figure shows that the growth of *E. coli* cells using fish peptone was higher than commercial peptone as well as a bit longer time to obtain maximum growth. On the other side, Figure 2 describes that the growth of *S. aureus* using fish peptone was lower as compared to commercial peptone. It indicated that fish peptone of *T. jarbua* was more selective for *E. coli* than *S. aureus*. Peptone from enzymatic hydrolysis by crude papain enzyme showed different performance in both bacteria growth profile. The hydrolysis process conducted the important roles in processing peptone which contributed to reveal the better performance of fish waste peptone in microbial growth [5]. It also might depend on the ability of bacteria to use the fish peptone to improve the cell growth. Fish peptone has a fairly high protein containing nitrogen easily for bacteria to absorp well [5][6]. The previous research showed that proteolytic enzyme papain for protein recovery from *Scomberomorus commerson* fish produced bioactive properties as well [10]. Bioactive compounds from fish protein hydrolysates also could be obtained by using base, acid, endogenous enzymes, bacteria, and digestive protease [11].

4. Conclusion

Peptone has been produced from from *T. jarbua* of which the optimum condition obtained with 0.2% papain enzyme, pH 7, at 65°C for 4 hr hydrolysis. The fish peptone was more selective for *E. coli* than *S. aureus*.

Acknowledgement

The authors greatly acknowledge the funding of the independently research project of LIPI 2016. The authors also thanked to the collaboration of staffs and supply the material.

References

[1] Saputra D and Tati N 2016 *ComTech.* **7 (1)** 11-18.
[2] Fallah M, Somayeh B and Seyed R J 2014 *Food Sci Nutr.* **3(2)** 153-157.
[3] Ghaly A E, Ramakrishnan V V, Brook M S, Budge S M and Dave D 2013 *J.Microb. Biochem Tech.* **5(3)** 107-129.
[4] Sabrah M M, Lamiaa I M, El-Sawy M and Emad H A E 2015 *J. Fish. Aquat. Sci.***11(2)** 147-162.
[5] Husin N, Siti M M K, Ling T C, Nurul F M and Nohani J 2015 *IOP Conf. Biomed. Engine. Technol.* **81** 54-57.
[6] Mawaddah R, Aninta S, Yunisha A, Fajar S, and Rahayu K R 2011 *IOP Conf. Chem.Biol.Env.Sci.* 334-336.
[7] Nicharee W and Sasithorn K 2015 *J.Med. Bioengine.* **4(6)** 466-470.
[8] Sesugh A, Leke L, Eneji I and Yakubu S 2012 *Elixir Food Sci.* **53** 11801-11803.
[9] Himonides A T, Anthony K D T and Anne J M 2011 *Food Nutr. Sci.* **2** 575-586.
[10] Abdulazeez S S, Baranitharan R and P. Ponnusamy 2013 *Int. J. Pharm. Bio. Sci.* **3(1)** 134-144.
[11] Kristisson H G and Barbara A R 2000 *Critical Reviews in Food Science and Nutrition: Fish Protein Hydrolysate: Production Biochemical and Functional Properties.* **40(1)** 43-81.

Bio-prospective of *Polyscias fruticosa* leaf extract as redactor and stabilizer of gold nanoparticles formation

Y Yulizar* **and Q Ayun**
Department of Chemistry, Faculty of Mathematics and Natural Sciences, University of Indonesia, Depok 16424, Indonesia Tel. +62-21-7270027, Fax +62-21-7863432
*Email : yokiy@ui.ac.id

Abstract. Metal nanoparticle is a great interest to researches due to its applications toward catalysis, sensors, and drug delivery. Biosynthesis of gold nanoparticles (AuNPs) using aqueous leaf extract of *Polycias fruticosa* (PFE) is reported in this article. PFE plays a role as reductor and stabilizer of AuNPs. The formation of PFE-AuNPs under radiation of natrium lamp for 15 min was monitored by UV – Vis spectrophotometer. The growth process and stability of PFE-AuNPs was observed from the colour and absorbance change in the wavelength range of 529-533 nm. The optimum synthesis condition of PFE-AuNPs was obtained at 0.06% (w/v) of PFE concentration. Size and its distribution of PFE-AuNPs were identified by particle size analyzer (PSA) as 35.02 nm and stable up until 21 days. The stable PFE-AuNPs was further characterized by Fourier transform infrared (FT-IR) spectroscopy to identify the functional group in phenolic compound of PFE interact with AuNps.

1. Introduction

Currently, nanoscience and technology have many applications for human life. Their applications can be found in biomedicine [1] agriculture, food [2], environment [3], electronics [4], and energy [5]. Metal nanoparticles have been used to study, such as gold nanoparticles (AuNPs). AuNPs has attracted an attention of researchers due to has a high stability, catalytic activity and its application in many fields [6] as biomedical material [7], electronics [8], catalyst [9], optical applications [10], sensors and biomedicine [11].

Synthesis of AuNPs has been successfully carried out by the physical and chemical methods. The chemical method was obtained using reductor such as sodium borohydride [12]; N,N-dimethylformamide, trisodium citrate, hydrazine, Tollens reagents and polyethylene glycol. However, synthesis of AuNPs using chemical methods has disadvantages: toxic and hazardous material for the environment [13]. There are alternative more eco-friendly methods of nanoparticle synthesis using microorganism, enzym, and plants or plant extracts. Hence, green synthesis method using plant extract is proposed in this research. The green synthesis method is a simple, low-cost, stable, and eco-friendly method [14]. It can be applied for nanoparticles synthesis from many metals such as Pt, Au, Ag, Cu, and Zn [15-16]. Au metal is prepared by green synthesis due to possess a distinctive characteristic. Synthesis of AuNPs using the plant has been reported such as fruit extract [17], shoots [18], flowers [19] and rods [20]. Synthesis of AuNPs using leaf extracts have been conducted such as *Terminalia catappa* [21], *Euphorbia hirta* [22], and *Plumeria alba* [23].

Polyscias fruticosa (PF) plant is very abundant in Indonesia and used as herbal plant. Previous study reported the plants can be used as a reductor and stabilizer of Ag [24-25] and Au [26]. Plants are used as reductor and stabilizer in AuNP synthesis due to contain secondary metabolites such as flavonoid. PF leaves have been studied contain flavonoid, steroid, triterpenoid, saponins, tannin, p-hydroxybenzoic acid, and vanillic acid [27]. Andarwulan N (2010) reported PF leaf contains a flavonoid (52.19 mg/100 g) higher than *Etlingera elatior* leaf (1.18 mg/100 g). PF leaf has potency as a reductor and stabilizer of AuNPs formation. Previous study has used a leaf extract through maceration in water, while in this research, maceration was conducted in methanol. Methanol extract was partitioned gradually using hexane, ethyl acetate and water. Water extract can be optimized to reduce and stabilize AuNPs. There is no report of PF leaf used to reduce and stabilize AuNPs. Therefore, the present study conduct the synthesis of AuNPs using water PF leaf extract as reductor and stabilizer of nanoparticle.

2. Experimental

2.1. Materials

Polyscias fruticosa (PF) leaves were collected from Sawangan, Depok West Java, Indonesia, and gold metal 99.9% purity obtained from PT Antam, Indonesia. Stock solution of $HAuCl_4$ was prepared from gold metal dissolved in aquaregia (HCl : HNO_3 = 3:1).

2.2. PF leaves extract preparation

PF leaves were washed and dried for one week in room temperature. Methanol of 500 mL was added to powder dried PF leaves, and macerated for 5 days in room temperature. Filtrate from maceration was then concentrated using vacuum rotatory evaporator. Methanol extract of PF leaves was gradually partitioned using hexane, ethyl acetate, and water solvents. Final fraction of water was concentrated using vacuum rotatory evaporator to get stock solution of PF extract (PFE) for AuNPs synthesis.

2.3. Preliminary phytochemical screening

Each extract was phytochemically tested for the presence of flavonoids, tannins, phenols, saponins, steroids, terpenoids and alkaloids [28]. Test of flavonoid was done using alkaline reagent. Test of phenols and tannins were done with addition of $FeCl_3$ solution in PFE. Test of saponins with addition of distilled water in PFE, and was shaken vigorously. Test of steroids with addition of chloroform, concentrated H_2SO_4 and acetic acid in PFE. Test of terpenoids with addition of PFE in 2 mL of chloroform, and evaporated to dryness. 2 mL of concentrated H_2SO_4 was added and heated for 2 min. Test of alkaloids with mixing PFE and 2 mL of 1% HCl, and heated gently. Mayer's and Wagner's reagents were then added to the mixture. Turbidity of the resulted precipitate was observed.

2.4. Synthesis of AuNPs using PFE and its stability

9.0 mL of $HAuCl_4$ 0,1 mM was added 1 mL PFE with various concentration 0,02 – 0,10% (v/w). The mixture was reacted for 15 min under Na-light exposure. The AuNPs stability was observed against reaction time using UV-Vis spectrophotometer and particle size analyzer (PSA).

2.5. Characterization of AuNPs

UV-Vis spectrophotometer (Shimadzu 2600) was used to get absorption spectra of surface Plasmon resonance (SPR) for AuNPs formation. Average particles diameter, particle size distribution were measured by particle size analyzer (PSA), and its nanoparticle colloidal stability observed from the potential Zeta charge (PZC) Malvern ZEN 1600. Identification of functional groups of biocompounds in PFE leaves extract before and after reaction was done with FTIR spectroscopy (Perkin-Elmer). Transmission electron microscopy JEM 1400 - selected area electron diffraction (TEM-SAED) was used to observe the shape, size, and crystalline of AuNPs.

3. Result and discussion

3.1. Identification of aqueous PFE

Identification of aqueous PFE was done by phytochemcal test. It is conducted to qualitatively analize the presence of secunder metabolites. Phytochemical test of methanol fraction PFE positively contained flavonoids, steroids, alkaloids, and saponnin, while of water fraction PFE positively contained flavonoids.

Identification of aqueous PFE and PFE stabilized AuNPs (PFE-AuNPs) were also measured using FTIR to investigate the interaction of functional groups in PFE and PFE-AuNPs colloids. Figure 1 shows the wavenumber shift from 3342 cm^{-1} (PFE) to 3385 cm^{-1} (PFE-AuNPs) as –OH stretch that identified from flavonoid. This wavenumber is suitable with the research of Yu J et al [29]. Figure 1 shows also the wavenumber shift of 1613 to 1629 cm^{-1} as a -C=O stretch. The significant wavenumber shifts of –OH group (alcohol) and -C=O (carboxylic acid and its derivative) from flavonoid were observed as reductor and stabilizer of AuNP formation. Based on phytochemical test, the water fraction PFE positively contained flavonoids, and this result of secondary metabolite in accordance with the FTIR data.

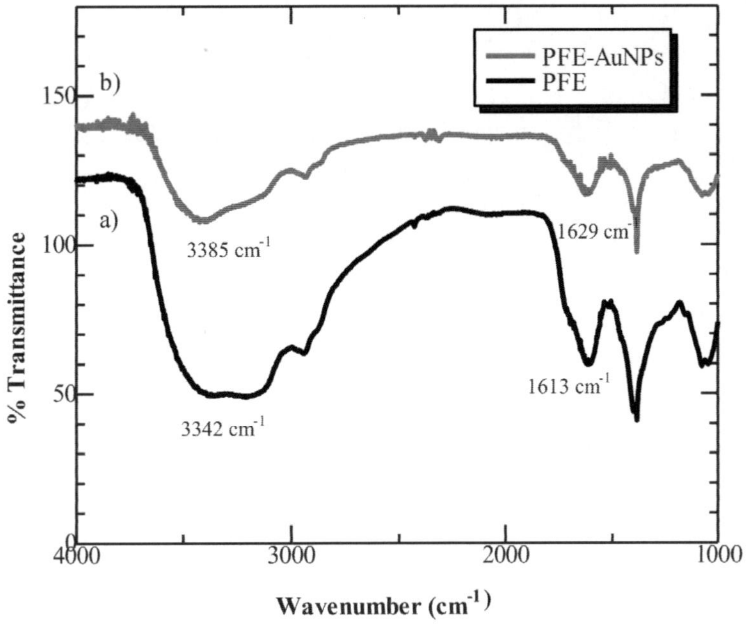

Figure 1. FTIR spectra of PFE and PFE-AuNPs.

3.2. Various PFE concentration of AuNPs synthesis

Synthesis of AuNPs was conducted by reducing the trivalent Au ions to uncharged Au atoms. In this research, synthesis of AuNPs was done with the reaction of aqueous PFE as reductor with $HAuCl_4$ solution under the exposure of Na-light. UV-Vis absorption spectra of PFE-AuNP were observed at various PFE concentration as shown in Figure 2 (a). AuNPs formation was confirmed by a colour change of aqueous PFE from pale yellow to bright red at 529-533 nm which was proved from SPR peak formation [30-31]. UV–Vis spectrophotometer was used to determine the optimum AuNPs formation through the change of wavelength and absorbance. Figure 2 (b) explains that optimum PFE concentration 0.06% was indicated by high absorbance, small maximum wavelength, λ_{max} value, and sharp peak. This is consistent with the research of Philip D (2010) which the optimum concentration for AuNPs formation was indicated by small λ_{max} and high absorbance [32]. The further experiment to confirm UV-Vis absorption spectra was conducted using TEM and PSA.

a

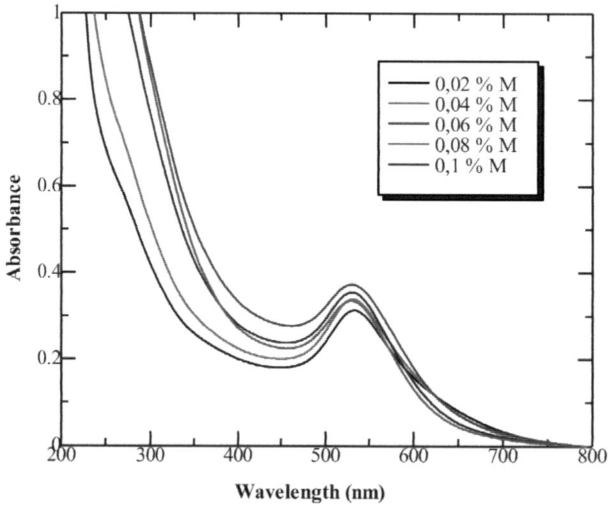

b

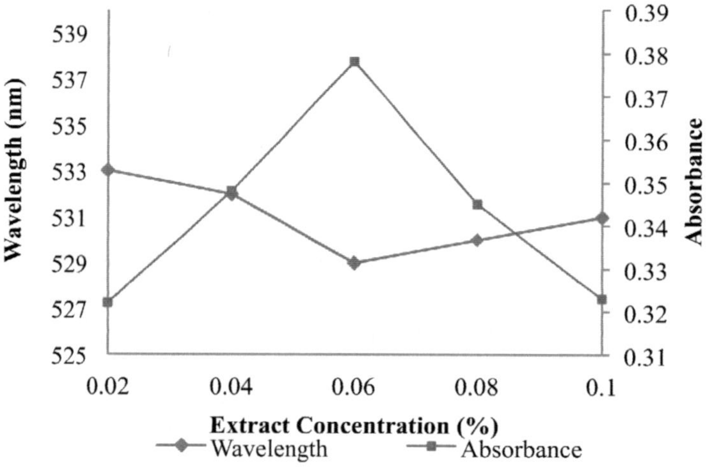

Figure 2. **(a)** UV-Vis absorption spectra of PFE-AuNPs and **(b)** relational curve between λ_{max} and absorbance against PFE concentration of PFE-AuNPs.

3.3. Stability of AuNPs

Synthesis of AuNPs will agglomerate due to existing of interparticle force which cause a strong interaction. Figure 3 shows the relation between absorbance value and λ_{max} against reaction time of PFE-AuNPs formation for 30 days. λ_{max} was observed to increase from 21 until 25 days, and shows that AuNPs was stable for 21 days. AuNP stability is affected by the active compounds in PFE which act as a stabilizer of AuNP. However, there is a trend of decrease after 21 days, which indicates AuNPs agglomeration.

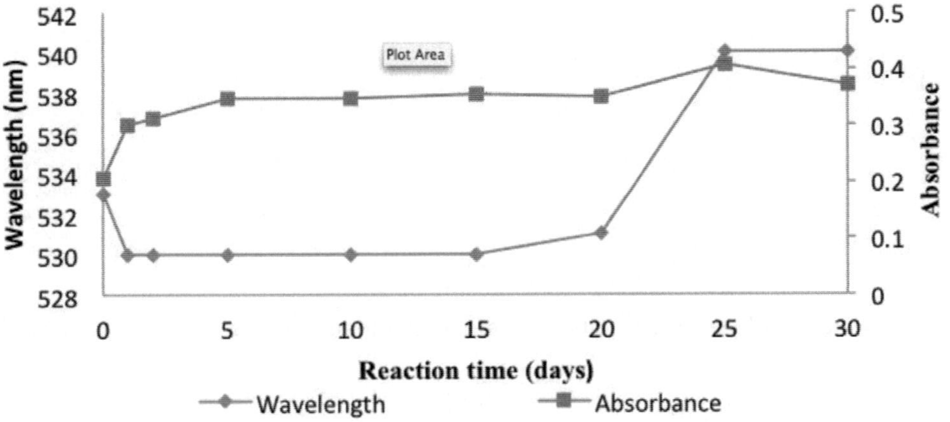

Figure 3. Curve of λ_{max} and absorbance value against stability time of PFE-AuNPs.

Figure 4 shows potential zeta charge (PZC) of PFE-AuNPs. PZC measurement was conducted to observe the stability of PFE-AuNPs and to measure the total charge between PFE interacted on AuNPs surface. PZC value of PFE-AuNPs is -12.3 mV, which means the particles are stable due to the negative charge prevent aggregation between particles as an effect of repelling force between the charges. In this study, PFE-AuNPs was observed stable for 21 days, after that it starts to agglomerate each nanoparticle.

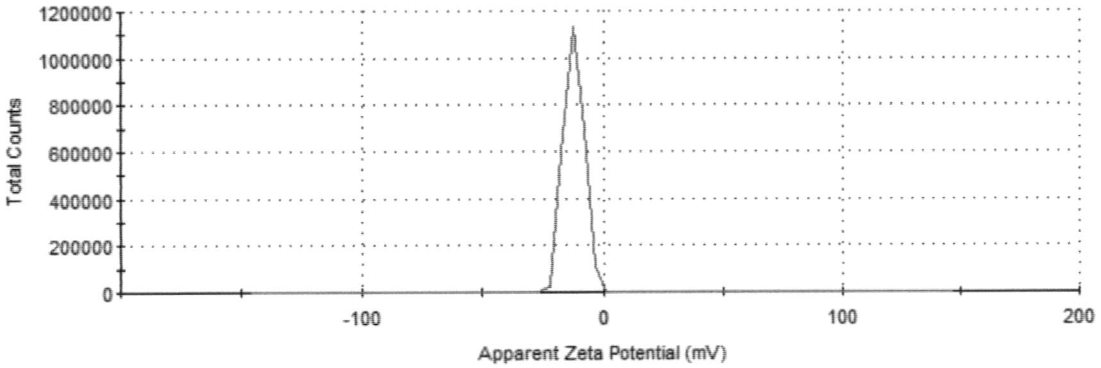

Figure 4. Zeta potensial charge of optimum PFE-AuNPs.

3.4. Particle size of AuNPs

Particel size of PFE-AuNPs was measured using PSA and TEM. Figure 5 explains PSA spectrum for intensity (percent) against particle size (nm) of PFE-AuNPs at optimum PFE concentration (0,06%). Intensity at Figure 5 indicates the amount of formed nanoparticles that scatter the light. Particle size and its distribution of AuNPs were polidispersed with size majority at 35.02 nm.

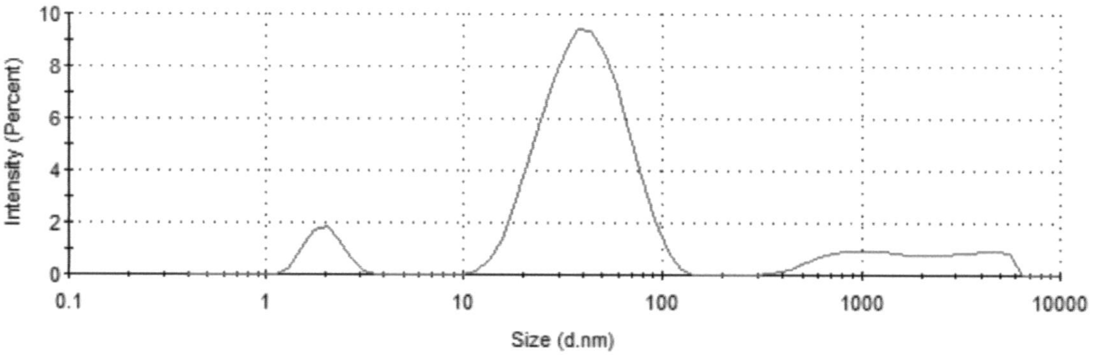

Figure 5. Particle size and distribution of PFE-AuNPs.

TEM characterization was conducted to identify morphology, shape, and size of nanoparticles. TEM measurement of PFE-AuNPs was conducted at optimum PFE concentration (0.06%). TEM image showed a homogenously dispered PFE-AuNPs and has an average diameter of 20 nm with relatively spherical shapes as shown in Figure 6. The homogenously dispersed and controlled nanoparticle size due to the aquoeous PFE played a role as a good stabilizer through capping on AuNPs surface.

The resulted data shows that AuNPs size from PSA was greater than TEM. The nanoparticle size using PSA measured PFE capped AuNPs through its scattering total, while TEM data only show image of AuNPs due to it has a high density.

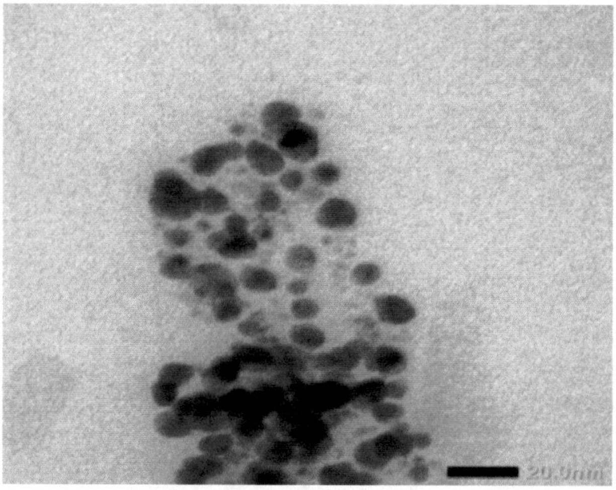

Figure 6. TEM image of AuNPs.

3.5. Analysis of AuNPs using TEM-SAED

TEM image showed spherical shapes with size of 20 nm. SAED pattern in Figure 7 shows that AuNPs formed a crsytalline structure of face center cubic (fcc) with Miller index (111), (200), (220), and

(311). SAED analysis of AuNPs indicates the existence of some ring that adjusted with JCPDS of Au on (111), (200), (220), and (311).

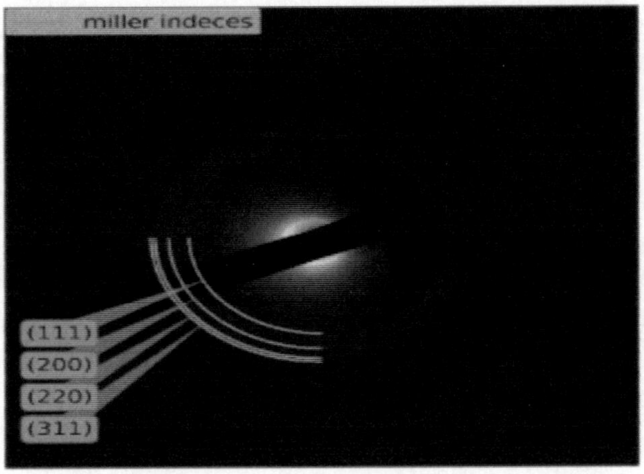

Figure 7. SAED pattern of AuNPs.

4. Conclusion

AuNPs was successfully synthesized using *Polycias fruticosa* leaf extract (PFE) in water fraction by green synthesis method. PFE played a role as reductor and stabilizer of PFE-AuNPs. Synthesis of AuNPs was done at optimum PFE concentration of 0.06%. Various PFE concentration has affected the maximum wavelength (λ_{max}), absorbance value, and spectrum peak shape of AuNPs. AuNPs was stable for 21 days, and has optimum λ_{max} 531 nm. Particle size of PFE-AuNPs based on PSA measurement is 35.02 nm, and the PZC value is –12.3 mV. Based on FTIR characterization, PFE contained a flavonoid that acted as reductor and stabilizer of AuNPs formation. The result of TEM image showed that AuNP has an average diameter of 20 nm with spherical shape. Analysis of SAED pattern of AuNPs indicated the presence of ring that adjusted with JCPDS of Au on (111), (200), (220), and (311). This explains that the crystaline characteristic of AuNPs is fcc (*face center cubic*).

Acknowledgment

This work was funded by hibah PITTA 2016, from the Universitas Indonesia through the directorate of research and community services, Universitas Indonesia (No.2040/UN2.R12/HKP.05.00/2016).

References

[1] Bansal A and Zhang Y 2014 *Acc. Chem. Res.* **47** (10) 3052-60.
[2] Sekhon B S 2014 *Nanotechnol.* **7** 31–53.
[3] Tang S C N and Lo I R M 2013 *Water Res.* **47** (8) 2613-32.
[4] Alshehri A H, Jakubowsa, Lozniak M M, Horaczek A, Rudka M, Free D, and Carey J D 2012 *ACS Appl Mater Interfaces.* **4** (12) 7007-10.
[5] Ahn S J, Kim K, Cho A, Gwak J, Yun J H, Shin K, Ahn S K, and Yoon K 2012 *ACS Appl Mater Interfaces.* **4** (3) 1530-6.
[6] Sadeghi B, Rostami A, Momeni S S 2014. *Facile green synthesis of silver nanoparticles using seed aqueous extract of Pistacia atlantica and its antibacterial activity*. Department of Chemistry, Islamic Azad University, Iran.
[7] Cao X L, Cheng C, Ma Y L, Zhao C S 2010 *J Mater Sci Mater Med.* **21** 2861-8.
[8] Mohan Y M, Lee K J, Premkumar T, Geckeler K E 2007 *Polymer.* **48** 158–64.

[9] Feng X, Qi X, Li J, Yang L W, Qiu M C, Yin J J, Lu F, Zhong J X 2011 *Appl Surf Sci.* **257** 5571- 5.
[10] Hayward R C, Saville D A, Aksay I A 2000 *Nature.* **404** 56–9.
[11] Sugihara K K, Sujitha M V, and Kannan S 2013 *Spec Act Part A.* **102** 15–23.
[12] Yulizar Y, Kadja G T M and Safaat M 2016 *Reac Kinet Mech Cat.* **117** 353-63.
[13] Joseph S, Mathew B 2014. *Microwave assisted facile green synthesis of silver and gold nanocatalysts using the leaf extract of Aerva lanata.* Department of Chemistry, Kerala, India.
[14] Ahmed S, Ahmad M, Swami B L and Ikram S 2016 *J Adv Res.* **7** 17-28.
[15] Mohanpuria P, Rana N K and Yadav S K 2008 *Nanopart Res.* **10** 507–17.
[16] Raveendran P, Fu J and Wallen S L 2003 *J Am Chem.* **125** 13940-1.
[17] Elumalai K, Velmurugan S, Ravi S, Kathiravan V and Adaikala Raj G 2015 J *Adv Pow Tec.* **26** 1639-51.
[18] Armendariz V, Herrera I, Peralta V J R, Jose Y M, Troiani H, Santiago P, and Gardea T J L 2004 *J Nanopart Res.* **6(4)** 377–82.
[19] Anand K, Gengan R M, Phulukdaree A, Chuturgoon A 2014 *J Ind Eng Chem.* **21** 1105-11.
[20] Suman T Y, Rajasree R S R, Ramkumar R, Rajthilak C and Perumal P 2013 *Spec Act Part A.* **118C** 11–6.
[21] Ankamwar B 2010 *E-J Chem.* **7** 1334-9.
[22] Annamalai A, Christina V L P, Sudha D, Kalpana M, and Lakshmi P T V 2015 *J Colloid interf sci.* **108** 60– 5.
[23] Mata R, Bhaskaran A and Sadras S R 2016 *Particuology.* **24** 78–86.
[24] Zuas O, Hamim N and Sampora Y 2014 *Mater Lett.* **123** 156-9.
[25] Kasthuri J, Veerapandian S and Rajendiran N 2009 *Colloid Surface. B* **68** 55-60.
[26] Kumar N R B, Kumar K H A and Mandal B K 2015 *J Mol Liq.* **211** 868-75.
[27] Andarwulan N, Batari R, Agustini D S, Bolling B and Wijaya H 2010 *Food Chem.* **121** 1231-35.
[28] Yadav R N S and Agarwala M 2011 *Phytology.* **3** (12) 10-4.
[29] Yu J, Xu D, Guan H D and Wang C 2016 *Mater Lett.* **166** 110-2.
[30] Yallapa S, Manjanna J, Dhananjaya B L 2015 *Spechtrochim Acta Part A.* **137** 236-43.
[31] Devi P R, Kumar C S, Selvamani P, Subramanian N, and Ruckmani K 2015 *Mater Lett.* **139** 241-4.
[32] Philip D 2010 *Spechtrochim Acta Part A.* **75** 1078-81.

IOP Conf. Series: Earth and Environmental Science **60** (2017) 012007 doi:10.1088/1755-1315/60/1/012007

Production of antioxidant compounds of grape seed skin by fermentation and its optimization using response surface method

D G S Andayani*, C Risdian, V Saraswati, I Primadona and P C Mawarda
Research Unit for Clean Technology, Indonesian Institute of Sciences (LIPI)
Jalan Cisitu 21/154 D, Bandung 40135, Indonesia
*Email : desakgedesa@gmail.com

Abstract. Skins and seeds of grape are waste generated from food industry. These wastes contain nutrients of which able to be utilized as an important source for antioxidant metabolite production. Through an environmentally friendly process, natural antioxidant material was produced. This study aimed to generate antioxidant compounds by liquid fermentation. Optimization was carried out by using *Schizosaccharomyces cerevisiae* in Katu leaf substrate. Optimization variables through response surface methodology (RSM) were of sucrose concentration, skins and seeds of grape concentration, and pH. Results showed that the optimum conditions for antioxidant production were of 5 g/L sucrose, 5 g/L skins and seed at pH 5, respectively. The resulted antioxidant activity was of 1.62 mg/mL. Mathematical model of variance analysis using a second order polynomial corresponding to the resulted data for the antioxidant was of $20.70124 - 3.86997\ A - 0.65996\ B - 1.88367\ C + 0.19634\ A^2 - 0.016638\ B^2 + 0.28848\ C^2 + 0.26980\ AB - 0.068333\ AC - 0.12367\ BC$. From the gained equation, the optimum yield from all variables was significant. Chemical analysis of the antioxidant was carried out using 2,2-Diphenyl-1-picrylhydrazyl (DPPH).

1. Introduction

Skins and seeds of grape are by-products of food industry. High content of nutrients in skins and seeds of grapes can be utilized as a source of nitrogen, vitamins, and minerals. Utilization of skins and seeds of grapes as a medium for antioxidant flavonoid production is an attempt to enhance its add value and improving the quality of living things [1, 2]. However, these natural materials are oxidized easily. Fermentation technology is able to maintain natural materials stability as well as increasing its antioxidant activity. Microbial fermentation is considered as a potential technology for producing new bioactive compounds. Several fermentation processes are applied in the production of secondary metabolite. However, its bioconversion yield varies, depending on the application of fermentation parameters for instance the kind of microorganisms, medium used, temperature, and pH [3].

Engineering of antioxidant biosynthesis in microorganisms by fermentation has significant scientific and economically importance. Response Surface Methodology (RSM) is one of the strategies for increase antioxidant production yields [1, 4]. The skins and seeds of grape contain fiber and resveratrol which is one source of antioxidants. Antioxidant consists of quercetin, procyanidins, catechins, and anthocyanins. The chemical compounds, namely: terpenes (geraniol, linalool, terpineol, nerolidol), norisoprenoids (damascenone ß, ß ionone), thiols (hexan-1-ol-L-cysteine) [5].

In this study, RSM was used to determine the optimum conditions for antioxidant production with various pH, sucrose concentration and concentration of grape skins and seeds. With choose antioxidant activity in response, optimum conditions antioxidant production process using yeast *S. cereviceae* as biocatalysts can be determined [6-9].

2. Experimental

2.1. Materials

Skins and seeds of grape were obtained from food industry whereas Katu leaves from traditional market. Commercial sucrose was purchased from chemicals agent whilst *S. cerevisiae* was taken from collection of Research Unit for Clean Technology, Indonesian Institute of Sciences.

2.2. Method

2.2.1. Cultivation of microorganisms

S. cerevisiae was cultivated in potato dextrose agar (PDA) at 30°C for 48 hr.

2.2.2. Inoculums preparation

Pure culture of *S. cerevisiae* was activated by putting it to PDA slant medium for 48 hr and inoculated to PDB medium for 4 days. The phase of inoculums preparation was began with the stage of microorganism breeding.

2.2.3. Pulverizing of skins and seeds of grapes

Skins and seeds of grape were oven-dried for a week at 50°C then crushed with a grinder to form a powder.

2.2.4. Blanching of Katu leaves

Katu leaves were boiled for 15 min at 80 °C then blended in distilled water (1: 4) w /v [2]. The resulted liquor was separated from the mixture.

2.2.5. Fermentation process

The liquor of Katu leafs was inoculated with 10% w/v of *S. cerevisiae* in fermentation media without the skins and seeds of grape, then incubated in the dark room for seven days at 30 °C, pH 7.

2.2.6. Fermentation process optimization

Optimum condition for antioxidant production was carried out with experimental design of Response Surface Methodology with Design Expert 6.0.6. software supported 20 experiments including six-center point designed in Central Composite Design. Each experiment was done in duplicate. Variables were varied at pH (5-8), concentration of skins and seeds of grape (5-10) g/L, and concentration of sucrose (5-10) %.

2.2.7. Antioxidant analysis

Inhibition of DPPH free radicals were measured in fermentation broth without grape skins and seeds in Katu leafs substrate and compared with the broth containing skins and seeds of grape. Inhibition of the fermentation was carried out in 1-7 days. Antioxidant activity test was conducted to the highest inhibition. A total of 2 mL of the fermentation broth at a concentration of 25, 125, 250 and 500 μg /mL in 500 μL of methanol were mixed with 0.1 mM DPPH in methanol. Homogeny solution was incubated in dark room at 30°C for 30 min. A blank solution was made for each sample solution by mixing 500 μL of sample and 1.5 mL of methanol. As a negative control 500 μL of 0.1 mM DPPH solution was added with 1.5 mL of methanol. Absorbance was measured at of 515 nm using a spectrophotometer Hitachi U-2800.

$$\text{Inhibition} = (\text{control absorbance-sample absorbance})/\text{control absorbance}) \times 100\%$$

IC_{50} values were determined on curve of inhibitions versus concentration obtain the regression equation. From the regression equation can be determined the amount of concentration of the fermentation broth that has the ability inhibition of DPPH activity by 50% [10].

IOP Conf. Series: Earth and Environmental Science **60** (2017) 012007 doi:10.1088/1755-1315/60/1/012007

3. Result and Discussion

3.1. Optimization of fermentation process by RSM

Observation of five points for optimization using CCD namely (-α), (-), (0), (+) and (+ α) is a factor of the variable pH, glucose concentration, and the concentration of yeast extract (Table 1). The optimum results for fermentation process by RSM that of activation of antioxidants as presented in the response and the rightmost column shows the average value of experiments in duplicate (Table 2). Variant Analysis (ANOVA) for the experimental results is presented in Table 3. From the analysis of the model fit summary of the results suggested that data is a quadratic model.

Table 1. Central Composite Design.

Parameter of optimization process	(-)	(+)	(-α)	(+α)	(0)
pH	5	8	3.98	9.02	6.5
Sucrose (%)	5	10	3.3	11.7	7.5
Grape seed skin (g/L)	5	10	3.3	11.7	7.5

The F value for the model 11.63 that shows the model is significant at 95% confidence level was showed in Table 3 whereas value of p (Prob> F) models was less than 0.05 (<0.0001) indicating that the overall variables was significant. In other words, variables in the model show significant effect on the response. In the above analysis, variables A, B, C, A2, B2, C2, AB, AC, BC were significant variables with p (Prob> F) values of less than 0.05. Meanwhile, AC and BC did not significantly affect towards antioxidant activity response, but still included in the model to get the model hierarchy. Mathematical equation model was presented in the equation I.1.

Table 2. Optimization of fermentation process by RSM.

Experiment	A:Sucrose (%)	B: grape seed skin (g/L)	C: pH	antioxidant (mg/mL)
1	10	10	8	9.44
2	7.5	7.5	6.5	2.65
3	7.5	7.5	6.5	2.52
4	7.5	3.3	6.5	2.18
5	7.5	7.5	3.98	3.76
6	7.5	7.5	6.5	2.33
7	5	5	8	4.82
8	7.5	7.5	6.5	2.77
9	**5**	**5**	**5**	**1.62**
10	7.5	7.5	6.5	2.72
11	11.7	7.5	6.5	8.53
12	7.5	7.5	6.5	2.68
13	5	10	8	3.39
14	3.3	7.5	6.5	3.05
15	7.5	11.7	6.5	1.87
16	7.5	7.5	9.02	4.55
17	10	10	5	9.12
18	5	10	5	1.86
19	10	5	5	1.95
20	10	5	8	4.31

The resulted models need to be examined its adequacy as well as the normal probability plots and residual plots. The normal probability plots of residuals and residuals show that in general are in a straight line, which means errors are normally distributed (Figure 1.a.). On the other hand, Figure 1.b. shows that the residuals do not show a specific pattern (random). It means the model becomes accurate for predicting the production of antioxidants to review pH, concentration of sucrose and concentration of the seeds and skins of grape.

Table 3. Varian analysis (ANOVA) for quadratic model by response of flavonoid antioxidant activities.

Variable	*Sum of Squares*	DF	*Mean Square*	*F Value*	*Prob > F*
Model	102.20	9	11.36	11.63	0.0003
A	36.56	1	36.56	37.44	0.0001
B	8.21	1	8.21	8.41	0.0159
C	5.59	1	5.59	5.73	0.0376
A2	21.70	1	21.70	22.22	0.0008
B2	0.16	1	0.16	0.16	0.6980
C2	6.07	1	6.07	6.22	0.0318
AB	22.75	1	22.75	23.29	0.0007
AC	0.53	1	0.53	0.54	0.4802
BC	1.72	1	1.72	1.76	0.2139
Residual	9.77	10	0.98		

Antioxidant activity by analysis of variance for Response Surface Quadratic Model is
Antioxidant = 20.70124 – 3.86997 A – 0.65996 B – 1.88367 C + 0.19634 A^2 – 0.016638 B^2 + 0.28848 C^2 + 0.26980 AB – 0.068333 AC – 0.12367 BC..(I.1)

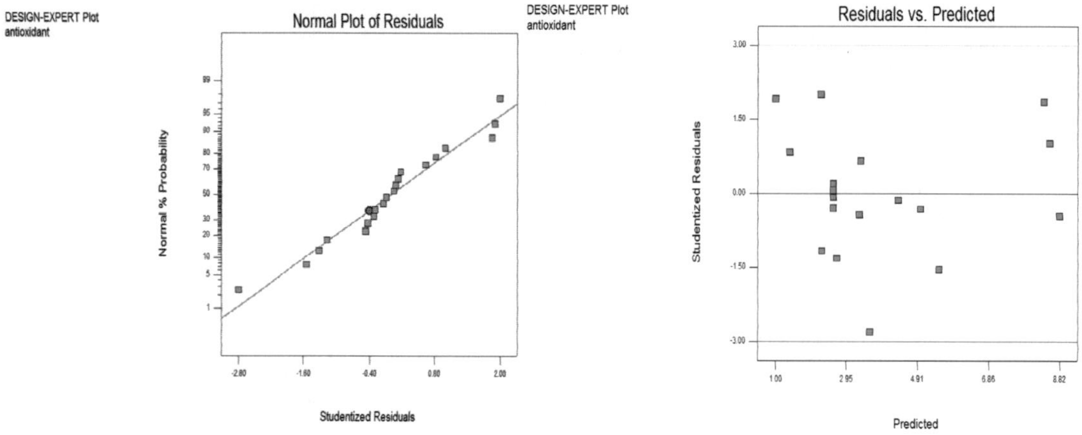

a. b.
b.

Figure 1. a. The normal probability for antioxidant activities.
b. Response for residuals and predicted for antioxidant activities.

The effect of pH and the concentration of sucrose against the production of antioxidants at concentrations of grape seed skin is presented in Figure 2.b. Antioxidant activity increased as rising pH to reach maximum activity. Optimum pH production for antioxidants was of 5. Based on ANOVA analysis, pH was a variable that influences significantly for antioxidants production. These results indicated that *S. cerevisiae* produces antioxidants in accordance with pH condition.

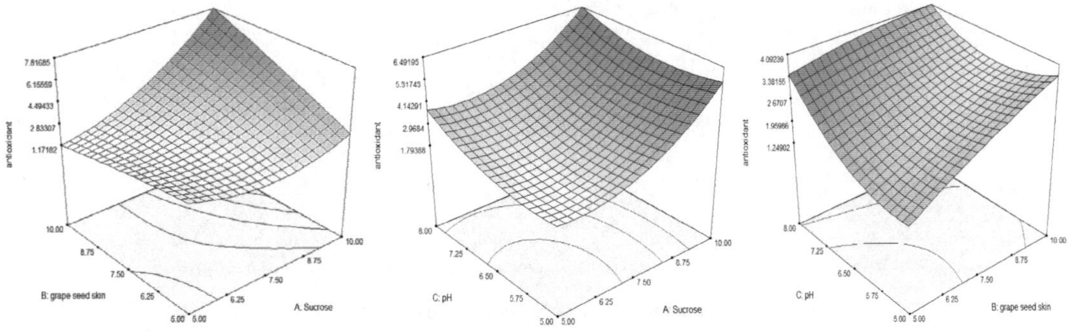

a. b. c.

Figure 2.Three dimension graphic of the relation among sucrose 5%, grape seeds and skin, and pH 5.
a. Three dimensions relation between sucrose and grape seed skin at pH 5
b. Three dimensions relation between sucrose and pH at 5 g/L grape seed skin
c. Three dimensions relation between grape seed skin and pH at 5% sucrose

The range of pH for *S. cerevisiae to* grows well is quite wide, of which its optimum around pH 5 [1], whereas others reported that the growth range *S. cerevisiae* was in the range 25-30°C and at pH of 5 [1,8]. Moreover, effect of sucrose and grape seeds and skins is shown in Figure 3.a. Optimum condition for antioxidant was achieved 5% of sucrose and of 5 g/L grape seed skin. Sucrose and grape seeds and skins are considered to represent carbon, and nitrogen of which provide a positive influence on the production of antioxidants [1,8]. Checking the accuracy for the optimum conditions generated by the software Design Expert was carried out through validation. Model validation was based on optimization as presented in Table 4. The tests showed that the predicted results and the experimental results differ greatly so the model is validated.

Table 4. RSM model validation between prediction and experiment.

Sucrose (%)	pH	Grape seed skin (g/L)	Antioxidant - prediction	Antioxidant - validation
5.07	5	5	1.70	1.62

3.2. Antioxidant analysis

The antioxidant activity of fermented liquid determined by DPPH. DPPH is purple free radical molecule. Antioxidant molecules will donate one hydrogen atom to be stable DPPH that marked with purple color changes to yellow. The antioxidant activity usually expressed as the concentration, which caused a loss of 50% DPPH activity.

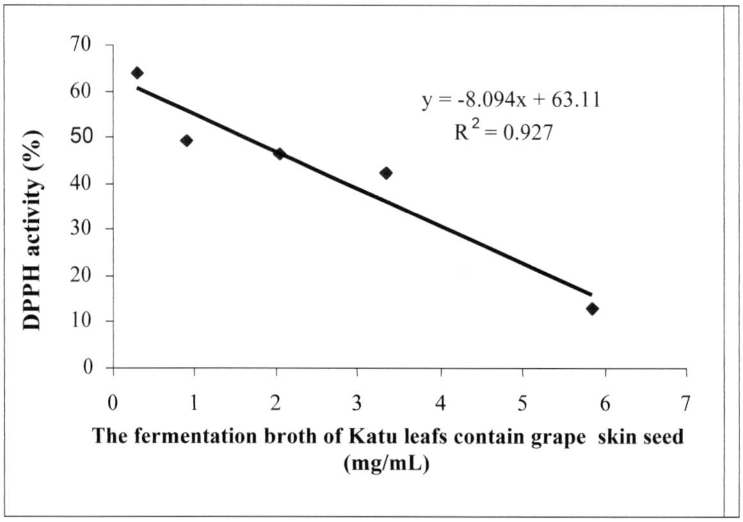

Figure 3. A regression equation to calculate the IC_{50} values by the broth fermentation in variation of 5g/L grape seed skin, 5% sucrose and pH 5.

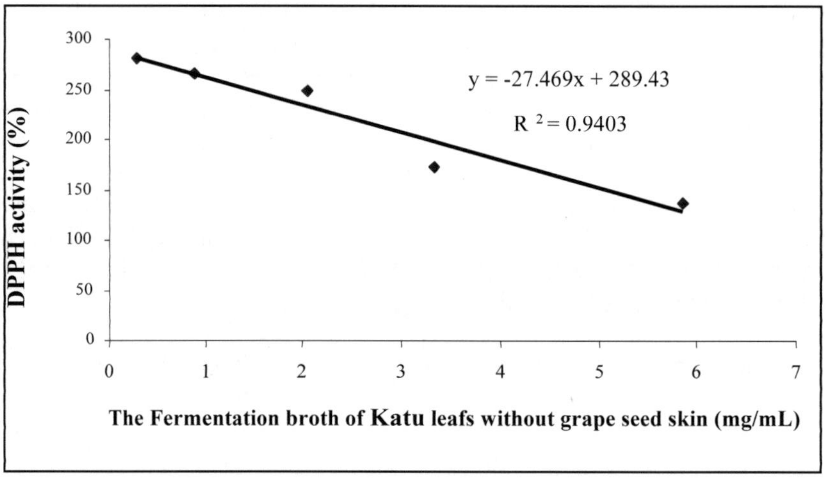

Figure 4. A regression equation to calculate the IC_{50} values by the broth fermentation in variation of 5% sucrose and pH 5.

Figure 3 and 4 show the DPPH radical activity decreases with increasing concentration of active ingredient. The optimum antioxidant activity containing 5 g/L grape seed skin, 5% sucrose and pH 5 with IC_{50} value was of 1.62 mg/mL whereas the fermentation broth that not contains grape seed skin was 8.72 mg/mL. The smaller the IC_{50} value, the greater the antioxidant activity in a test sample in reducing free radicals.

4. Conclusion

Antioxidant has been successfully produced from skins and seeds of grape. Antioxidant activity with IC50 value was of 1.62 μg/mL, increased 5 times higher than that of without skins and seed of grape.

Acknowledgement
The research was funded by DIPA of Indonesian Institute of Sciences (LIPI), Research Unit for Clean Technology, 2016.

5. References

[1] Abbas C A 2006 *Production of Antioxidants, Aromas, Colours, Flavour,and Vitamins by Yeast.* (Berlin: Springer-Verlag) pp 285-334
[2] www.FAO.org. 2015 *Vegetable specific processing technologies*
[3] Wang Y, Chen S, and Yu O 2011 *Appl. Microb. Biotechnol.* **91** 949
[4] Panda B P, Javed S, Ali M 2010 *Food and Bioproc. Technol.* **3** 373–378
[5] Miller HF, Rigelhof F, MarquartAP, and Kanter M 2000 *J. of The American Coll. of Nutr.* **19** 3125-3195
[6] Montgomery D 2009 *Design and Analysis of Experiments*, 5th Ed, (New York: John Wiley & Sons Inc.)
[7] Wu Y, Chen Q, Ruan H, and Guoqing H 2013 *Adv. J. of Food Scie. and Technol.* **5** 217-224.
[8] Jeppsson H, Yu S, and Gerdal H B 1996 *Appl. Environ. Microb.* **62** 1705–1709
[9] YaoweiFS A, Shu L, Shujun W, Mingsheng L, and Yuliang J 2013 *Carbohy. Polym.* **98** 1377–1382
[10] Molyneux P 2004 *Songklanakarin J. Sci.Technol.* **26** 211-219

IOP Conf. Series: Earth and Environmental Science **60** (2017) 012008 doi:10.1088/1755-1315/60/1/012008

Utilization of sulphurized palm oil as cutting fluid base oil for broaching process

Sukirno* **and Y R Ningsih**
Department of Chemical Engineering, University of Indonesia, Depok, Indonesia
*E-mail: sukirnos@che.ui.ac.id

Abstract. Broaching is one of the most severe metal cutting operation that requires the use of cutting fluids formulated with extreme pressure (EP) additives to minimize metal-to-metal contact and improve tool life. Enhancement of EP performances of the cutting fluids can be achieved by addition of sulphur containing compounds that will allow the formation of metal sulfide film that has low shear strength and good antiweld properties and acts as protection layer from wear and seizure. Most of the cutting fluids are mineral oil based. However, as regards to health and environmental issues, reseach on vegetable oil based cutting fluid have been increased recently. This paper reports a study on the sulphurization of palm oil derivatives and its usage as broaching oil. Sulphurization of the palm oil derivative was conducted via non-catalytic sulphurization using elemental sulphur at various composition and under heating of 150-160°C for 3 hr. Broaching oil was made by blending the sulphurized palm oil and additive packages. The performance parameters of the broaching oil that has been observed including load carrying capacity, wear scar diameter, corrosion protection, oxidative stability, and surface finish of workpiece. From this research, it was found that sulphurized FAME based broaching oil has excellent EP properties. The optimum formulation was obtained on composition of sulphurized FAME-mineral oil with 6% wt of sulphur. The result from the test showed that kinematic viscosity of sulphurized palm oil was about 25.3 cSt (at 40 °C), load carrying capacity was 400 kg_f, and wear scar diameter was 0.407 mm. In addition, it can be concluded that the class of corrosion protection of modified palm oil was 1.b (slight tarnish category), oxidative stability at 160 °C was obtained for 0.11 hr, and the surface roughness of workpiece was about 0.0418-0.0579 μm. These performances are comparable to commercial broaching oil. By this result, it indicates that sulphurized palm oil is applicable for industrial cutting fluids formulation.

1. Introduction

Metal working fluids consumption is increasing as the consequence of machining industry development. Kline & Company Market Analysis (2012) has reported that 38 million tones of lubricants were globally used and 2.2 million tones of it were allocated for metalworking fluid requirements. Asia pacific was the largest market of metalworking fluid with 42% of total world's consumption. North America, Europe, and another region contributed with 28%, 26%, and 4% of total world's consumption, respectively [1]. Metal cutting in manufacturing industry requires cutting fluids for lubricating and cooling the contact surfaces between the cutting tools and the workpieces, so it will generate a product with good quality and good surface texture [2].

Broaching is one of the metal cutting processes where a toothed tool called a broach is run linearly or rotary and pressed into a surface of the workpiece to effect the cut and to remove material.

Broaching process is categorized as a heavy machining process because it works in extreme pressure and relatively high temperature [3]. The heavy machining processes generally require middle or heavy cutting oil, such as neat cutting oils that contain 90% base oil and have better EP performances [4]. Broaching process requires a cutting fluid that contains highly active sulphur. Most of broaching oils is formulated by blending a low viscosity of mineral base oils, i.e paraffinic, and a highly contained sulphur additive [5]. Since it contains mineral oil, the spills dan the waste oil of the broaching oils are considered as pollutant that can possibly contaminate soil and water. Moreover, directly disposal of the waste oils to the environment is environmentally unsafe [6].

The utilization of vegetable oils as the green raw material for cutting fluid application has been studied by several researchers. One of the main objective research is to increase the poor thermo-oxidative stability properties of the vegetable oil due to the presence of bisallylic protons at its structure, to meet the stability requirement during various tribochemical processes. Another objective is to pursue a cutting fluid that has excellent boundary lubricating properties through strong physical and chemical adsorption with the metal surface.

Elimination of the bisallylic protons at the chemical structure of the vegetable oil by incorporating sulphur element into the structure of vegetable oil is considered to be able to improve EP performances and also thermo-oxidative stability. The polar group of sulphur is known as active site of molecules that can increase its adsorption on the metal surfaces to form protection layer [7]. EP performances of vegetable oil may also be obtained by the incorporation of phosphate, chlorine, nitrogen and zinc, however, compounds or additives that contain those elements have been considered as the cause of numerous health and environmental concerns [7,8,9]. Meanwhile, sulphur contaning compounds or additives are well-known for their excellent antiwear properties and EP characteristics, low toxicity, and categorized as a non-carcinogenic additive [9,10].

Incorporation of sulphur to the oils can be done by sulphurization method such as sulphur flower reaction (with and without H_2S, aminic, and other suitable catalysts), sulphur chlorination with S_2Cl_2, organic halides with alkalipolysulfides, and mercaptan route. Today's large-scale production technology in general avoids any halogen-containing reaction steps because there are low limits (max 30 ppm chlorine) in the final lubricants that may not be exceeded [11]. Black sulphurization is the simplest sulphurization method. The manufacturing equipment needs to withstand pressure above 1–2 bars (it may even be pressureless).

In this research, palm and its derivatives are sulphurized with sulphur flower/elemental sulphur. The mixture of oil and elemental sulphur is heated above the melting point of sulphur. An uncatalyzed reaction starts to become exothermic above 150–160°C, with the evolution of substantial amounts of H_2S [11]. This research was aimed to produce sulphurized oil that will be use for formulating broaching oil with maximum viscosity of 107 cSt at 40°C [12] using sulphurized palm oil for steel-made process machines. For broaching process, the desire viscosity of broaching oil is low to allow a good penetration of lubricants. This process will lead into the elimination of chips and contaminants from workpiece and working tools rapidly by fluid current. The high level of viscosity will decrease the efficiency of working tools due to the trapped chips [4]. The formulation of broaching oil include the addition of additive packages, such as antioxidant and anticorrosive additives, into the sulphurized oil and also the blending with other oil. The characterization of sulphurized oil are viscosity, sulphur contents, and the spectrum of FTIR. Performance tests for the formulated broaching oil to workpiece and worktool include EP performance, surface finish, corrosion protection level, and thermo-oxidative stability. Furthermore, these performance tests were also done to a broaching oil of commercial products as comparison.

2. Experimental

This research consists of three main parts: preparation of base oil, synthesis of broaching oil, and performance test of broaching oil. The process flow diagram of the research is shown in the Figure 1.

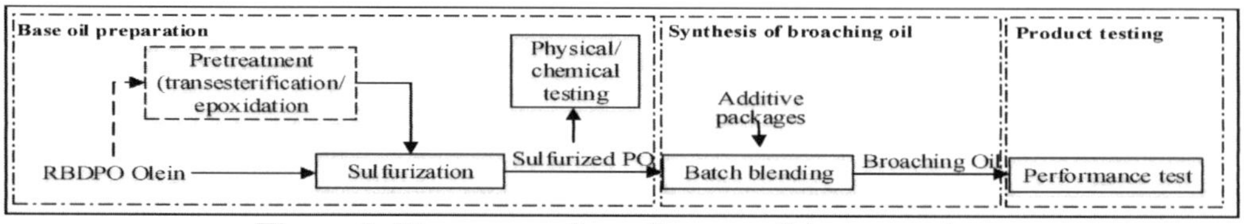

Figure 1. The process flow diagram of the research.

2.1 Base oil preparation and synthesis of broaching oil
Base oil are the products of sulphurization process of RBDPO olein, epoxydized of RBDO olein, FAME and epoxydized FAME. The FAME was prepared by process of esterification of RBDPO olein using methanol and NaOH as catalyst. Epoxydized of RBDO olein and epoxydized FAME were obtain from epoxydation reaction using oxidator hydrogen peroxide formic acid as the catalyst. The sulphurizations were conducted using elemental sulphur of technical grade, which was added into the oil, heated at 150-160°C, stirred at 500 rpm for 3 hr. The condition of temperature, time, and speed rate were remained constant while the quantity of the elemental sulphur was varied at 2, 4, 6, 8, and 10% wt, respectively. The desire sulphurized oil was the high sulphur content with viscosity less than 86 cSt which was considered as the proper viscosity for broaching oil. The selected sulphurized oil was then used in formulation of broaching oils by blending it with additive package which contain sodium tolytriazole as anticorrosion and butylated hydroxy toluen as antioxidant. Mineral base of parafinik HVI 60 was also used as base oil for the purpose of comparison.

2.2 Performance test of the product
Each of the sulphurized oils was measured its viscosity by ASTM D-445 method and the sulphur content by ASTM D-2622 method. Fourier transformation infra-red (FTIR) analysis was carried out to examine the change of functional group of the oil's chemical structure. To examine the degree of unsaturation of the oils was determined by iodine value measurement.

Several parameters of broaching oil performance were examined to obtain the quality of broaching oil. The examinations of EP performance of the broaching oil were conducted by 4-ball test for obtaining wear scar diameter (ASTM D-4172 method) and load carrying capacity-weld point (ASTM D-2783 method). The examination of workpiece's surface finish was carried out by surface roughness tester tool kit. Morphology of the workpiece by scanning electron microscopy (SEM) equipped with energy dispersive x-ray (EDX). Corrosion protection level performance was analyzed by copper strip corrosion test (ASTM D-130), and oxidative stability performance was carried out by induction time measurement with Rancimat-892.

3. Result and discussion
3.1. Sulphurized palm oil characteristic
Sulphurization of the palm oil and its derivatives including RBDPO olein, epoxydized olein, FAME, and EFAME was darkened their color from goldish yellow into brown-blackish proprotional with the addition of sulphur. The rise of sulphur content also caused the increase of the viscosity of the sulphurized palm oil products. Sulphur incorporation of various amount into olein, epoxydized olein, FAME, and EFAME results in various viscosity of sulphurized palm oil as shown in Figure 2. It is apparent that the highest sulphur content with the lowest viscosity can only be obtained by sulphurized FAME. The sulphurization with up to 10 wt% of sulphur content resulted in the sulphurized FAME with viscosity of 33 cSt.

To function as a broaching oil, its viscosity l has to be remained low to allow a good penetration of lubricants. This process will lead into the elimination of chips and contaminants from workpiece and working tools rapidly by fluid current. The high level of viscosity will decrease the efficiency of

IOP Conf. Series: Earth and Environmental Science **60** (2017) 012008 doi:10.1088/1755-1315/60/1/012008

working tools due to the trapped chips [4]. Base on viscosity consideration, this sulphurized FAME still meet requirement of broaching oil viscosity for effective operation. Another configuration of sulphurized products, especially the sulphurized olein and sulphurized epoxy olein have too high viscosity for effective broaching oil

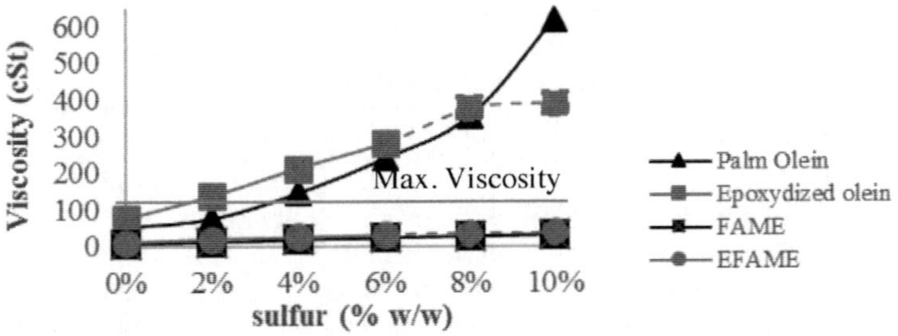

Figure 2. Various viscosity obtained from sulphur incorporation into several palm oil configuration.

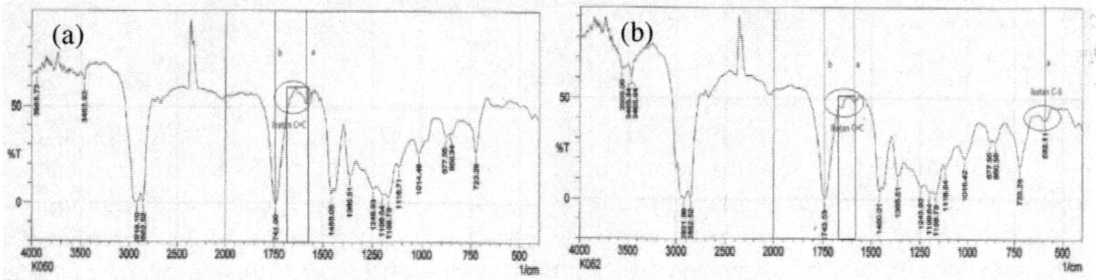

Figure 3. Fourier transform infrared spectrum of FAME: (a) before sulphurization and (b) after sulphurization.

Figure 3 shows the FTIR spectrum of FAME before and after sulphurization. The sulphurized FAME is characterized by the appearance of C-S group (592 cm^{-1}). This confirming the incorporation of sulphur into FAME at the reactive site C=C group. Therefore, theoritically the peak height of C=C group (1620-1680 cm^{-1}) should decrease or disappear after sulphurization. From the spectrum, the ratio peak height of C=C to peak height of C=O (1750 cm^{-1}) after sulphurization is lower than before sulphurization, but C=C peak does not disappear after sulphurization. This fact implied that not all C=C group had been replaced by C-S group. This is in agreement with the result of iodine value measurement shown on table 1, where the iodine value is decrease with increasing the sulphur content, but never reach value of zero

Table 1. Quantitative analysis result of sulphurized FAME.

Theoritical Composition		Actual Analysis Result		
FAME (% wt)	S Content (%wt)	S Content (%wt)	Viscosity (cSt)	Iodine Value (g Iod/100 g)
100	0	0	5.7	47.1
98	2	1.9	11.0	42.6
94	6	6.0	23.4	31.8
90	10	10.0	32.6	30.8
85	15	14.1	39.7	25.9
80	20	18.1	46.3	25.0

3.2 The performance of formulated broaching oil

The formulated broaching oil sample are mixture of sulphurized FAME containing sulphur 6% wt (S FAME 6) plus additive package, mixture of sulphurized FAME containing sulphur 10% wt (S FAME 10), mixture of sulphurized FAME containing 6% wt sulphur plus mineral base oil (paraffinic HVI 60)

IOP Conf. Series: Earth and Environmental Science **60** (2017) 012008 doi:10.1088/1755-1315/60/1/012008

as combined base oil (S FAME+Min) plus additive package. Commercial product of broaching oil that used mineral oil as base oil (Min Commercial), was used as the comparison. Table 2 shows the matrix of performance test.

Table 2. Matrix of broaching oil performance test.

Configuration of Product	Parameter of Performance Test			
	Extreme Pressure	Surface Finish	Corrosion Protection Level	Oxidative Stability
S FAME 6	4 Ball-scar diameter 4 Ball-weld point	Surface rouhgness Workpiece temperature	Copper strip corrossion	Induction time
S FAME 10	4 Ball-scar diameter 4 Ball-weld point	Surface rouhgness Workpiece temperature	Copper strip corrossion	Induction time
S FAME+Min	x x	Surface rouhgness Workpiece temperature	Copper strip corrossion	Induction time
Min Commercial	4 Ball-scar diameter 4 Ball-weld point	Surface rouhgness Workpiece temperature	Copper strip corrossion	Induction time

3.2.1 *Extreme pressure (EP) performance test.*

The EP performance of the formulated broaching oil sample were assesed by the wear scar diameter and weld point obtained from 4-ball test method. EP performance describes the ability of broaching oil to protect surface by avoiding metal-to-metal contact between workpiece and working tool.

As shown in Figure 4, the EP performance of S FAME 6 resembled the Min Commercial with the average scar diameter value of 0,407 and 0,384 mm, respectively, while the scar diameter value of S FAME 10 was at 0,689 mm.

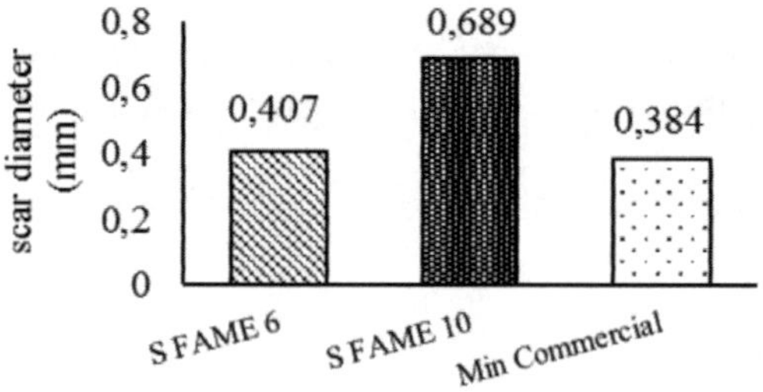

Figure 4. The comparison of wear scar diameter between the samples.

Broaching oil sample using S FAME 10 gave the highest wear scar diameter, while the lowest wear scar diameter is mineral base broaching oil of Min Commercial. This results agree with the experiment's result reported that higher sulphur content increased the load carrying capacity decreased the wear protection level of broaching oil [13]. The exessive sulphur content of broaching oil will accelerate both the wear scar rate and corrosion rate [14]. Another factor considered to have contribution to the rate of wear (scar diameter) is oxygen containing compounds in the broaching oil. Wear scar diameter reflects the oxidation rate during the examination. Oxygen containing compounds contribute the oxidation of the ball's surface. This process cause the surface to brittle. Thus the wear scar rate becomes higher. Mineral based broaching oil does not have O bonds on its structure. Therefore, the oxidation can be minimized [15].

The morphology of scar diameter from the samples is shown in Figure 5. It shows that the formation of scar are generally similar, but the diameter size are different. The wear scars are affected

IOP Conf. Series: Earth and Environmental Science **60** (2017) 012008 doi:10.1088/1755-1315/60/1/012008

by the chemical reaction between the metal and sulphur or so called corrosive wear effect. The excessive sulphur content and the metal-to-metal friction will trigger the corrosion on the metal surfaces, leading to the brittleness of the suraface. The most common corrosive compounds that generally attack metals are hydrogen sulfide (H_2S), sulphur dioxide (SO_2), chlorine (Cl_2), nitrogen dioxide (NO_2), as well as oxygen (O_2) [16]. Among those elements, sulphur is the most corrosive compound. The result shows that the commercial broaching oil may use EP additives of non-sulphur compound, allowing the product to form the lowest wear scar diameter.

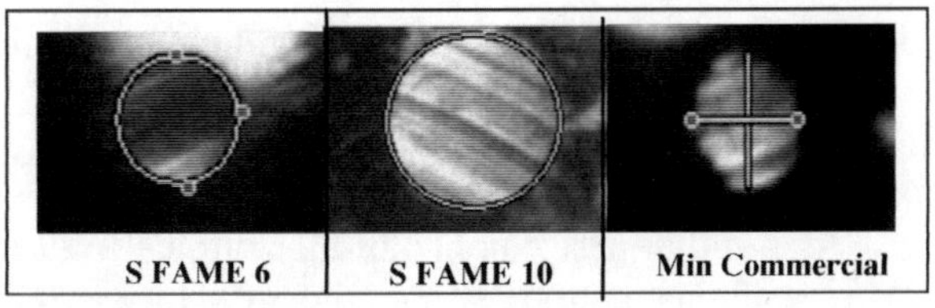

Figure 5. The comparison of wear scar diameter and morphology among the samples.

Load carrying capacity performance test was carried out by weld point examination. The higher the weld point, the better the load carrying capacity of the lubricants. Figure 6 shows the picture of weld phenomena of the lower balls of 4-ball tester after the weld test. Figure 7 depicts the result of weld point of broaching oil sample whereas the sample of S FAME 10 shows the highest weld point load of 800 kg_f. On the other hand, both sample of S FAME 6 and Min Commercial has the same value of 400 kg_f. The best load carrying capacity performance was obtained by broaching oil having highest sulphur content (10% wt sulphur), since the formation of salt-metal layer sulphur containing compound was thought to enhance the anti-weld properties and reduce the friction coefficient. The oil with a high load carrying capacity would be applicable for heavy duty metal cutting process.

Figure 6. Weld phenomena.

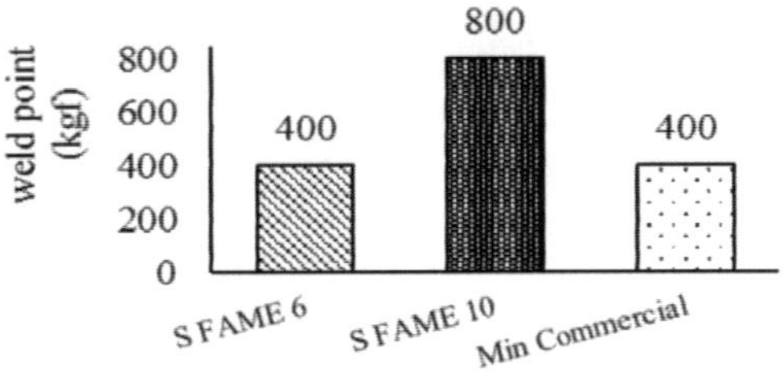

Figure 7. Weld point of the samples.

3.2.2 Surface finish of workpiece performance

The performance of surface finish can be observed from the surface roughness, the cooling level of lubricated workpiece, and the morphology of workpiece after cutting process. The surface roughness is an arithmetic average (R_a) of the peak height and the valley depth at the surface profile, commonly expressed in unit μm. Figure 8 shows the roughness of the workpieces after cutting process at certain cutting speed.

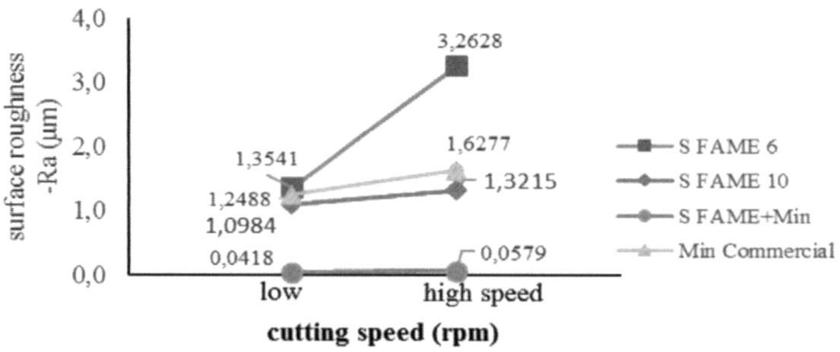

Figure 8. Surface roughness profile of workpiece compared to cutting speed level.

After heavy duty machining process, the surface roughness of workpiece increases as the cutting speed rises. In general, broaching operation at lower lower speed cutting speed would result in better surface finish quality. Figure 8 shows that performance of all samples were relatively the same on the lower cutting speed. However, at higher cutting speed, the sample wf S FAME 6 has the poorest performance. The finest surface roughness value was given by sample of S FAME+Min at both lower and higher cutting speed. These results demonstrated that cutting speed level has a significant influence on the heavy duty metal cutting process, where the lower cutting speed level will generate better surface roughness value than the higher cutting speed level. The surface finish is affected by lubricating properties and cooling performance of the cutting oil. The poor lubricating properties of cutting fluid will accelerate wear of the working tool, thus resulting in the low quality of product [4,17]. Cooling performance of oil is needed to lower temperature or remove heat generated when cutting at high speed. Figure 9 shows the workpiece final temperature during cutting process plotted versus cutting speed level.

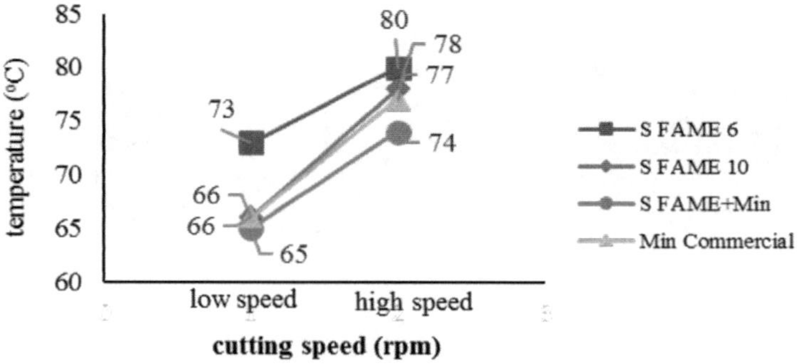

Figure 9. The final temperature profile of workpiece compared to the cutting speed level.

It is clearly seen from Figure 9 that the higher cutting speed level will increase the temperature of workpiece. Most of the sample show the final temperature are about the same in the range of 75-77 C^{o}, with the lower temperature given by the sample of S FAME+Min. It concluded that the surface finish performance of the formulated broaching oil resembles the commercial product.

Micro-structure observation of the surface of the cut workpiece by SEM is shown in Figure 10. Samples of S FAME 10 and Min Commercial shows relatively deep crater-like scratch lines. The crater-like scratch on the surface of S FAME 10 sample is a bit deeper than that of Min Commercial sample. This is might be due to the effect of corrosive wear stimulated by the excessive sulphur of the S FAME 10 sample.

IOP Conf. Series: Earth and Environmental Science **60** (2017) 012008 doi:10.1088/1755-1315/60/1/012008

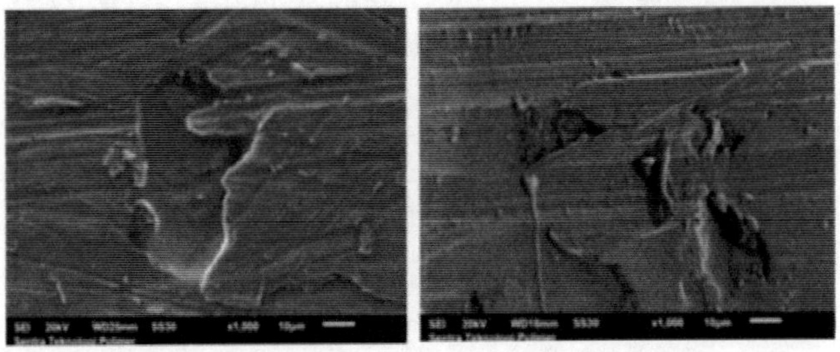

a) b)

Figure 10. Morphology of the workpieces lubricated (1000x magnification)
(a) Formulated broaching oil (S FAME 10) and
(b) Commercial broaching oil (Min Commercial)

To evaluate the elemental composition of the workpiece cut surface, the EDX analysis was done. The results of EDX analysis are shown in Figure 11 and the summary is available in Table 3.

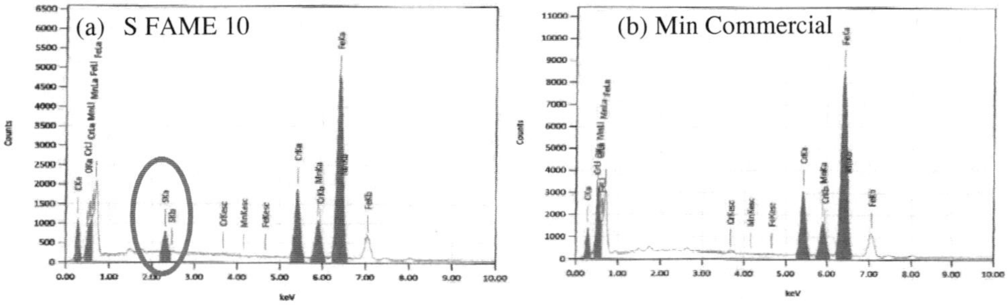

Figure 11. EDX spectra of the workpieces lubricated by (a) formulated broaching oil (S FAME 10) and (b) commercial broaching oil (Min Commercial).

Table 3. EDX results of cutted workpiece.

	Elemental composition (wt%)							
	C	O	Al	S	Cr	Mn	Fe	Mo
Formulated broaching oil (10% w/w S)	14.82 ± 5.08	2.02 ± 0.57	0.3 ± 0.53	0.53 ± 0.91	12.82 ± 0.83	7.22 ±0.53	60.13 ± 4.22	2.16 ±2.16
Commercial broaching oil	6.83 ± 4.37	6.17 ± 4.39	-	-	13.21 ± 1.31	7.34 ±1.06	66.46 ±5.87	-

As shown in Table 3, elemental sulphur is found on the cut workpiece surface lubricated by the S FAME 10 oil. On the other hand, it is not found in the surface lubricated in Min Commercial oil. This result confirmed that the sulphur containing compound in the formulated broaching oil adsorbed in the surface of workpiece and the adsorbed sulphur upgraded the surfcace's load carrying capacity performance whereas, Min Commercial oil contain non-sulphur additives.

3.2.3 Corrosion protection level performance

Elemental sulphur of sulphur EP additives will degraded and form sulfide metals on boundary lubrication condition, when the rise of local temperature is high [18]. Under boundary lubrication, the metal-to-metal contact between the working tool and the workpiece increase the temperature around the surface, leading to the degradation of sulphur containing compound and the formation of highly reactive metal sulfide. The sulphur will be highly active and then be able to stimulate metal corrosion, color alteration, and layer formation on the surface of metals, especially the yellow metals, such as copper and its alloys. Table 4 shows the corrosion test result of copper strip that lubricated by various broaching oils.

Table 4. Broaching oil corrosivity level.

Property	S FAME 6	S FAME 10	S FAME+Min	Min Commercial
Corrosivity level	1.b (dark orange)	4.a (dark grey)	1.b (dark orange)	1.a (light orange)

As shown in Table 4, S FAME 10 sample is the most corrosive broaching oil, as it caused the sample strip to darken into class 4.a, a corrosion category that is equal to dark grey-colored copper strip. On the other hand, the corrosivity of S FAME 6 and S FAME+Min are class 1, equal to dark orange-colored copper strip. The Min Commercial is in class 1.a, slight tarnish with the light orange-colored copper strip. Except the S FAME 10, the 3 samples meet the requirements of lubricant corrosivity level for general use at maximum class 2.a, moderate tarnish category with the red wine-colored copper strip [19].

The activity of sulphur depends on the sulphur chain in the molecules, where the activity of monosulfide and disulfide are moderate to soft metals than the pentasulfides [11]. Sulphurization of FAME would possibly produce the sulphur bridges structure (polysulfides) that will affect the corrosion protection level performance of yellow metals. The elemental sulphur will be more effective on the extreme pressure condition where the high corrosive metals do not exist. The higher content of sulphur EP additives in the lubricants will generate the highly active type of sulphur as same as the elemental sulphur [20]. The higher the ontent of sulphur in S FAME 10 originated from elemental sulphur will make it as one of the sulphur-active type lubricants. Therefore, the formulation of broaching oil that exceeds 6% wt is not recommended for yellow metal-contained cutting process as it will cause corrosion on the workpiece.

3.2.4 Oxidative stability performance.

Oxidative stability properties are essential to the product's storage and handling for it relates to the shelf life and the usage of broaching oil commercially. The results of oxidative stability performance test are shown in Figure 12.

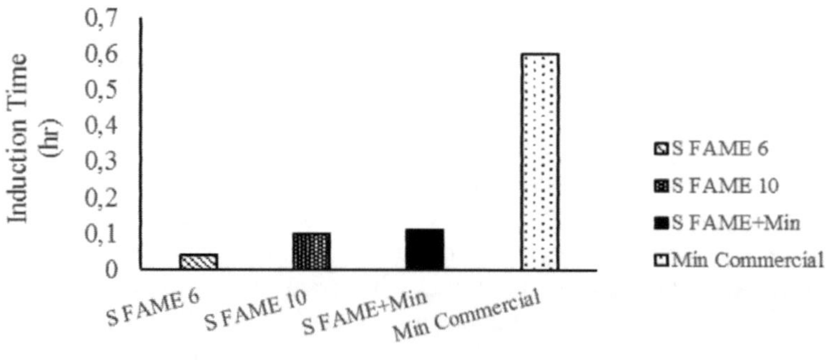

Figure 12. The oxidative stability.

It is clearly showed in Figure 12 that the oxidative stability of the formulated broaching oils are inferior to the Min Commercial broaching oil, with the their inducton time is still far shorter than that of the Min Commersial oil. These results indicate that the oxidative stability problem of the green oil based lubricants has not been overcome yet. This is might be due to the partial conversion of bisallylic protons on the sulphurized lubricant could not be carried out completely. This uncomplete convertion of C=C was confirmed by the result of iodine value that is still in the range of 25–30 g/100 g sample, even with the sulphur addition of 10% wt. Mineral oil based lubricants has more thermo-oxidative stability performance at elevated temperature compared to the green oil based lubricants, as it contains less unsaturated than the green base oil lubricants which contains unsaturated and polar group, such as C=C group, carbonyl group (C=O) and hydroxyl group (O-H) that are susceptible to oxidation.

From Figure 12, it is also interesting to notice that the variation of sulphur in the samples generates various oxidative stability, where the higher sulphur content of broaching oil S FAME shows better oxidative stability compared to the lower sulphur content of broaching oil S FAME 6. This might be due to the more conversion of C=C bonds to C-S bonds or S-S bonds, the higher the stability of the oil. However, excessive sulphur can also attack antioxidant agent in the lubricants [21]. Therefore, the optimum amount of sulphur that will not inhibit antioxidant performance should be passed over. The antioxidants have a good synergy with the sulphurized products that contain inactive type of sulphur. Meanwhile, a long chains of sulphur bridges in the polysulfides are thermally less stable than short sulphur bridges, where sulphur is linked to the carbon atom of the raw material [11]. From the test results, it might be possible that the polysulfide chains formed as a result of sulphurization reaction, and leading to the poor oxidative stability performance.

4. Conclusion

From the result obtained, it concluded that sulphurized FAME from palm oil is the best for broaching oil formulation with low viscosity and high EP performance or high content of sulphur. Addition of sulphur into palm oil or its derivatives by sulphurization produced sulphurized FAME containing more than 6%wt of sulphur. The higher the sulphur content the better the EP performance of the sulphurized products. Excessive sulphur content cause corrosion and britleness of metal surface and accelate wear rate. Broaching oil formulated by sulphurized FAME containing 6%wt sulphur and mineral oil EP additive showed the EP performance as good as the broaching oil of commercial product.

Acknowledgement

We acknowledge to Departement of Chemical Engineering, Faculty of Enginering, University of Indonesia for financial support for the research.

References

[1] Phadke M. 2014 *Tribol. Lubr. Technol.* **70**(3):38-40
[2] Diniz A E and Micaroni R 2007 *Int. J. Mach. Tools Manuf.* **47**: 247-254
[3] Walsh R A and Cormier D 2006 *McGraw-Hill Machining and Metalworking Handbook* (United States of America: The McGraw-Hill Companies, Inc.)
[4] Trent E M and Wright P K 2000 *Metal Cutting 4th Ed* (London: Butterworths & Co Ltd) p. 311-337
[5] Kammann K P and Phillips A I 1985 *J. Am. Oil Chem. Soc.* **62**: 917–923
[6] Pop L, Puscas C, Bandur G, Vlase G and Nutiu R 2008 *J. Am. Oil Chem. Soc.* **85**: 71
[7] Sharma B K, Adhvaryu A and Erhan S Z 2009 *Tribol. Int.* **42**: 353-358
[8] Canter N 2007 *Tribol. Lubr. Technol.* **63**(9): 10-18
[9] Datta S, Bhatnagar P, Saxena D, Mooken R T, Naithani K P, Malhotra R K and Kumar A 2010 *Proc. Int. Conf. on Petrotech* (India) Paper ID: 20100991
[10] Kumar B S, Padmanabhan G and Krishna P V 2015 *J. Manuf. Sci. Technol.* **3**(1): 1-9

[11] Rossrucker and Fessenbecker 2009 *Sulphur Carrier : Lubricant Additives Chemistry and Applications* 2nd Edition L.R Rudnick (New York: CRC Press) Chapter 9 pp 251-269
[12] Oberg E, Jones F, Horton and Ryffel H 2012 *Machinery's Handbook 29th ed.* (New York: Industrial Press) pp 1185 – 1186
[13] Marmerstein L, Gober V and Vence J 2014 *Cost Effective Formulating: Sulphurized Additives* (Ohio: Elco Corporation)
[14] Scibbe H W, Townsend D P and Aron P R 1984 *NASA Technical Paper*
[15] Syahrullail, Hariz and Hamid 2013 *Procedia Eng.* **68**: 166-171
[16] Song J, Wang L, Zibart A and Koch C 2012 *J. Metals* **2**: 450 -477
[17] Cookson J O 1977 *Tribol. Int.* **10**(1): 5-11
[18] Fibria M C, Yuliani and M. Haniffudin 2012 *Lembaran Publikasi Minyak dan Gas Bumi*. **46**(1): 1-7
[19] Costello M T 2009 *Lubricant Additives Chemistry and Applications* 2nd Edition) ed L.R Rudnick (New York: CRC Press New York) Chapter 17 pp 421-444
[20] Lincoln H B and Byrkit G D 1945 Lubricant *US Patent No. 2,386,222*
[21] Shreir L L, Jarman R A and Burstein G T 1963 *Procedia Eng*. **68**: 166-171

IOP Conf. Series: Earth and Environmental Science **60** (2017) 012009 doi:10.1088/1755-1315/60/1/012009

Production of peptone from boso fish (*Oxyeleotris marmorata*) for bacterial growth medium

S Priatni*, W Kosasih, T A Budiwati and D Ratnaningrum
Research Unit for Clean Technology, Indonesian Institute of Sciences
Jalan Sangkuriang Bandung 40135, Indonesia
*Email: sripriatni@yahoo.com

Abstract. Underutilized *Oxyeleotris marmorata* fish is abundant and widespread in Indonesia. The study aimed to use *O. marmorata* fish for peptone production using papain from dried latex of papaya fruit. The peptone was applied as nitrogen sources for bacterial growth. The resulted peptone was optimized at 50-65°C for 5-8 hr, using 0.1% of papain at pH 6.0. Characterization of peptone was based on the soluble protein content, N-amino content, % degree hydrolysis (DH), SDS PAGE profile and growth of bacteria *Escherichia coli* and *Staphylococcus aureus*. The results indicated that the optimum condition for hydrolysis was at 50°C for 7 hr ($p < 0.05$). Fish peptone soluble protein content was of 8.6 mg/mL, α-amino was 0.59%, and AN/TN 5.47%. Sodium dodecyl sulphate-polyacrylamide gel electrophoresis (SDS PAGE) profile of peptone showed a major band with molecular weight between 17-28 kDa. Fish peptone effectiveness for *E. coli* and *S. aureus* growth was similar with commercial bacterial peptone.

1. Introduction

Peptone is a water-soluble mixture of polypeptides and amino acids which are widely used in many biological and biotechnological applications. Peptone is primarily obtained from bovine or porcine origin, such as meat, internal organs, gelatin, and milk, as well as from plants and yeasts [1]. Peptone is a partially digested protein, and prepared by the enzymatic or acidic hydrolysis of proteinaceous material. The resulted hydrolysate contains secondary protein derivatives such as polypeptides, dipeptides, and amino acids. They provide a source of nitrogen which is water soluble, does not coagulate on heating, and is therefore particularly suitable for inclusion in microbiological culture media [2]. Peptone is one of the good supplements and replacement of serum for the growth of the mammalian cells and production of the biological products. The benefits of peptone supplementation in cell culture applications have been well documented for many years as it contains low molecular weight nutrients, peptides and free amino acids which are good for cell culture [3].

Fish peptone is one of the alternative approaches for protein hydrolysate production for nitrogen source in microbial growth. The use of fish is not only an eco-friendly but also as a cheaper raw materials for media formulation. The utilization of sardine and mackerel wastes from fish processing industry has been studied for peptone recovery. The recovered peptones performance in microbial growth was found to be affected by hydrolysis method [4].

Indonesian marine biodiversity provides significant biological potential and is able to be exploited for its fish, seaweed, and other biological potential. Boso fish (*Oxyeleotris marmorata*) is a widely distributed species of sleeper goby native to fresh and brackish waters in Indonesia [5]. *O. marmorata* fish was potential for peptone production because of its abundant and cheap due to it less favored for people. Indonesia still imports peptone to fulfill its demand since there is no peptone industry available in Indonesia. Imported peptone within the last two years increase by 3,296 tones with the value of USD 12,5 million in 2012 and become 5,102 tones with the value of USD 20,76 million in 2013 [6]. The study aimed to utilize *O. marmorata* fish for peptone using papain from dried latex of papaya fruit.

2. Experimental

2.1. Materials

Boso fish (*O. marmorata*) was obtained from north beach of Indramayu, Indonesia. Papaya latex was harvested directly from papaya fruit at a California papaya farm in Bandung, Indonesia. *Escherichhia coli* and *Streptococcus aureus* strains were obtained from Research Unit for Clean Technology collection. Oxoid as commercial peptone was used for comparison.

2.2. Methods

2.2.1. Enzyme preparation

Papain enzyme was prepared from papaya latex by drying it at 50°C for overnight. Enzyme activity of dried papain was characterized using casein as substrates. Yield of papain enzyme was of ±30% w/w.

2.2.2. Determination of papain activity

Papain activity was determined by modification of Afaq, S. and Iqbal, J. (2001) methods [7]. One mL of 2% casein (w/v) was mixed with 2 mL of phosphate buffer 0,1 M pH 7,0 and incubated at 50°C for 5 min. This solution was added with 0.5 mL of papain sample and incubated at 50°C for 20 min. The reaction was stopped with addition 1 mL of TCA 20% (v/v) and incubated at 50°C for 20 min. Control sample solution was prepared by the addition of 50 µl L-cystein chloride with 2 mL of phosphate buffer 0.1M, pH 7.0, 0.5 mL of papain, 1 mL of TCA 20%, and incubated at 50°C for 20 min. This solution was added with 1 mL of casein 2% and incubated at 50°C for 20 min. Tyrosine (0-450 µg/mL) was used as the standard solution. Sample and its control were centrifuged at 4°C, 6.000 rpm for 10 min. Absorbance of supernatant was measured by a spectrophotometer UV-Vis at 280 nm. The unit activity of papain enzyme was calculated based on µg tyrosine per gr of enzyme.

2.2.3. Peptone preparation

Peptone preparation was carried out by using a modification of Fahraniah et al. (2009) methods [8]. The frozen of boso fish was thawed and mixed with distilled water with a ratio 1:4. Samples were blended and pH adjusted to 6.0. The optimization of peptone production was carried out in a water bath by using 0.1% of papain at 50-65°C for 5-8 hr. Enzymatic hydrolysis process was stopped at 85°C and allowed to stand for 15 min. Peptone extract was vacuum-filtrated and filtrate was stored at -20°C. For molecular weight assay, the selected peptone extract was freeze dried.

2.2.4. Analysis of soluble protein content

Soluble protein content was analyzed by using a modification of Lowry (1951) method [9]. A 0.5 mL of peptone extract was added to a reaction tube and mixed with 5 mL of solution C (50 mL of Lowry A and 1 mL of Lowry B solution), then incubated at room temperature for 30 min. The sample solution was mixed with 0.5 mL of 1N Folin reagent. Absorbance was measured by a spectrophotometer UV-Vis at 500 nm. A series concentration of bovine serum albumin (BSA) was used for standard curve of protein.

Determination of Degree Hydrolysis

Degree hydrolysis (DH) of peptone extract was calculated using relationship between α -amino nitrogen (AN) and total nitrogen (TN) according to equation [10]:

$$\%DH = \frac{\alpha - amino\, Nitrogen\; (AN)}{Total\; Nitrogen}\; x\; 100$$

Total nitrogen was determined by Kjeldahl method [11]. α-Amino nitrogen was analyzed using a modification of Sudarmadji (1997) method [12]. Five mL of peptone extract was mixed with 10 mL of distilled water and 0.2 mL of saturated K-oxalate (K-oxalate : water = 1:3) and 5 drops of 1% phenolphtalein, then stirred the solution for 2 min. The mixture was titrated with 0.1 N NaOH to pink color and added with 2 mL of formaldehyde. Titration was continued with 0.1 NaOH to its final end-point. The concentration of α-AN was calculated using the following equation:

$$\alpha - AN\,(\%) = \frac{V}{W\,x\,10}\,x\,N\,NaoH\,x\,14.008$$

V: titration volume, W: weight of sample.

2.2.5. Statistical analysis
The data of soluble protein content and % DH was evaluated by General Linear Model with univariate analysis. Duncan's multiple range tests were used to compare the means to identify which groups were significantly different from others. Significance was defined at $P < 0.05$.

2.2.6. *Application of peptone produced as bacterial growth media*
Bacterial growth assay was determined by Rahayuningsih and Wiranti (2014) method [6]. The strains of *E. coli* and *S. aureus* were used to test the microbial growth performance. The inoculum was prepared by inoculating one loops of strains to 100 mL of Nutrient Broth (NB) media and incubated on a shaker incubator at 30°C for 18 hr. Furthermore, 1 mL of inoculum was inoculated into 100 mL of NB media contains 5% (v/v) of peptone extract. Both strains were grown at 30°C and cultivation was performed for 24 hr on a shaker incubator. The assay was compared with a commercial peptone. The growths of both microbes were monitored every 2 hr by measuring optical density at 600 nm.

2.2.7. Molecular weight analysis
Sodium dodecyl sulphate-polyacrylamide gel electrophoresis (SDS-PAGE) was done on all samples and protein marker on a discontinuous buffered system according to LaemmLi in Bollag et al. (1996) method [13]. Twenty μl of each sample was added with 40 μl of sample buffer. Samples were placed in a foam rack and placed in a beaker of boiling water for 4 min. The prepared of polyacrylamide gel was placed in an electrophoresis unit. Running buffer was filled to the upper buffer chamber of the gel until the buffer reaches halfway between the tops of the short and long glass plates. Five μl of standard protein markers and 25 μl of each samples were added to the polyacrylamide gel. Electrophoresis was conducted at a constant 200 V for 30 min. The gel was removed and placed in staining solution.

3. Results and discussion
Fish peptone has been produced by enzymatic hydrolysis process using papain enzyme from dried papaya latex. The proteolytic activity of papain was of 68.46 ± 4.95 unit, of which unit activity was calculated based on μg tyrosine production per gram of enzyme. Latex obtained from unripe papaya (*Carica papaya* L.) fruits is a mixture of proteolytic enzymes including papain, chymopapain A and B (EC 3.4.22.6), endopeptidase papain III, endopeptidase papain IV and endopeptidase papain omega [14]. Catalytic activity of papain involves hydrolysis of proteins with broad specificity for peptide bonds, but preference for an amino acid bearing a large hydrophobic side chain [15]. Meanwhile, protein and fat content of boso fish were 67.32% ± 1.86 and 4.02% ± 0.22, respectively.

Generally, enzymatic hydrolysis process influenced by some conditions such as temperature, time, pH, substrate and enzyme concentrations. In this study, the optimization of fish peptone production was carried at 50-65°C for 5-8 hr. Other conditions, such as pH, substrate and enzyme concentrations have been determined. In this study, optimum pH of substrate was 6.0 with 25% (b/v) substrate and 0.1% of enzyme concentration. Protein soluble and ratio of α-amino (AN) to total nitrogen (TN) were analyzed to evaluate the optimum condition of fish peptone production. The effect of hydrolysis temperature on soluble protein content of fish peptone extract was monitored for 4-8 hr. The data is presented on Figure 1.

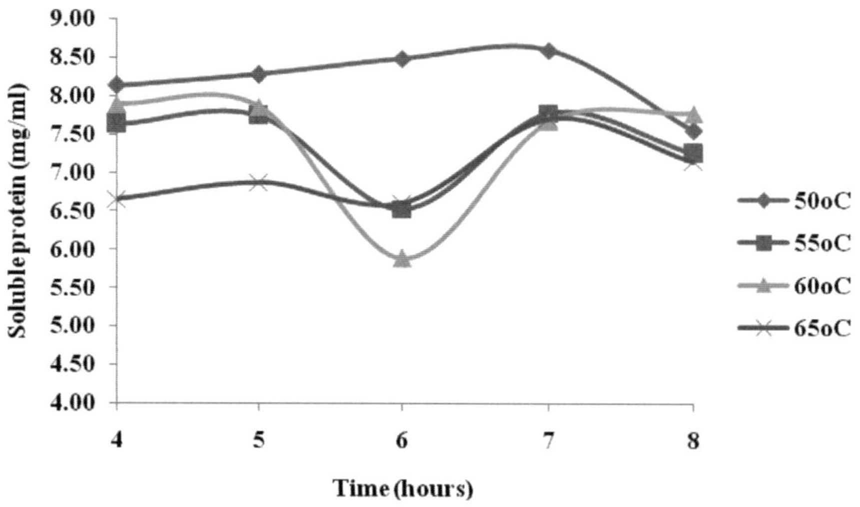

Figure 1. The effect of hydrolysis temperature on soluble protein content of peptone.

Figure 1 shows that the highest soluble protein content was obtained at 50°C after 7 hr of enzymatic hydrolysis. At this condition, fish peptone had the soluble protein content of 8.6 mg/mL. Statistical analysis followed by Duncan test ($p < 0.05$) showed that hydrolysis temperature of 50°C was significantly different with other hydrolysis temperatures. Rahayuningsih and Wiranti (2014) reported the best temperature for peanut meal peptone production using 0.4% of crude papain was 55°C for 4 hr. Moreover, the production of soybean meal peptone production was at 60°C for 5 hr [8]. Based on statistical analysis, hydrolysis time was also significant to soluble protein content. The data showed that the soluble protein content of fish peptone at hydrolysis temperature 55-65°C, after 5 hr hydrolysis was decreased and increased after 6 hr hydrolysis. Temperature correlates with the chemical reaction due to the changes of kinetic energy. The increase of enzymatic reaction will be performed on certain temperature. When hydrolysis temperature is too high, protein enzyme denatures, tertiary structure of enzyme disrupts and the catalytic activity of this enzyme decreases [6]. In this study, protein soluble of fish peptone was increased after 7 hr hydrolysis. This phenomena was occurred at hydrolysis temperature 55-65°C, but was not occurred at 50°C. This assumed that the stability of protein was disrupted at temperature more than 50°C. Degree hydrolysis was generally used for monitoring the hydrolysis process in peptone production. Peptones are partially digested protein; they are prepared by the enzymatic or acidic hydrolysis of proteinaceous material. These hydrolysates contain secondary protein derivatives such as polypeptides, dipeptides, and amino acids. Degree hydrolysis of proteins was measured by the number of peptide bonds cut, divided by the total number of peptone bonds [2]. The effect of hydrolysis temperature to AN/TN of fish peptone extract was also monitored for 4-8 hr (Figure 2).

Figure 2 shows that degree of hydrolysis (AN/TN) of fish peptone production at 50-65°C for 4-8 hr was not significantly different ($p > 0.05$). Degree of hydrolysis of protein depends on three basic principles namely the quantification of the nitrogen released by protein hydrolysis, the determination of free amino groups, and the titration of protons released [10]. Based on the data in Figure 1, the optimum condition for enzymatic hydrolysis was reached at 50°C for 7 hr. At this condition, fish peptone product had α -amino content 0.59% and AN/TN 5.47%. Study of fish peptone using enzymatic hydrolysis of silver carp by-products in *S. aureus* media showed that degree of the protein hydrolysate from alcalase and trypsin was of 4.94 and 4.6%, respectively [1].

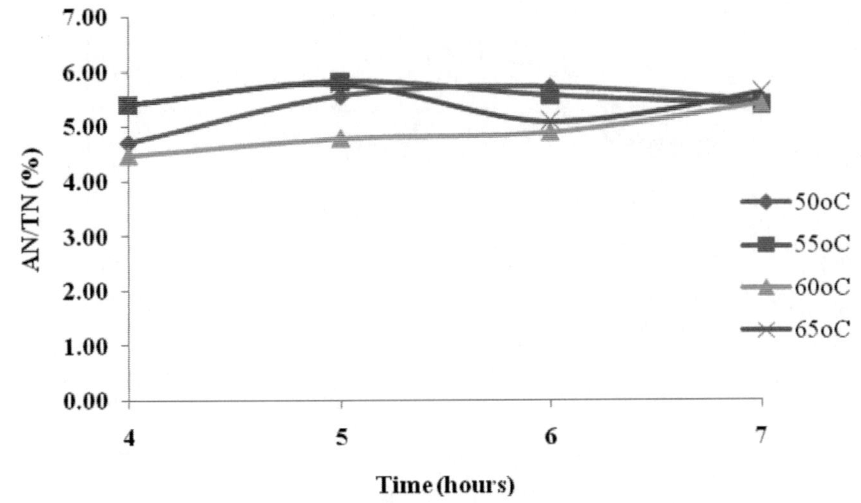

Figure 2. The effect of hydrolysis temperature on AN/TN of fish peptone extract.

SDS PAGE profile shows that the molecular weight was of between 17 -28 kDa (arrow sign) whereas commercial peptone (CP) was ≤ 6 kDa (Figure 3). Commercial peptone used in this study was produced from soya bean as raw material. Beside raw materials, parameters such as time, pH, temperature and enzyme concentration was also influences to the enzymatic activity co-operatively [16]. The molecular weight of peptone from fish waste which hydrolyzed by alcalase was < 6.5 kDa and by alkaline was < 21 kDa [4]. The hydrolysis of a protein molecule is a gradual process by which the gigantic molecule of protein is converted in to products of successively lower molecular weight such as peptones [2].

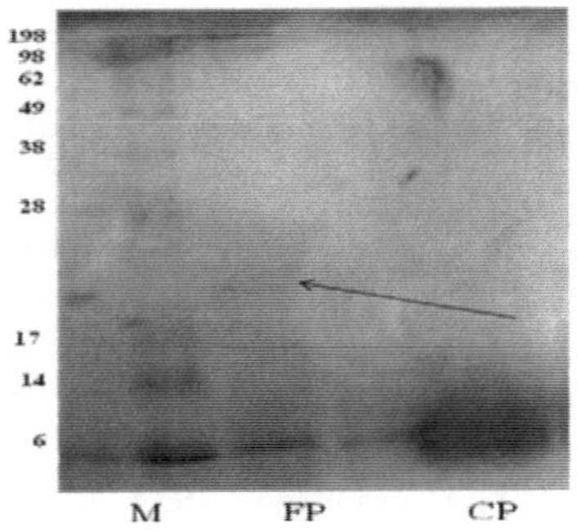

Figure 3. SDS PAGE profile of fish peptone extract from boso fish. M: protein marker, FP: fish peptone extract, CP: commercial peptone.

Fish peptone extract was tested for bacterial growth media and compared with commercial peptone. In this study, the concentration of fish peptone extract in media was calculated based on the protein content which equal with the protein content in commercial peptone. The growth of *E. coli* and *S. aureus* was monitored every two hr until 24 hr. The growth curves of *E.coli* and *S. aureus* is presented in Figure 4.

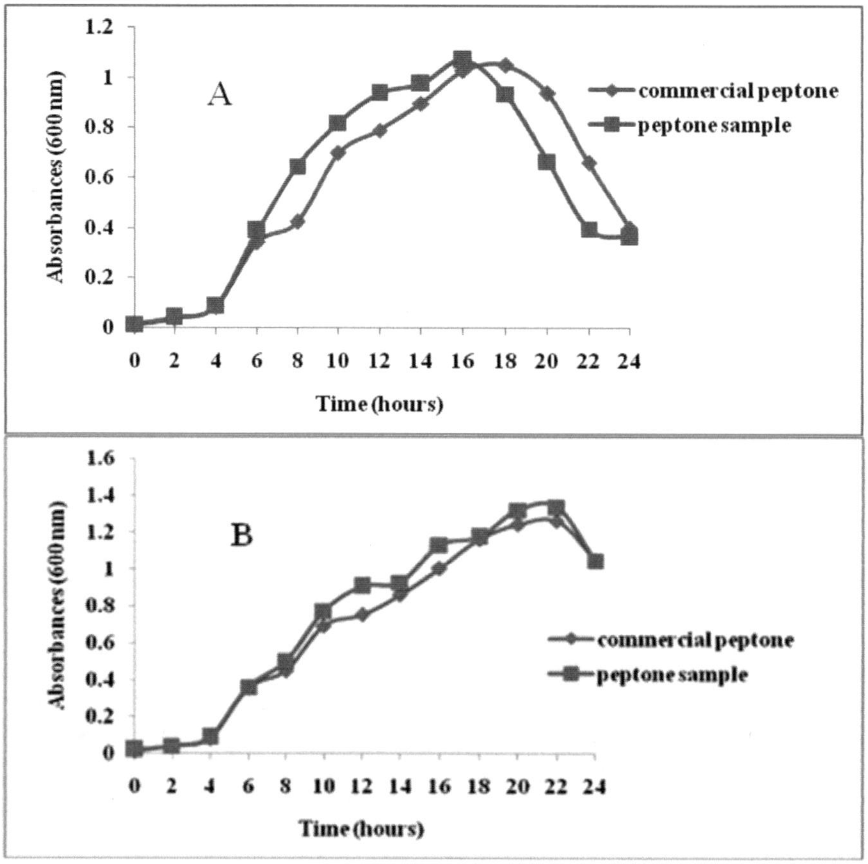

Figure 4. Growth curves of *E. coli* (A) and *S. aureus* (B).

Cell density for the peptones from boso fish showed superior performance in *E.coli* compared with commercial peptone. The data shows that lag phase of *E. coli* in peptone was about 4 hr. Log phase of *E.coli* in fish peptone sample was occurred for 12 hr with the maximum absorbance was A_{600} =1.0275 whereas commercial peptone was occurred for 16 hr with the maximum absorbance was A_{600} =0.9350, respectively. Furthermore, the absorbance of *E. coli* decreased, indicating that the death phase was occurred after 16 hr for peptone sample and 18 hr for commercial peptone. Vieira et al. 2005 [17] reported the performance of peptones produced from different fisheries by-product for *E. coli* compared with commercial peptone. The better performance of fish peptone in comparison with commercial media was also reported by other researches [1]. As shown on Figure 4B, the growth curves of *S. aureus* in fish peptone sample was similar with commercial peptone. *S. aureus* showed the same lag phase with *E.coli*. However, *S.aureus* has longer log phase than *E. coli*. The maximum absorbances of *S. aureus* in both fish peptone sample (A_{600} =1.2450) and commercial peptone (A_{600} = 1.323) were reached after 22 hr incubation. Rahman et al (2004) reported peptone from waste beer industry gave the same effectiveness for *S. aureus* as that from commercial Bacto peptone and yeast extract. The maximum absorbances of *S. aureus* in peptone from waste beer industry reached after 16 hr incubation. The lag phase is defined to transition to the exponential phase after the initial population has doubled. The lag phase is thought to be due to the physiological adaptation of the cell to the culture conditions. This may involve a time requirement for induction of specific messenger RNA (mRNA) and protein synthesis to meet new culture requirements. Meanwhile, during log phase or exponential growth the rate of increase of cells in the culture is proportional to the number of cells present at any particular time and the death phase is characterized by a net loss of culturable cells [18].

4. Conclusion

Peptone has been successfully produced from Boso fish (*O. marmorata* of which the optimum condition for hydrolysis process was obtained at 50°C for 7 hr ($p < 0.05$). Soluble protein content of the resulted peptone was of 8.6 mg/mL whereas α-amino content and AN/TN was of 0.5 and 5.47%, respectively. SDS PAGE profile of fish product showed a major band with molecular weight between 17-28 kDa. The effective of fish peptone for *E. coli* and *S. aureus* growth was similar with commercial bacterial peptone.

Acknowledgement

This research was supported by DIPA Project 2016 - Research Unit for Clean Technology - Indonesian Institute of Sciences (LIPI).

References

[1] Fallah M, Bahram S, Javadian SR 2015 *Food Sci & Nutrition.* **3**(2) 153–157
[2] Al Bahri MBAG, Al Naimi SA and Ahammed SH 2009 *J. Al-Khwarizmi Eng.* **5**(1) 1-19
[3] Durrani A Ali A, Durrani S, Shaikh JB, Upadhyay A, .Khan ZH 2011 *J. Int Biol.* **3**(1) 140-145
[4] Husin N, Kamal SMM, Chuan LT, Muhammad NF, Jusoh N 2015 5th *ICBET-IPCBEE* **81**
[5] Froese, Rainer, Pauly, Daniel 2013 Fish Base.
[6] Rahayuningsih M and Wiranti NG 2014. *Proceeding ASEAN COSAT* 85-92
[7] Afaq S and Iqbal, J 2001 *Electronic J. Biotech.* **4**(3) 120-124
[8] Fahraniah, Fardiaz D, Idiyanti T 2002 *Jurnal Teknol dan Industri Pangan* XIII **3** 260-267
[9] Rahman, Fardiaz D, Idiyanti T 2004 *Jurnal Teknol dan Industri Pangan* XV **2** 159-165
[10] Silvestre MPC, Silva MC, Silva VDM, Silva MR, Amorin LL 2011 WPC *J. Asian Sci Research.* **4**(4) 302-314
[11] AOAC, 1995, Official Methods of Analysis of The Association Analytical Chemistry, Washington DC
[12] Sudarmadji S 1997 *Prosedur Analisa Bahan Makanan dan Pertanian* Liberty Yogyakarta
[13] Bollag DM, Rozycki MD, Edelstein SJ 1996 Protein Methods p 107-154 Wiley Liss Inc Publication
[14] Mahecha MMA, Rodrigu OM, Hugo A, Correa M 2011 *Acta Agron.* 60 (3)
[15] Amri E and Mamboya F 2012 *American J. of Biochem. Biotechnol.* **8**(2) 99-104
[16] S. Salwanee, W.M. Wan Aida, S. Mamot, M.Y. Maskat& S. Ibrahim 2013 *Sains Malaysiana.* **42**(3) 279–287
[17] Vieira, G H, Vieira R H, Macrae A, Sousa O V 2005 *J. Sci. Food Agric.* **85**:1235–1237.
[18] Maier RM 2009 Environmental Microbiology Chapter 3 p 38. Academic Press. Inc.

Influence of pH on Cr(VI) ions removal from aqueous solutions using carboxymethyl cellulose-based hydrogel as adsorbent

L Anah[*] and N Astrini
Research Unit for Clean Technology, Indonesian Institute of Sciences (LIPI)
Jalan Cisitu 21/154 D, Bandung 40135, Indonesia
[*]E-mail: likanah@yahoo.com

Abstract. The major problem in heavy metal pollution is that these metals are not biodegradable and accordingly accumulate in the bodies of living organisms, causing dangerous diseases and serious cell disorder. According to World Health Organization (WHO), the long term exposure of Cr(VI) levels of over 0.1 ppm causes respiratory problems, liver and kidney damage, and carcinogenicity.Due to its easy operation and of various cheap adsorbents development, adsorption has been proved to be efficient and most economically attractive technique and feasible to the removal of toxic heavy metal from wastewater. The study aimed to report the removal of Cr(VI) ions from aqueous solutions through adsorption process using carboxymethyl cellulose-graft-poly(acrylic acid) (CMC-g-PAA) hydrogel as adsorbent.Effect of pH was studied to remove hexavalent chromium.Graft copolymerization of poly(acrylic acid) onto carboxymethyl cellulose was carried out in the presence of benzoyl peroxide redox initiator and methylenbisacrylamide as crosslinker agent. Batch experiments were carried out to investigate the effects ofinitial pH.The adsorption of Cr(VI) ions as a function of pH was conducted in the initial pH range of 1 to 8. The results indicated that acidic pH strongly favored the adsorption. The optimum pH for adsorption of Cr(VI) ranged from 1 to 3, and the maximum uptake of Cr(VI) from the solution was 6.53 mg/g at pH 1 and 30°C. FTIR spectroscopy, SEM analyses were performed on the adsorbent before and after Cr(VI) binding. All analyses confirmed the complexation of Cr(VI) ions on the adsorbent.

1. Introduction

Heavy metals ion such as chromium, copper, lead, zinc, mercury, and cadmium are present in wastewater from various industries such as electroplating, tanning factories, steel works, wood preservation, plastic procession, metallurgical and mining activities, atomic power plants, dyes and pigments, chromate preparation, ceramic paints, and fertilizer industries [1-2]. The introduction of these harmful pollutants into aquatic systems is due to various industrial operations, which increase annually to cover the continuous needs from the rapid increase in the global population. The major problem in heavy metal polution is that these metals are not biodegradable and accordinglyaccumulate in the bodies of living organisms, causing dangerous diseases and serious cell disorders. Among of the very complicated environmental pollution problem, the pollution of water resources with different agricultural and industrial effluent is a matter of great importance. According to World Health Organization (WHO), the long term exposure of Cr(VI) levels of over 0.1 ppm causes respiratory problems, liver and kidney damage, and carcinogenicity. Therefore it is necessary to treat effluents to reduce the Cr(VI) concentration in water and wastewater prior to its transport and cycling in environment [1].

Chromium species presence mainly in two different oxidation states in environmental water, namely Cr(VI) and Cr(III), of which possesses contrasting physiological effects. Cr(III) is considered as an essential trace element for the maintenance of an effective glucose, lipid, protein metabolism in mammals. On the other hand, Cr(VI) can be toxic for biological systems and cancerogenic in humans. Therefore speciation of Cr(VI) and Cr(III) is necessary to evaluate the toxicological behavior. Recently, chromium speciation was reported [4].

The removal of metal ions from aqueous solutions by adsorption depends on the solution pH since the acidity of the solution affects the ionitation of the metal ions and concentration of the counter H^+ ion of the surface groups. To maximize the removal of heavy metals by adsorbents, knowledge of an optimum pH is important [10].

When Cr(VI) comes in contact with organic substances or reducing agent (adsorbent), Cr(VI) is easily or spontaneously reduced to Cr(III) in a acidic medium due to its high redox potential value. The mechanism of Cr(VI) removal by cellulose adsorbent is an electrostatic attraction process in which Cr(VI) are present in the form of anions, such as $HCrO_4$ and Cr_2O_7, while the surface of cellulose adsorbent has positive charges [1].

Techniques used for removal of heavy metal from polluted wastewater include chemical precipitation, ion exchange, adsorption, solvent extraction, flocculation, membrane separation, filtration, electrolysis and reverse osmosis. Due to its easy operation and of various cheap adsorbents development, adsorption has been proved to be efficient and most economically attractive technique and feasible to the removal of toxic heavy metal from wastewater [4-6].

Various adsorbent has been developed to remove heavy metal ions from aqueous solutions, such as clay, ion exchange resins, activated carbon, carbon nano-tubes, and metal oxides [1]. A variety of biomaterial have been exploited as adsorbent for the removal of Cr(VI) from industrial effluents [5-6]. Cellulose, the main constituent of plants, is the most abundant biomass in the world. However, the compact and inactive molecular structure of cellulose requires to be modified to improve is hydrophilicity as an adsorbent [7]. It is renewable, cheap, biodegradable, and can be chemically modified for enhancing its effectiveness by introduction of new functional groups. There is intensive research nowadays for the application of cellulosic material for hazardous contaminant adsorption from aqueous wastes. To date, the cellulose-based material are prepared by carboxymethylization, grafting and cross-linking [7]. Among of the cellulose-based materials, carboxymethyl cellulose (CMC) is a representative water-soluble cellulose derivative, biocompatible, nontoxic and low cost. CMC is used as an adsorbent for heavy metal and dye sorption from various wastewaters systems which have binding sites for divalent cations due to the presence of various functional group within its [2,7].

This paper aims to report the removal of chromium ions from aqueous solutions using carboxymethyl cellulose-acrylic acid graft hydrogel as adsorbent. Effect of pH has been studied to remove hexavalent chromium.

2. Experimental

2.1. Material

Potassium dichromate ($K_2Cr_2O_7$) analytical reagentgrade was used as a source of Cr(VI). Sodium hydroxide (NaOH) and hydrochloric acid (HCl) were used for pH adjustment. All chemicals were purchased from E-Merck Company Inc., German and were used without further purification. Deionized water was used in this work.

2.2. Method

2.2.1. Preparation of the adsorbent

The graft copolymerof CMC-g-PAA was synthesized by free radical and solution polymerization. CMC was dissolved in distilled water then stirred and purged by passing nitrogen gas. The solution was slowly heated up to desired temperature then added with benzoyl peroxide (BPO) initiator and methylenbisacrylamide (MBA) crosslinker agent. The reaction was continued until desired time and process of polymerization was completed. The hydrogel adsorbent was oven-dried, then purified by washing with distilled water and acetone [8].

2.2.2. Preparation of Cr(VI) solution

The stock of Cr(VI) ions solution at concentration of 1000 mg/L was prepared by using $K_2Cr_2O_7$ as a source of Cr(VI) ions. Experimental solutions of the desired concentrations were obtained by successive dilution. 0.5 M NaOH or 4 N HNO_3 was used for pH adjusment and controlled by pH meter [9].

2.2.3.Effect of pH

In this study, the pH on adsorption of Cr(VI) was varied of 1 – 8 at room temperature by shaking dryCMC-based hydrogel (100 mg) with Cr(VI) solution (100 mL, 10 mg/L) for 10 hrs at 300 rpm. HCl 4 N was applied to decrease the pH and NaOH 0.5M to increase pH. Absorbance of the standards and samples of chromium was recorded on atomic absorption spectrophotometer (Varian Type).

2.2.4. Adsorption

Adsorption was carried out in a 250 mL volumetric flask using 100 ml Cr(VI) ions solution. The flasks were placed on a Gallenkamp Orbital Incubator Shaker running at 100 rpm, 30°C for 10 hr.
The amount of Cr(VI) adsorbed by the CMC-g-PAA hydrogel adsorbent (Q_e) and removal percentage (*R%*)of Cr(VI) were calculatedfor each run by following equations,

$$Q_e = \frac{(C_0 - C_e)V}{W} \quad (1)$$

$$R\% = \frac{C_0 - C_e}{C_0} \times 100 \quad (2)$$

Where Q_e (mg/g) is the adsorption capacity, R (%) is removal percentage, C_0 (mg/L) is the initial concentration of Cr(VI), C_e (mg/L) is the equilibrium concentration of Cr(VI), V (L) is the volume of the solution of Cr(VI), W (g) is the weight of adsorbent added, and C_0 and $C_e$are initial and f respectively [11-12].

3. Result and discussion

3.1. Effect of pH

Result of the adsorption of Cr(VI) as a function of pH is shown in Figure 1. It shows that the adsorption of the cellulose adsorbent was highly pH dependent. Cr(VI) removal efficiency decreased with increasingthe initial pH. The optimum pH for Cr(VI) removal was found at 1.0, and a maximum adsorption capacity of Cr(VI) 6.53 mg/g was obtained. About 21.5% removal was achieved at a pH of 2, which increases to about 64.3% at pH 1. The improved removal of Cr at low pH is probably due to reduction of hexavalent Cr to trivalent Cr [15].
At low pH :

$$Cr_2O_7^- + 14H^+ + 6e^- \longrightarrow 2Cr^{3+} + 7H_2O$$

At moderate pH :

$$HCrO_4^- + 7H^+ + 3e^- \longrightarrow Cr^{3+} + 4H_2O$$

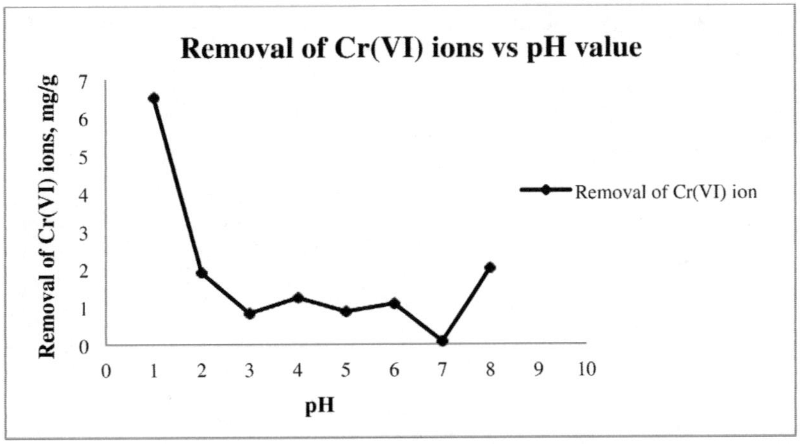

Figure 1. The amount of Cr ion using CMC-g-PAA hydrogel adsorbent at different pH.

At a lower pH (pH ≤ 3) value, Cr(VI) exist mainly is $HCrO_4^-$, CrO_7^{2-} , which have negative charge, while the surface of adsorbent is charged positively. The stable forms of chromium such as H_2CrO_4 and CrO_3 exist as polynuclear species at high Cr(VI) concentration, and hence the low pH value results in a higher percentage removal of Cr(VI). High adsorption at acidic pH condition could be explained by the chemical character of the chromium species and the adsorbent surface. In the acidic pH range (1 – 2), the prominent formation of the chromium are $Cr_2O_7^-$, $HCrO_4^-$, and above pH 7 the primary stable species is CrO_4^-. Therefore, at lower pH, $Cr_2O_7^-$ and $HCrO_4^-$were adsorbed onto the adsorbent, resulting in a high percentage of adsorption. At basic pH condition, the supression of the hydrolysis of Cr(VI) may be the reason for the decreased adsorption [16]. At low pH, there is presence of a large number of H^+ ions, which in turn neutralizes the negatively charged adsorbent surface thereby reducing hindrance to diffusion of dichromate ion. It is anticipated that the effect of pH on adsorption is also governed by the development of an electric double layer on the adsorbent. The polarity of the double layer at the adsorbent surface may be the changed from positive to negative as the H^+ion concentration changes from acidic to basic with the increase of pH. It was found that at lower pH the system attain equilibrium faster and also the percentage of chromium adsorbed increased[15].

At pH ≥ 1.0, the adsorption efficiency of Cr(VI) decreases as pH increases. This may be explained on the basis of the lower extent protonated hydroxyl groups with rising pH or due to the dual competition between both the anions (CrO_4^{2-} and OH^-) to be adsorbed on the surface of the adsorbent of which OH^- predominant [1,3].

pH solution plays an important role during the adsorption process and affect the surface charge of cellulose adsorbent, the degree of ionization, and speciation of adsorbate. The hydrogen and hydroxyl ions are adsorbed quite strongly, and therefore the adsorption of other ions is affected by the pH of the solution [1,13].The pH of the aqueous solution affects the solubility of the metal ions, concentration of the counter ions on the functional groups on the sorbent and the degree of the ionization of the sorbent during the reaction. The active sites on the sorbent can either be protonated or deprotonated depending on the pH. Concurently, the sorbent speciation in a solution also depends on the pH [14].

3.2. FTIR spectrum

Figure 2 shows the FT-IR spectrum of the CMC-g-PAA hydrogel adsorbent before and after adsorptionof Cr(VI). The peak at 2920 cm^{-1} and 2922 cm^{-1} were due to the stretching vibration of aliphatic C-H. The peak at 1548 cm^{-1} and 1544 $cm^{-1}$was due to asymmetrical stretching vibration of COO^- groups. Similar peak at 1398 cm^{-1} and 1402 cm^{-1} are assigned to symmetrical stretching vibration of COO^- groups whereas peak appeared in 1017 cm^{-1} and 1021 cm^{-1}were attributed to the C-O-C stretching vibration[2].

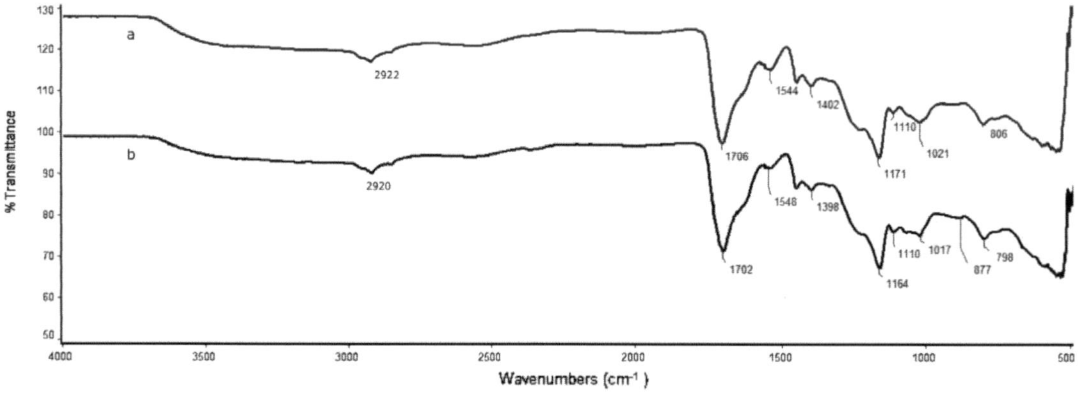

Figure 2. IR spectra of CMC-g-PAA hydrogel adsorbent (a) and Cr(VI) loaded adsorbent (b).

To identify the possible functional groups on carboxymethyl cellulose adsorbent involved in the binding of Cr(VI), FTIR spectra were obtained before and after metal adsorption. If a ligand coordinates to a metal, the energy of the ligand material will most likely be perturbed leading ultimately to subtle shifts in the absorption peaks in the FTIR spectra. These FTIR absorption bands are usually shifted to lower or higher frequencies. Figure 2 shows the FTIR spectra of carboxymethyl cellulose adsorbent before and after Cr(VI) adsorption. There was a small shift of the ring vibration peak at 1706 cm^{-1}, 1402 cm^{-1}, 1171 cm^{-1}, 1021 cm^{-1} in carboxymethyl cellulose adsorbent to 1702 cm^{-1}, 1398 cm^{-1}, 1164 cm^{-1}, 1017 cm^{-1} for Cr(VI) loaded adsorbent [1].

3.3. SEM Image

SEM images of carboxymethyl cellulose adsorbent before and after adsorption are shown in Figure 3.

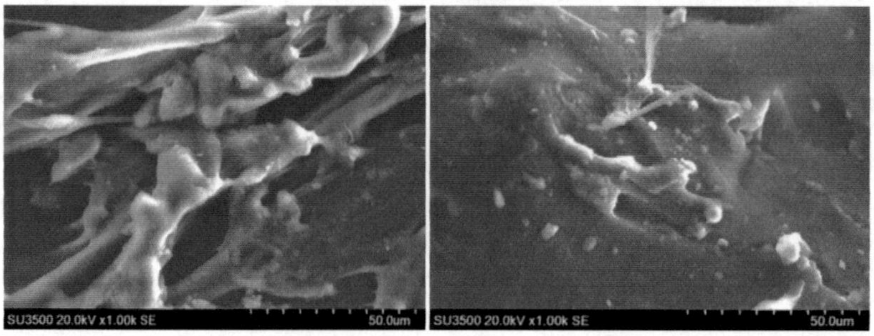

(a) (b)

Figure 3. SEM micrograph of CMC-g-PAA adsorbent before (a) and after (b) adsorption.

The SEM micrograph of hydrogel adsorbent showed fractured and rough surface morphology providing better adsorption sites. It can be observed that the surface of the adsorbent after adsorption changed, and the visible pores of the adsorbent had disappeared or had been filled up [1].

4. Conclusion

Current study concluded that the adsorption of Cr(VI) was pH dependent. The removal efficiency of adsorbent increased with the decreasing in pH. At pH of 1 about 64,3% Cr(VI) removal was possible.The present study showed that CMC-g-PAA hydrogel waseffective adsorbent for Cr(VI) metal ions removal from aqueous solution. The adsorption of Cr(VI) ions waspH dependent. CMC-g-

PAA hydrogel was found to be a favourable adsorbent. Characterization with FTIR and SEM confirmed the binding of Cr(VI) onto the carboxymethyl cellulose adsorbent.

Acknowledgement
The authors gratefully acknowledge the Research Unit for Clean Technology-LIPI, Bandung, Indonesia, for providing financial support and facilities required to carry out this work.

References

[1] Chunxiang L, Sha Q, Wei L, Yifan L, Danhui L, Xiaojuan L, and Minghua L 2014 *BioResources*. (**9**) 4 6998-7017.
[2] Ali J Salim 2015 *Journal of Al-Nahrein University*. **18** (4) 40-48.
[3] Mack C, Wilhelmi B, Duncan J R, Burgess J E 2007 *Biotechnol. Adv*. **25**264-271.
[4] Jun Dai, Ren F L, Tao C Y 2012 *Int. J. Environ. Res. Public Health.* **9** 1757-1770.
[5] Li C, Zhang Y, Peng J,Wu H, Li J, Zhai M 2012 *Radiation Phy. and Chem.* **81** 967-970.
[6] Abdel-Halim E S, and Al-Hoqbani A A 2015 *BioResources.* **10** (2) 3112-3130.
[7] Genlin Z, Lijuan Y, Hui D, Ping S 2013 School of Chemistry and Chemical Engineering, Key Laboratory for Green Processing of Chemical Engineering of Xinjiang Bingtuan, Shihezi University, Shihezi 832003, China.
[8] Lik A, Nuri A, Agus H 2015 *Macromol. Symp.* **353**178-184.
[9] Selvarani M, Prema P 2012 *Int. J. of Env. Scie*. **2**(4).
[10] Sanna Hokkanen 2014 *Thesis, Lappeenranta University of Technology, Finland.*
[11] Jian H C, Hai T X, Hong X G, Guo P L, Wen W, Shi R H 2013 *J. of Hazardous Mat.* 285-294.
[12] Gad, H M H, Nabila A M 2014 *Int. J. of Adv. Scie. Tech. Res.* **4** (4) 184-201.
[13] Govindarajan C, Ramasubramaniam S, Gomathi T, Sudha P N 2011 *Archives of Appl. Scie. Res.* **3** (5):572-580.
[14] Abdul-Reheim M A R, Farag R, El-Saeed S M, and Abdel-Raouf M E 2015 *Res. J. of Pharmac. Biol. Chem. Scie.* **6** (6) 1197-1212.
[15] Verma A, Chakraborty S, Basu J K 2006 *Separation Purification Technol.* **50** 336-341.
[16] Xuemei H, Huidong X, Hui L 2015 *World J. of Eng. Technol.* **3**, 234-240.

Physical and mechanical of breadfruit leaves-polyethylene composites

R Satoto[1*], S Rohmah[2] and A Suhandi[2]
[1]Research Unit for Clean Technology, Indonesian Institute of Sciences (LIPI)
Jalan Cisitu 21/154 D, Bandung 40135, Indonesia
[2]Deparment of Physics Education, UniversitasPendidikan Indonesia
Jalan Dr.Setiabudhi 229, Bandung, Indonesia
*Email : rsatoto@yahoo.com

Abstract. A degradable polymeric composite was prepared from polyethylene (PE) and breadfruit leaves tree powders. Breadfruit leaves tree powders were mixed with PE in an internal mixer at a temperature above the glass transition of PE without additives. Formulations were based on PE/leaves ratio of 100/0, 90/10, 85/15, 80/20 and 75/25 on a dry weight basis. The effect of leaves powders of 140 and 100 meshon the composite mechanical properties was evaluated by means of universal testing machine (UTM) and hardness Rockwell tester. The results showed increases of elastic modulus and flexural modulus of the composites with increasing leaves powders. Tensile strength, yield strength, yield strain, and hardness decreased with increasing percentage of leaves powders in the composites. Flexural strength was slightly decreased with the presence of the leaves powders but independent on the percentage of the leaves powders. In general, the composite properties of 140 mesh leaves powders were more enhanced compare with that of 100 mesh. PE/leaves composites increased water absorption and caused the bulk composite surface more porous. Hence, oxo-biodegradation processes will more easily take place in the PE/leaves composites.

1. Introduction

Single-use articles and packaging waste materials such as food packaging, small container, wrap, shock absorberare produced every day. In municipal region, the quantity of packaging waste is increasing from day to day. Waste removal, using an incinerator may be a way to solve the problem on a long term, but it sometimes produces air pollution and it requires high investment and operational cost. Recycling is an environmentally attractive solution. However, only a minor portion of plastics is recyclable, and most plastics end up in municipal land fill sites. Natural decomposition and/ or natural degradation are believed the best way to solve the plastics waste problem. Packaging material should be designed appropriately to meet this purpose.A scientific solution can form a more significant contribution to this municipal problem.

Polylactic acid (PLA) is known as a biodegradable plastic with good mechanical properties [1-2]. The disadvantage of this plastic is too expensive as a packaging material as well as needs relative long time to be biodegraded. Pure PLA showed no weight and tensile strength change over 1 year in a soil buried treatment and looked like new or no discoloration, pitting [3].While blends of starch/PLA for the same treatment showed significant change in weightandtensile strength after 3 months.

The role of starch was very significant for degradation initiation of the starch/polycaprolactone composites [4]. In previous papers, starch/PEcomposites [5-20], lignin/PE [21-24], husk/PE [25-27], cellulose/PE [28-32], and wood flour/PE [33-34] showed that all of the biomass were being interest for research.Preparation of biodegradable films by acetylated acid modified rice starches and the thermal properties have been extensively studied [35].Thermoplastic starch could be an ideal matrix for

IOP Conf. Series: Earth and Environmental Science **60** (2017) 012011 doi:10.1088/1755-1315/60/1/012011

packaging material or single-use articles, because it is a renewable resource material and biodegradable [36]. Since water is a plasticizer for starch, the performance of thermoplastic starch may be unstable and depending on the moisture contents, being in equilibrium with the ambient relative humidity [37]. A product made from thermoplastic starch is sensitive to water and inclined to retro-gradation which limits its application. A product composed from pure thermoplastic starch will become unstable in a moist environment and its mechanical performance will decrease strongly. For a practical extension of the utilization of starch-based products, reinforcement or modification of starch is essential. Efforts[38-39] have been made to provide suitable alternatives to upgrade starch-based product performance. However starch is consumed as food and still relatively expensive to be used for starter of degradability of the plastic composites.

Leaf-tree is one of biomass which has not been reported yet for producing polymer composites. A fall leaf treeis good availability and the lowest economically comparing to the starch, lignin, cellulose and wood. Besides, preparation of leaf flour needs the lowest energy compare to that of biomass flour in question.Since no reported data regarding leaf/plastic up to date, we would utilise the fall leaf as a degradable starter in our composites.

Biomass polymer and agricultural production are enviromentally easier degradable. In this context, polymers from biomass have been advocated as environmentally friendly alternatives to unmodified polyethylene. On a few practical application fields of packaging either blending or composites, it is strongly desired that these materials naturally degradable or at least compostable after their servises. As per standards [41-45], a minimum level of mineralization needs to be exhibited for the material to be classified as ‘‘compostable’’. Since during lifetime and after become waste packaging materials can be exposed to sunlight, heat, oxygen, moisture and micro organisms, it is of great importance to investigate the durability of these materials when exposed to these conditions. Firstly, only the effect of water on degradability will be investigated.

This study aimed to investigate the optimal formula and operational conditions for preparing the leafs-plastics composites, to characterize the physical and mechanical properties of the composites. Blending leaf-tree and plastics is one of the alternatives to upgrade the natural degradability properties.

2. Experimental

2.1. Materials

The sample used in this experiment was Polyethylene (PE) produced by PT. Chandra Asri Petrochemical, Tbk. Indonesia. Lot number MI-LLD 45-12. It contained a conventional thermal stabilizer of undisclosed composition.The falled breadfruit leafs were washed and sun-dried, crushed and shaker sieved for 100 and 140 mesh. Powder violet leafs were dried in Nabertherm oven at 105°C for at least 24 h. Water content was kept below 1% before melt blending.

2.2. Breadfruit leaf-based composites preparation

Variousleaf concentration of0, 10, 15, 20 and 25 wt% were melt blended with PE using an *Internal Mixer* (Labo Plastomill Model 30R150, Toyo Seiki Ltd. Japan, Capacity of 60 ml) at the set up temperature of 135°C. The screw rotation was kept of 50 rpm. The mixing was stopped for mixing time of 8 min.

The composite sheets were prepared from the melt blended samples using a hand-press Gonno (ram stroke diameter 152 mm, capacity 37 ton). Pressing temperature was set at 130°C. The melt blended samples were pressed in such a way till the end of the processing pressure of 50 kg/cm^2. After pressing for around 8 min, the sheets were dried quenching using a cold press.

2.3. Characterization techniques

Tensile properties were measured using Universal Testing Machine, Orientec Co. Ltd, Model UCT-5T. The composite sheet was cut into dumbbell according to ISO 572-2 type 5A. Dumbbell specimens were conditioned at 23°C and 50% relative humidity for not less than 48 h before testing. The speed of testing was 50 mm/min.

Flexural properties were measured using the same machine with that of tensile properties. Loading *nose* and *supports* have cylindrical surfaces with radii of 5 mm. Conditioning before testing and measurement following ASTM D 790 – 02. Testing was performed by procedure B in ASTM D 790 – 02, since the samples did not break after the strain more than 5 %. The rate of crosshead motion was 13 mm/min.

In order to evaluate the hardness of the leaves composites,Rockwell Hardness Testing Machine, Matsuzawa Mrk-M was used to measure the hardness of samples. Testing procedure following ASTM D 785 – 98, using ball indenter of 12.7 mm diameter. The minor and major loads were 10 and 60 kg, respectively. The Rockwell hardness scale R (HRR) was determined 15 sec after releasing major load to minor load.

The rate of water absorption was measured by weighing the composites every day. The composites sample was cut into dumbbell ISO 527-2 type 1A, 7 specimens for each sample were prepared and weight before soaking in aquadest, then kept in room temperature. For water absorption measurement, samples were dried using absorbing toilet tissue paper. The relative weight increase (Wr) was determined with respect to the initial average weight according to equation (1).

$$Wr = ((Wd - Wi) / Wi) \times 100\% \quad \ldots\ldots\ldots\ldots\ldots\ldots \quad (1)$$

whereas Wd and Wi were average weight at the soaking duration and the initial average weights, respectively.

Degradability of the leaves composites was investigated by measuring tensile properties after water absorption for 8 weeks, without conditioning at room temperature.

3. Results and discussion

To describe the mechanical properties of different PE composites, their elastic modulus, yield strength and yield strain were determined. Figure 1 shows the elastic modulus of 100 and 140 mesh PE composites. Elastic modulus is in general increased with increasing leaves addition, and they were higher than elastic modulus of pure PE. The elastic modulus of 140 mesh PE composites was lower than that of 100 mesh.

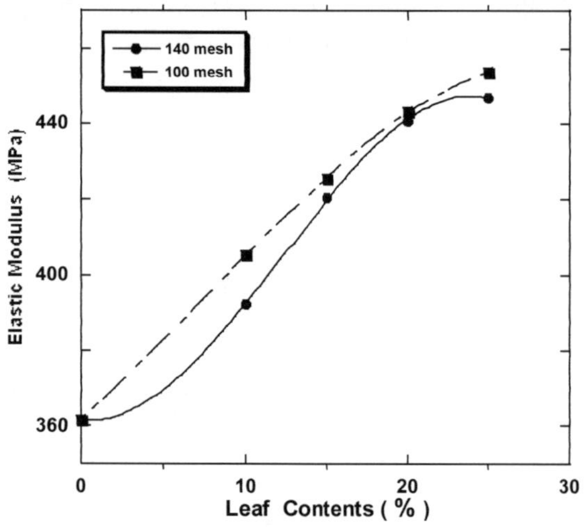

Figure 1. Tensile elastic modulus as a function of leaf contents.

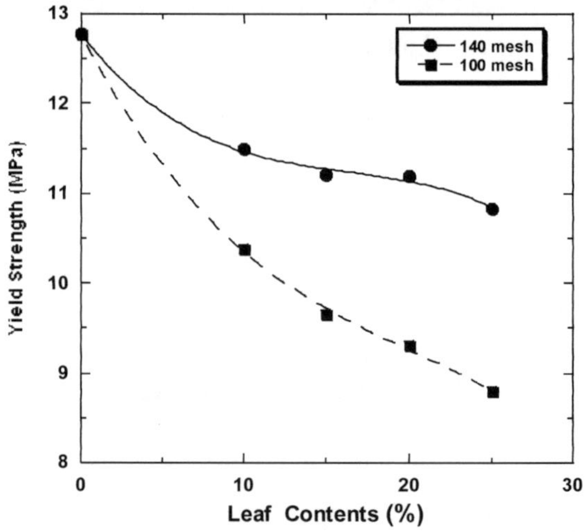

Figure 2. Yield strength as a function of leaf contents in tensile measurement.

Figure 2 shows the tensile strength of PE composites decreases with increasing leaves composition. The decreasing tensile strength was more severe in the PE composites of 100 mesh leaves. This can be explained by taking into account the adhesion of leaves powders-PE was smaller compared with the cohesion among PE molecules. For 100 and 140 mesh PE composites, tensile strength of PE composite with 10% leaves was of 84 and 90% of the tensile strength of pure PE, respectively. Composite with 100 mesh-25% leaves, shows the lowest tensile strength of 8.8 MPa.

These PE/leaves composites have the greater tensile strength compared with that of natural additive starch/ lignin LDPE blends prepared by others [46]. The tensile strength of the blends with natural additives is 6.8 – 7.2 MPa or 64% of the tensile strength of LDPE alone.

The elongation at breaking of PE/leaves composites was shown in Figue 3. Elongation at breaking for all composites was shorter than pure PE and decreased with increasing leaves content in the composites. The elongation at breaking of PE/leaves composites 100 mesh was lower than that of 140 mesh for all leaves composition. This suggests that characteristic of composites was more brittle with the higher leaves powders. The higher the leaves powders the less homogeneous composites/blends. The stress-strain curve of the PE/leaves composite did not show clear yield point.

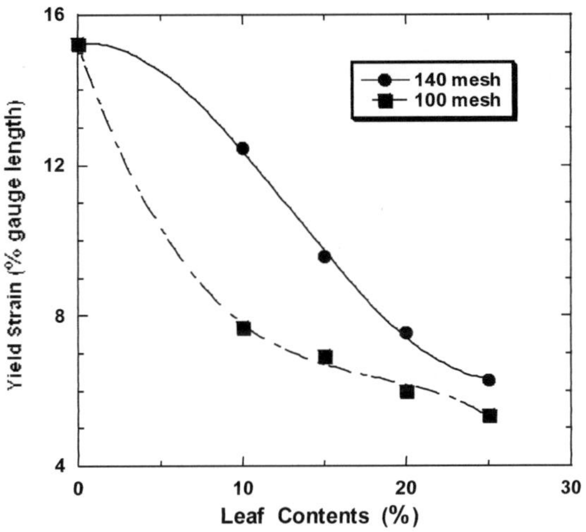

Figure 3. Yield Strain as a function of leaf contents in tensile test.

Figure 4 shows the flexural modulus of both PE/leaves composites of 100 and 140 mesh, obtained by the tangent at 0.4% of the strain. Their flexural modulus was higher than that of pure PE. The flexural modulus increased with increasing of leaves content in the composites. The highest flexural modulus was the 670 MPa, obtained from the composite with 25% leaves content of 140 mesh leaves powders. For all PE/leaves composites of 140 mesh showed slightly greater flexural modulus than that of 100 mesh for the same leaves powders content.

Flexural strength of PE/leaves composites was shown in Figure 5. For these composites samples did not yield or break within the strain limit when tested by procedure A of ASTM D 790 – 02, the testing were performed by increasing the strain rate to 0.1 mm/mm/min. However, the samples did not yield or broken within the required 5% strain limit. According to the standard, the flexural strength should be determined as the flexural stress at the strain of 5%, and the testing or loading specimen were terminated. The flexural strength of PE/leaves composites for all leaves composition was lower than that of pure PE, but seemed to be independent of the leaves contents. The flexural strength of PE/leaves composites of 140 mesh were greater than that of 100 mesh for the same leaves contents. These flexural data show that the composite are still flexible. Even though the yield strain of the tensile test show significant decreased (Figure 3), the composite was not brittle.

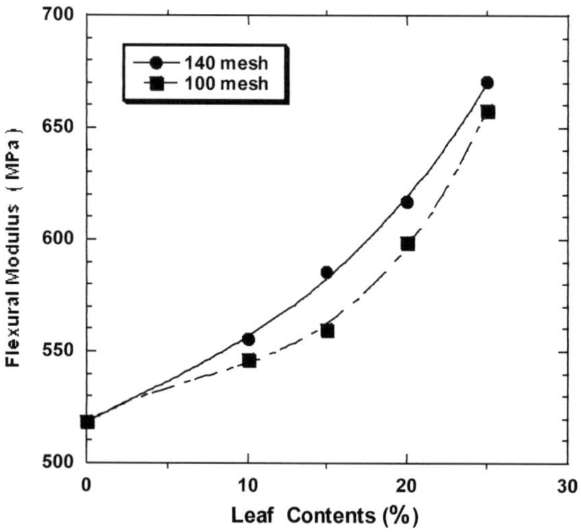

Figure 4. Flexural modulus as a function of leaf contents.

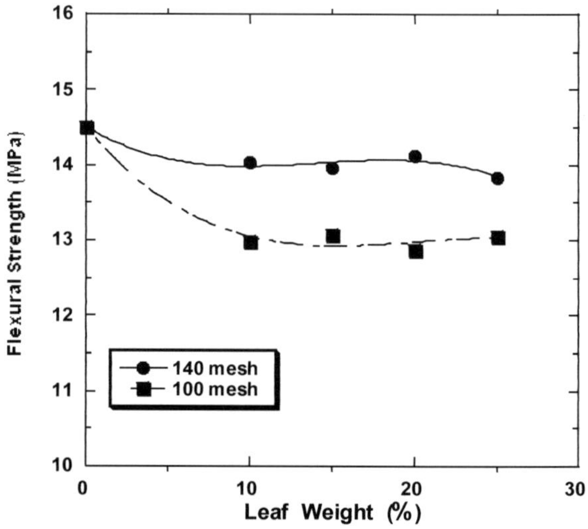

Figure 5. Flexural strength as a function of leaf contents.

Figure 6 shows the average hardness of the leaves composites as a function of leaves contents. The average hardness decreased with increasing of leaves contents. The average hardness of leaves composites of 140 mesh is higher than that of 100 mesh for the same leaves contents. The finer powder

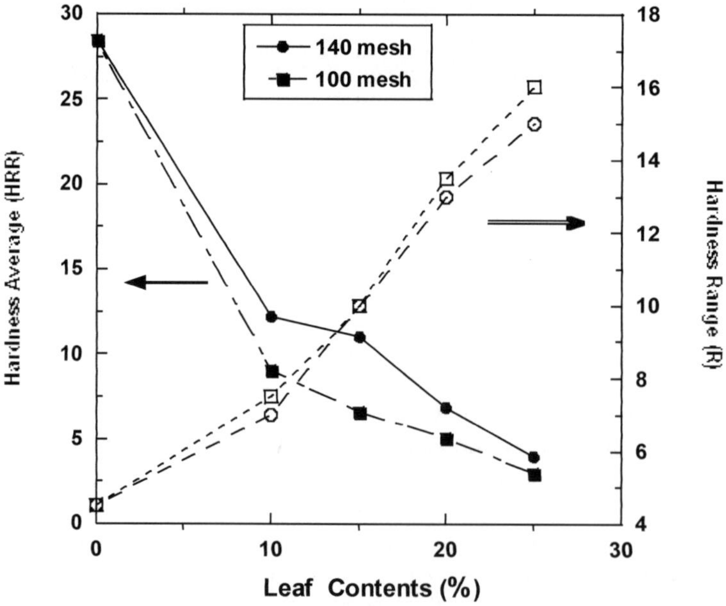

Figure 6. Rockwell hardness average and hardness range as a function of leaf contents.

leaves may resulting composites mixture more homogeneous. In hardness measurement, the indentation applied in small area, so that inhomogenity of compound was easily identified from each hardness value in a set of measurement. The range of hardness was determined from the difference of the highest and lowest hardness in a set measuring hardness for n =17, according to ASTM E18-2011. Figure 6 shows the hardness range (R) increased significantly with the increasing of leaves content in both 100 and 140 mesh composites. These facts may be attributed to less homogeneity in mixing the compounding by increasing the leaves content. To get a homogeneous mixture, the mixing time should be longer for higher leaves content.

The relative weight increased during water absorption is shown in Figure 7. For each composite shows a significant absorption after 3 days.

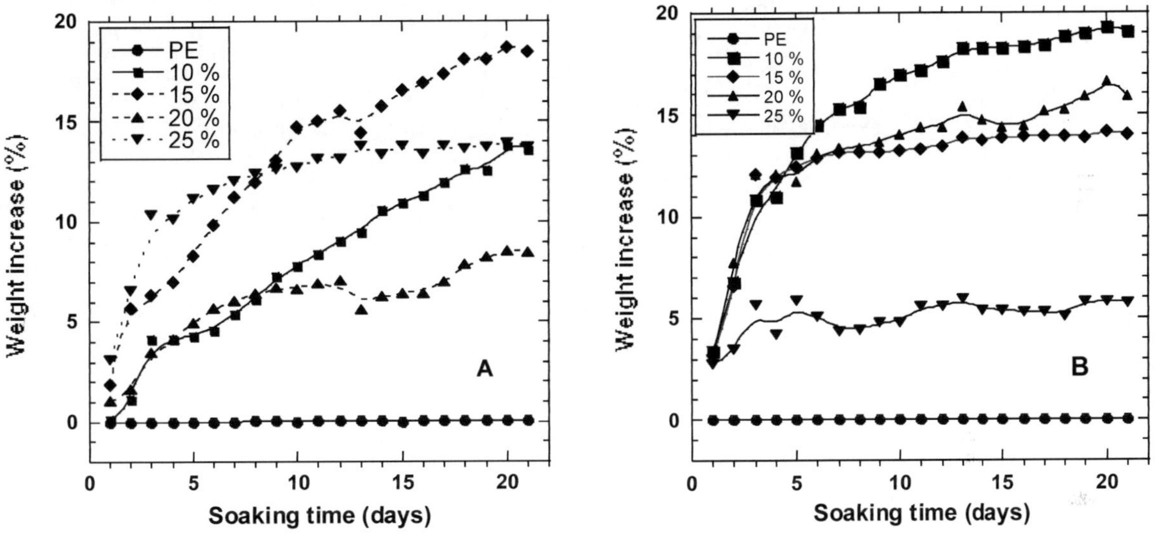

Figure 7. Relative weight increse during soaking in aquadest at room temperature.
A: 140 mesh and B: 100 mesh.

Water absorption increased with the increasing time, although the increasing fluctuation was not clear correlation with the leaves contents. The water absorption for composite 100 mesh-25% leaves increased from the 1st- 5th days, then almost stable until the following 3 weeks. A consequence of the water absorption was preferable by wild microorganism [47] and certain microorganism in the nature [48]. Further, biodegradation could be expected.

Figure 8 shows the elastic modulusof the leaf composites after 10 weeks soaking in destiled water. The elastic modulus of all the composites drops to around 160 MPa, whereas the elastic madulus of pure PE remain unchanged (360 MPa). Water absorption of pure PE was almost zero percent for 10 weeks (Fig. 7), so that its modulus was not affected. Upon the soaking of the composites, water absorption occured in the leaf powders. The water absorption affects interfacial adhesion between grain leaf powder and PE. The absorbed water in the leaf composite plastics deteriorated the elastic modulus of the composites and independent of the leaf contents.

Figure 9 shows the yield strength of the leaf composites after soaking in destiled water for 10 weeks. As shown in this figure, the yield strength of PE remained unchanged. The yield strength of composites decreased with increasing leaves contents. For the same leaves contents, yield strength of 140 mesh was slight greater than that of 100 mesh composites. This fact indicated that leaf composites are easy to deteriorated even only in water. The absorbed water will induces stress to PE molecules and may stimulate the PE matrix cleavage. Water molecules also disturbed the micro-structure of the PE

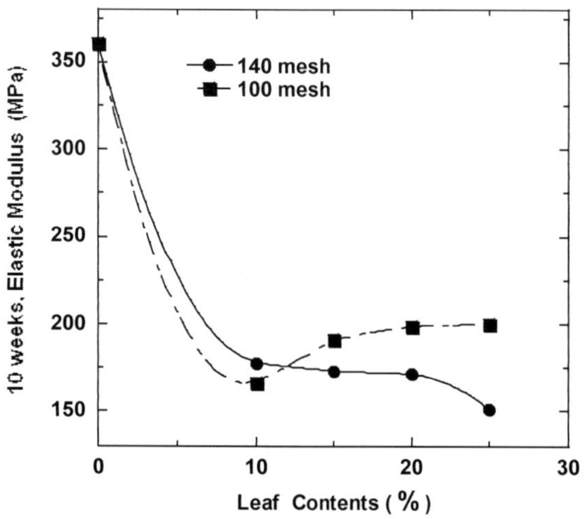

Figure 8. Tensile elastic modulus as a function of leaf contents after 10 weeks in water.

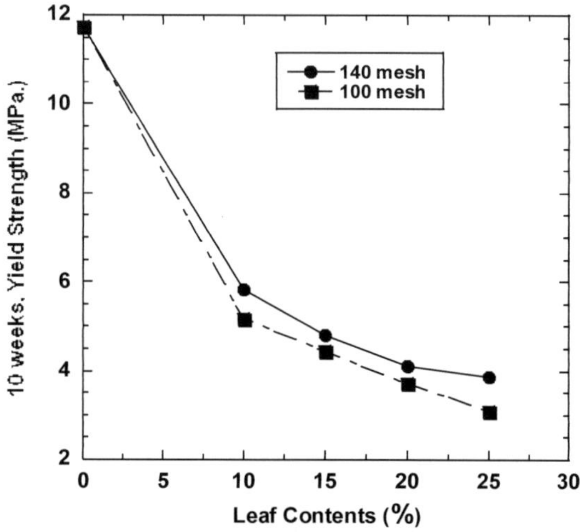

Figure 9. Yield Strength as a function of leaf contents after 10 weeks in water.

and oxo- biodegradation being started. So that the PE is further fractioned. In shortly, the absorbed water accelerate oxo-biodegradation processes [48].

4. Conclusion

Results of current work allow the comparation of PE and bio-composites characteristic. The mechanical and physical properties of the bio-composites are different fairly. Modulus elastic and the flexural modulus of the composites increased compare with that of pure PE. Whereas tensile strength, yield strength, yield strain and hardness were deceased compare to that of pure PE. The mechanical properties of the composite are changed by every kind of the leaves composition and the powder size. Leaves composites absorbed water quickly and raised stress or tension on PE molecules of which stimulated fragmentation.

Acknowledgement

The Authors are grateful to the Indonesian Institute of Science through Mandiri Project 2015, Research Unit for Clean Technology LIPI for financial support.

References

[1] Hakkarainen M 2002 *Adv.Polym. Sci.* **157** 113-38.
[2] Tsuji H, Echizen Y and Nishimura Y 2006 *Polym. Degrad. Stab.* **91** 1128-37.
[3] ShogrenaR L, Doane W M, Garlotta D, Lawton J W, Willett J L, 2003 *Polym. Deg. Stab.* **79** 405-11.
[4] Wu C S 2003 *Polym. Deg.Stab.* **80** 127-34.
[5] Seong I Y, Tae Y L, Jin-San Y, Ik-Mo L, Mal-Nam K and Han S L 2002 *J. Appl. Polym. Sci.* **83** 767-76.
[6] Baldev R, Annadurai V, Somashekar R, Madan R and Siddaramaiah S 2001 *Europ. Polym J.* **37** 943-8.
[7] Matzinos P, Bikiaris D, Kokkou S and Panayiotou C 2001 *J. Appl. Polym. Sci.* **79** 2548-57.
[8] Suda K, Prodepan T and Manit S 2001 *Polym. Deg.Stab.* **73** 363-75.
[9] Thakore I M, Desai S, Sarawade BD and Devi S 2001 *Europ. Polym J.* **37** 151-60.

[10] Zuchowska D, Hlavata D, Steller R, Adamiak W and Meissner W 1999 *Polym. Deg.Stab.* **64** 339-46.
[11] Bikiaris D, Aburto J, Alric I, Borredon E, Botev M. Betchev Cand Panayiotou C 2001 *J. Appl. Polym. Sci.* **71** 1089-100.
[12] Ioannis A, Costas G B, Hiromasa O and Noriaki K 1998 *Carbohydr. Polym.* **36** 89-101.
[13] Stephanie S and Edwin LT 1998 *Polymer***39(23)** 5588-99.
[14] Bikiaris D, Prinos J and Panayiotou C 1997 *Polym. Deg.Stab.* **58** 215-28.
[15] Bikiaris D, Prinos J and Panayiotou C 1997 *Polym. Deg.Stab.* **56** 1-9.
[16] Eleni P, Ioannis A, Costas G B, Hiromasa O and Noriaki K 1997 *Carbohydr. Polym.* **33** 227-42.
[17] Chandra R and Rustgi R 1997 *Polym. Deg.Stab.* **56** 185-202.
[18] Krisna S P, Satyanarayana D and Mohan-Rao D V 1998 *J. Appl. Polym. Sci.* **70** 2251-57.
[19] Bikiaris D and Panayiotou C 1998 *J. Appl. Polym. Sci.* **70** 1503-21.
[20] Zuchowska D, Steller R and Meissner W 1998 *Polym. Deg. Stab.* **60** 471-80.
[21] Feng L, Nan-ying N, Long C, Run S, Jing C, Qin Z and Qiang F 2009 *Chinese J. Polym. Sci.* **27(6)** 833-42.
[22] Sangram K S, Fernandes E G, Andrea C and Emo C 2014 *J. Polym. Environ.* **22** 58-68.
[23] Shuwei W, Na F, Jiulai Z, Keun-Byoung Y, Dongho L, Minjie Q, Xuequan Z and Hexin Z 2016 *Polym. Adv. Technol.* **27(10)** 1351-58.
[24] Hasan S and Dimitris S A 2016 *ACS Sustainable Chem. Eng.* **4(10)** 5160-66.
[25] Rahman M R, Islam M N, Huque M M, Hamdan S and Ahmed A S 2010 *BioResources***5(2)** 854-69.
[26] Atuanya C U, Olaitan S A, Azeez T O, Agaku C C, Onukwuli O D and Menkiti M C 2013 *Int. J. Cur. Res. Rev.* **5(15)** 111-8.
[27] Bilal A, Lin R J T and Jayaraman K 2014 *Appl. Mech. Mater.* **575** 223-6.
[28] Farao Z, Takashi E,Wulin Q, Liqun Y and Takahiro H 2002 *J. Appl. Polym. Sci.* **84** 1971-80.
[29] Kosaka P M, Kawano Y, Petri H M, Fantini M C A and Petri D F S 2007 *J. Appl. Polym. Sci.* **103** 403-11.
[30] Priscila M K, Kawano Y, Fantini M C A and Petri D F S 2006 *Macromol. Mater. Eng.* **291(5)** 531-9.
[31] Eve S, Sami L, Marja R and Jukka S 2012 *Cellulose***19** 661-9.
[32] Nenad D, Aleksandar D M, Jasmina B N, Sasa Z D, Milica R, Danijela V B and Petar S U 2016 *J. Serb. Chem. Soc.* **81(5)** 589-605.
[33] Hengchong S, Bin L, Cheng C and Yonglin J 2007 *J. Appl. Polym. Sci.* **104** 3161-70.
[34] Baojie D, Quan W, Wei Y, Zhuang L and Weihong G 2016 *J. Appl. Polym. Sci.* **133** 4295-8.
[35] Singh V, Isobe S, Okadome H, Othtsubo K 2007 Trends in App. Sci. Res. **2(2)** 103- 14.
[36] Averous L, Fringan C and Moro L 2001 Starch/Starke **53** 368- 71.
[37] Victoria G and Patricia T 2005 European Polym. J. **41** 715- 26.
[38] Baldev R, Annadurai V, Somashekar R, Madan R and Siddaramaiah S 2001 *Europ. Polym J.* **37** 943-8.
[39] Teruo N, Hiraku I and Masaru M 2002 *J. Macromol . Sci. Part B, Physics* **41(1)** 85-98.
[40] Muhammad P, Philip O and Muhini S 2014 *Mater. Sci. Appl.* **5** 845-56.
[41] ASTM-D6954 (2004) Standard guide for exposing and testing plastics that degrade in the environment by a combination of oxidation and biodegradation.
[42] ASTM-D6400 (2004) Standard specification for compostable plastics.
[43] EN13432:2000 (2000) Packaging. Requirements for packaging recoverable through composting and biodegradation. Test scheme and evaluation criteria for the final acceptance of packaging.
[44] ISO-17088 (2008) Specifications for compostable plastics.
[45] Thomas N, Clarke J, McLauclin A, Patrick S (2010) EV0422: assessing the environmental impact of oxo-degradable plastics across their life cycle; Report for the Department for Environment, Food and Rural Affairs.

[46] Koroleva A, Huebner M, Lukanina Y, Khvatov A, Popov A and Monakhova M 2012 Chemistry & Chemical Tech. **6(4)** 405-13.
[47] Emo C, Andrea C and Salvatore D A 2007 *Polym. Deg. Stab.* **92** 1378-83.
[48] Mara Z, Barbara C Z and Ana Maria C G 2014 *Materials Research* **17(1)** 121-6.

IOP Conf. Series: Earth and Environmental Science **60** (2017) 012012 doi:10.1088/1755-1315/60/1/012012

Effect of temperature, time, and milling process on yield, flavonoid, and total phenolic content of *Zingiber officinale* water extract

R Andriyani*, W Kosasih, D R Ningrum and S Pudjiraharti
Research Unit for Clean Technology, Indonesian Institute of Sciences (LIPI)
Jalan Cisitu 21/154 D, Bandung 40135, Indonesia
*Email : rina008@lipi.go.id

Abstract. Several parameters such as temperature, time of extraction, and size of simplicia play significant role in medicinal herb extraction. This study aimed to investigate the effect of those parameters on yield extract, flavonoid, and total phenolic content in water extract of *Zingiber officinale*. The temperatures used were 50, 70 and 90°C and the extraction times were 30, 60 and 90 min. *Z. officinale* in the form of powder and chips were used to study the effect of milling treatment. The correlation among those variables was analysed using ANOVA two-way factors without replication. The result showed that time and temperature did not influence the yield of extract of Powder simplicia. However, time of extraction influenced the extract of simplicia treated without milling process. On the other hand, flavonoid and total phenolic content were not influenced by temperature, time, and milling treatment.

1. Introduction

Recent trends in extraction techniques have largely focused on finding solutions that minimize the use of solvents or to find alternatives for petroleum solvents [1]. Based on The Conductor -like Screening Model for Real Solvent (COSMO-RS) approach, water is categorized as a green polar solvent due to renewable and non-toxic properties [2].

Extraction is often used for the separation of medicinally active compounds that contain in plant or animal tissues apart from the inactive or inert components. For this purpose, a suggestion temperature for drying process should not be more than 60°C and a selective solvent should be employed in standard extraction procedures[3,4].

The bioactivity analysis of water and ethanol extracts of *Zingiber officinale* has been carried out in previous study [5]. Bioactivity depends on the solvent and temperature used in extraction proces [5]. The polar solvent is used in extraction to get numerous secondary metabolites from the plant tissues such as flavonoid and phenolic compounds. Polyphenols are widely distributed in nature and found in terrestrial and marine plants or animals [6]. Among others, flavonoids are the most important component from a nutritional point of view [6]. Flavonoids are polyhydroxy chemical compounds synthesised in various plants[6]. Some of them possess medicinal properties [6]. Meanwhile, the effect of simplicia size is described in the current study. Hence, the present study aimed to investigate the effect of variable temperatures and times on extraction process of *Z. officinale* using water as solvent. Powder and chips simplicia were used in this study.The extraction method used was adapted from the methods of decoction and infusion.

2. Experimental

2.1. Material

Z. officinale was obtained from Bandung, West Java – Indonesia. The rhizome of *Z. officinale* was used for extraction. *Aqua pro distillate* was used for extraction, whereas pro analyses solvent was used for analysis.

2.2. Extraction

Z. officinale was peeled, washed, thinly sliced, and dried at 50°C for three days to get Z. officinale chips [5]. Z. officinale chips were then milled by using an electric food blender. Z. officinale of 10 g was put in 250 mL aqua pro distillate. The decoction was carried out at 90°C for 30 min whereas infusion was carried out at 90°C for 15 min. Temperature was varied at 50, 70 and 90°C whereas time was varied for 30, 60 and 90 min. The resulted extract was filtered, evaporated at 50°C into slightly viscous, and oven-dried at 50°C.

2.3. Total Flavonoid content

Total flavonoid assay was adapted from Saravanakumar and Andriyani methods [5,7]. A sample of 400 μL was put into a test tube, added with 1600 μL aquadest, 120 μL $NaNO_2$ 5%, and kept for 5 min. Further, added with 120 μL $AlCl_3$ 10% in ethanol, 800 μL NaOH 1M and 960 μL aquadest. The absorbance of the reaction mixture was measured at 420 nm. Rutine Quercetin-3-rutinoside hydrate (Sigma) was used as positive control.

2.4. Total phenolic content

Total phenolic assay was adapted from Saravanakumar and Andriyani methods [5,7]. A sample of 300 μL was put into a test tube, added with 1500 μL of Follin 10% and Na_2CO_3 75% . The mixture was incubated at 50°C for 15 min. and measured its absorbance at 760 nm. Gallic acid was used as a positive control.

3. Result and discussion

3.1. Effect of temperature and time on yield extract

As presented in Table 1, the optimal temperature and time for extraction was 70°C and 60 min respectively, either for *Z. officinale* that was treated by milling process (powder) or not (chips). At higher temperature, some of volatile compounds might have evaporated and other compounds degraded, thus the yield extract decreased. The highest yield extract was resulted from powder simplicia ranged from 8.8 to 18.9 % compared to that from chip simplicia with a range from 8.7 to 9.6 %. Milling process caused surface area of simplicia getting much wider and made it easier for the solvent to diffuse into the tissue leading to the active compound transfer. However, statistically, time and temperature variables did not give any significant influence on the yield as indicated in Table 2 and Table 3. Statistical analysis was used to evaluate the interaction between time and temperature parameters. ANOVA two factors for milling treatment is shown in Table 2 whereas without milling treatment is presented in Table 3.

Table 2 shows that residual (error) means square (MS) in extraction of simplicia treated with milling process was lower than MS of variable time. Therefore it was concluded that there was no interaction between variables temperature and time on the yield of extract. The same result was also shown by simplicia without milling process.

As shown in Table 3, since residual (error) MS in extraction of simplicia without milling treatment was smaller than MS of variable temperature, then it was concluded that there was no interaction of variable temperature on the yield of extract. However, as residual (error) was higher than MS of variable time, thus variable time influenced the yield extraction.

Table 1. Effect of temperature and time on yield extract.

Time (min)	Temperature (°C)					
	50	70	90	50	70	90
	Yield Extract of *Z. officinale* Chips (%, w/dw)			Yield Extract of Powder *Z. officinale* (%,w/dw)		
30	8.7 ± 0.6	9.6 ± 0.4	8.9 ± 0.8	17.0 ± 0.4	16.1 ± 0.9	12.0 ± 0.5
60	9.2 ± 0.4	9.6 ± 0.4	9.1 ± 0.5	16.8 ± 1.5	18.9 ± 1.8	15.3 ± 1.2
90	8.9 ± 0.6	9.5 ± 0.6	9.6 ± 0.4	16.9 ± 0.2	18.4 ± 1.6	8.8 ± 0.2

w: weight; dw: dry weight

Table 2. ANOVA two factors for powder *Z. officinale*.

Source of Variation	Mean Square
Time	4.8
Temperature	29.2
Error	4.1

Table 3. ANOVA two factors of *Z .officinale* chips.

Source of Variation	Mean Square
Time	0.06
Temperature	0.30
Error	0.07

3.2. Effect of milling treatment on total flavonoid and phenolic content

The results of milling treatment on total flavonoid and total phenolic content of *Z.officinale* extract are presented in Table 4. It shows that total flavonoid and total phenolic content of the extract from powder simplicia was higher than that from the chips form. The highest total flavonoid and phenolic content was obtained at 90°C for both form of simplicia, but a longer time was needed for the chips. These results indicated that extraction of bioactive compounds from simplicia in the form of powder was more effective compared to that from the chips form due to the small size and the wide surface of simplicia. The result of ANOVA two factors for flavonoid is presented in Table 5.

The same as yield extract, the residual (error) MS value was also smaller than either MS of time and temperature or milling treatment and without milling treatment. This result indicated that the effect of time-temperature did not interact with the milling treatment. Moreover, time and temperature variables as well as milling treatment have no effect on flavonoid content. Further, the result of ANOVA two factors for total phenolic content is presented in Table 6.

Table 6 shows that since residual (error) MS was smaller compared to both of the MS variables, therefore the effect of time-temperature did not interact with the milling treatment. Moreover, time and temperature variables have no effect on total phenolic content. In addition, milling/without milling did not affect the phenolic content.

Table 4. Effect of milling treatment on total flavonoid and phenolic content of extract.

Temperature (°C)	Time (min)	*Z. officinale* Chips		Powder *Z. officinale*	
		Total Flavonoid (%)	Total Phenolic Content (%)	Total Flavonoid (%)	Total Phenolic Content (%)
50	30	3.3 ± 0.12	0.6 ± 0.03	5.9 ± 0.13	1.3 ± 0.02
	60	3.9 ± 0.04	1.2 ± 0.18	5.4 ± 0.04	1.2 ± 0.04
	90	2.9 ± 0.02	0.8 ± 0.01	5.6 ± 0.17	1.3 ± 0.01
70	30	3.8 ± 0.20	1.2 ± 0.01	5.2 ± 0.08	1.1 ± 0.05
	60	4.0 ± 0.01	1.2 ± 0.01	6.0 ± 0.04	1.6 ± 0.03
	90	5.1 ± 0.03	1.3 ± 0.00	4.9 ± 0.03	1.4 ± 0.01
90	30	4.3 ± 0.11	1.0 ± 0.01	7.6 ± 0.01	1.8 ± 0.05
	60	5.9 ± 0.65	1.4 ± 0.08	7.0 ± 0.55	1.8 ± 0.03
	90	5.3 ± 0.07	1.4 ± 0.04	6.9 ± 0.09	1.9 ± 0.03

Table 5. ANOVA two factors of total flavonoid of powder/chips.

Source of Variation	Mean Square
Time and temperature	1.26
Powder and Chips	14.09
Error	0.56

Table 6. ANOVA two factors of total phenolic content of powder/chips.

Source of Variation	Mean Square
Time and temperature	0.12
Powder and Chips	0.67
Error	0.04

4. Conclusion

Z. officinale has been extracted in water through milling and without milling treatment. Even though statistically all the variables did not give any interaction, the optimum condition to obtain high yield extract was achieved at 70°C for 60 min. However, the highest total flavonoid and phenolic content was obtained at 90°C for 30 min for powder simplicia and 60 min for the chips.

Acknowledgement

The authors are grateful to Indonesian Institute of Sciences for financial support for this research.

References

[1] Chemat F, Vian MA 2014 *Alternative Solvents for Natural Products Extraction* 5.
[2] Laurianne Moity, Morgan Durand, Adrien Benazzouz VM, Aubry and J-M 2014 *Alternative Solvents for Natural Products Extraction: In Silico Search for Alternative Green Solvents* 1–24.
[3] Handa 2008 *International Centre For Science And High Technology: An Overview of Extraction Techniques for Medicinal and Aromatic Plan* 21–54.

[4] Anonim 2008 *Farmakope Herbal Indonesia I*. Departemen Kesehatan Republik Indonesia XXV.
[5] Andriyani R, Budiati TA and Pudjiraharti S 2015 *Procedia Chemistry* **(16)**149–54.
[6] Somasundaram SG and Oommen B 2013 *Bioactive Food as Dietary Interventions for Arthritis and Related Inflammatory Diseases: Antioxidant Flavonoids for Arthritis Treatment Human and Animal Models* 1-16.
[7] Saravanakumar A 2009 *Pak. J. Pharm. Sci.* **22(3)**:282–6.

Pineapple peel wastes as a potential source of antioxidant compounds

V Saraswaty[1]*, C Risdian[1], I Primadona[1], R Andriyani[1], D G S Andayani[1] and T Mozef[2]

[1]Research Unit for Clean Technology, Indonesian Institute of Sciences (LIPI)
Jalan Cisitu 21/154 D, Bandung 40135, Indonesia
[2]Research Centre for Chemistry, Indonesian Institute of Sciences, Kawasan Puspiptek Serpong, Tangerang Selatan, Banten, Indonesia, 15314
*Email : vsaraswaty@gmail.com

Abstract. Indonesia is a large pineapple (*Ananas comosus*) producing country. Food industries in Indonesia processed this fruit for new products and further resulted wastes of which cause an environmental problems. Approximately, one pineapple fruit total weight is 400 gr of which 60 g is of peel wastes. In order to reduce such pineapple peel wastes (PPW), processing to a valuable product using an environmentally friendly technique is indispensable. PPW contained phenolic compound, ferulic acid, and vitamin A and C as antioxidant. This study aimed to PPW using ethanol and water as well as to analyze its chemical properties. Both dried and fresh PPW were extracted using mixtures of ethanol and water with various concentrations ranging from 15 to 95% (v/v) at room temperature for 24 h. The chemical properties, such as antioxidant activity, total phenolic content (Gallic acid equivalent/GAE), and total sugar content were determined. The results showed that the range of Inhibition Concentration $(IC)_{50}$ value as antioxidant activity of extracts from dried and fresh PPW were in the range of 0.8±0.05 to 1.3±0.09 $mg.mL^{-1}$ and 0.25±0.01 to 0.59±0.01 $mg.mL^{-1}$, respectively, with the highest antioxidant activity was in water extract. The highest of total phenolic content of 0.9 $mg.g^{-1}$ GAE, was also found in water extract.

1. Introduction

Pineapple (*Ananas comosus*) is one of the main agricultural commodities from Subang, West Java, Indonesia. The city produces over 135.000 tons of pineapple in the year of 2014 [1]. As the impact of pineapple food industries, pineapple peel wastes (PPW) are the important issue of waste management of which urgently to be overcame. PPW is therefore converted into highly valuable product, since contains considerable content of antioxidant property, sugar, phenolic compound, high fiber, and protein. PPW also provides high potential bromelain enzyme as functional material [2-4].

Generally, PPW has been used for animal feed or land fertilizer [5], as reducing agent or antioxidant, and substrate for production of bio-ethanol [6-7]. Antioxidant or reducing agent plays an important role both in health sector and chemical industry. Unfortunately, chemically synthesis antioxidant such as butylated hydroxyanisole (BHA) and butylated hydroxytoluene (BHT) are not safe for health. It is found that at high dose both BHA and BHT induced tumor growth [8]. Therefore we aimed to explore PPW as a potential of natural antioxidant. One of interesting pineapple applications as reducing agent is in biosynthesis of metal nanoparticles. It is found that biosynthesis of metal nanoparticles can be carried out by aqueous pineapple extract [9-10]. The work had successfully produced silver and cooper nanoparticles. Chlorogenic acid and ferulic acid in aqueous pineapple extract is considered having an important role as reducing agent or antioxidant [9].

Study of antioxidant activity showed that methanol was the best organic solvent for pineapple extract [2,11]. However, methanol is a toxic organic solvent which caused several health problems [12]. Mixtures of ethanol and water are an environmentally preferable solvent [13]. Study on the pineapple as antioxidant by ethanol was considerable aginst DPPH radical and achieved at concentrations of 50 and 70% v/v. , and is considerable against DPPH radical [11]. The study aimed to investigate the effect of water-ethanol at various concentrations for PPW as antioxidant activity.

2. Experimental

2.1. Material

PPW was collected from UKM Alam Sari, Subang, West Java Indonesia. PPW was washed under tap water and dried in a blower oven at 55°C for 48 hr. The dried PPW was grinded mechanically and further used for extraction. Meanwhile for fresh material, PPW was cut into small pieces, then directly macerated in organic solvent. Chemical reagents sulphuric acid, anthrone, methanol, Na_2CO_3, Follin Ciocalteu are pro analysis grade from Merck, meanwhile DPPH radical, ascorbic acid, gallic acid are from SIGMA-ALDRICH, and ethanol is technical grade from Bratachem.

2.2.Method

2.2.1. Extraction

Ten g of fresh and dried PPW were macerated in 100 mL of ethanol:water at various concentration (0:100, 15:85, 35:65, 55:45, 75:25 and 100:0 for 24 hr at ambient temperature). The filtrates were evaporated under rotary vacuum evaporator (Heidolph, Germany). The yield of extraction was determined as percentage of concentrated extract compare to weight of sample before extraction.

2.2.2. Total sugar content

Total sugar content was determined according to anthrone method [19] with a slight modification. One g of anthrone was dissolved in 500 mL 72% v/v sulphuric acid. One mL of sample was put in a test tube and added by 5 mL of anthrone reagent. The reaction was incubated in a water bath at 80°C for 11 min. The absorbance was then measured at 630 nm by a UV-VIS Spectrophotometer (Hitachi, U2800) with glucose as standard. Total sugar content was calculated based on glucose standard curve. The analysis was performed triplicate.

2.2.3 Antioxidant activity

Free radical scavenging effect was performed according to the following protocol: 50 µl of samples at various concentrations (0.5, 1, 2, 4 and 8 $mg.mL^{-1}$) was put into 96-micro wells plate and added with 200 µl of 0.077 mM DPPH. The reaction was incubated at room temperature for 15 min. The absorbance was then measured at 520 nm by ELISA reader (Multiskan, BioRad). Percentage of inhibition was calculated by using the following formula:

$$\%\ \text{Inhibition} = \frac{[C-S]}{C} \times 100\% \qquad (1)$$

C is absorbance of DPPH, and S is absorbance of sample. Ascorbic acid was used as positive control. The procedure was performed triplicate [14].

2.2.4. Total phenolic content

Total phenolic content was determined spectrophotometrically according to Follin-Ciocalteu method. Sample of 300 µl of at concentration of 1000 µg/mL was mixed with 1200 µl Folin Ciocalteau (10% v/v) and 1500 µl Na_2CO_3 (7.5% w/v). The reaction was incubated at 55°C for 15 min. Absorbance was then measured at λ 760 nm by UV-VIS Spectrophotometer (Hitachi, U2800). Gallic acid was used as standard and total phenolic content was calculated based on gallic acid standard curve. The total of phenolic content was expressed as gallic acid equivalent (GAE). The procedure was performed triplicate [15]

3. Result and discussion

3.1 Yield of extraction

PPW is highly containing water of which will cause a rapid rottening process and significantly pollutes environment. Air contamination occures due to putrid odor whereas land contamination due to microbial growth. Water content of PPW is shown in Table 1 It shows that water content of fresh PPW is higher than dried PPW. This results revealed that high water content of PPW was reduced through drying process.

Yield of PPW in various ethanol concentration is shown in Table 1. It shows that yield of dried PPW extraction is higher than fresh PPW. This is due to the effect of drying process in which significantly reduced water content. Moreover, water content in fresh PPW directly diluted ethanol concentration and increased sample weight. Meanwhile, effect of ethanol concentration on either dried or fresh PPW extraction is similar. This result indicated that compounds of PPW were majorly soluble in mixture of ethanol and water, in which 55% v/v of ethanol/water is the most approriate solvent for extraction. This process of extraction followed the principle of like-dissolves-like [16]. Total sugar content determination is important since sugar plays important role in promotion of microbial growth as the carbon source. However, since microbial growth was not found, total sugar content in dried PPW was not evaluated. In addition, water content of dried PPW was less than 10% for which is not promote for microbial growth. Table 1 shows that PPW water extract of PPW reveals lower sugar content compared with ethanol extract. This is due to the sugar in PPW was more easily extracted by the mixture of ethanol-water rather than by water solely. Fresh PPW extract contains sugar so that easily promote the growth of microbes. Therefore, mixtures of ethanol and water has an advantage. In extraction process, presence of ethanol will reduce the growth of microbes due to that ethanol is an organic solvent of which exhibits a rapid and broad spectrum antimicrobial activity [17].

Table 1. Yield of extraction, antioxidant activity and total sugar content.

	Water content (% w/w)	[Ethanol] (% v/v)	Yield (% w/w)	Antioxidant activity (IC_{50}, mg.mL^{-1})	Total Sugar* (% w/w)
Fresh PPW	85.5	0	5.16	0.2±0.009	3.69
		15	6.48	0.4±0.004	7.51
		35	6.57	0.4±0.003	7.16
		55	6.94	0.5±0.020	7.40
		75	5.70	0.6±0.040	6.23
		95	4.18	0.6±0.001	5.65
Dried PPW	5.43	0	24.67	0.8±0.050	n.d
		15	30.95	1.3±0.090	
		35	28.93	1.2±0.001	
		55	45.38	0.9±0.070	
		75	30.09	0.9±0.070	
		95	30.37	1.2±0.110	

n.d = not determined, PPW : Pineapple peel wastes, *based on dry extract.

3.2 Antioxidant activity of PPW

Antioxidant activity of PPW extract in various ethanol concentrations was evaluated. The lower the IC_{50} value the more the active extract. Table 1 shows, the lowest IC_{50} value of antioxidant activity was found in water extract, both for fresh and dried PPW. These results indicated that water extract contains polar compound as phenolic compound and serves as antioxidant. Moreover, the increase of total sugar content in mixture of ethanol:water affected the decrease of PPW antioxidant activity. On the other hand, IC_{50} value of fresh PPW was smaller than dried PPW. This result indicated that drying of PPW at 50°C affected antioxidant activity in which significantly increased IC_{50} value. The lower antioxidant activity in dried extract may due to bioactive compound degradation which is sensitive to heat. Other antioxidant activity of different variety of pineapples have been carried out of which IC_{50} value of antioxidant activity (DPPH free radical scavenger) were in the range of 12.4 – 29.0 mg.mL^{-1} [11]. It was found that current IC_{50} of PPW extract was low both for fresh and dried PPW (Table 1). This result indicated that currently, PPW shows as high potential source of antioxidant for DPPH free radical scanvenger.

3.3 Total phenolic content of PPW

Phenolic compounds has strong correlation with antioxidant activity. Therefore, total of phenolic content from both dried and fresh PPW was evaluated. The results are shown in Figure 1. It shows that in ethanol 75% fresh PPW contained higher total phenolic content than dried PPW. We assumed, the drying process of PPW would degrade compounds and significantly affect the antioxidant activity.

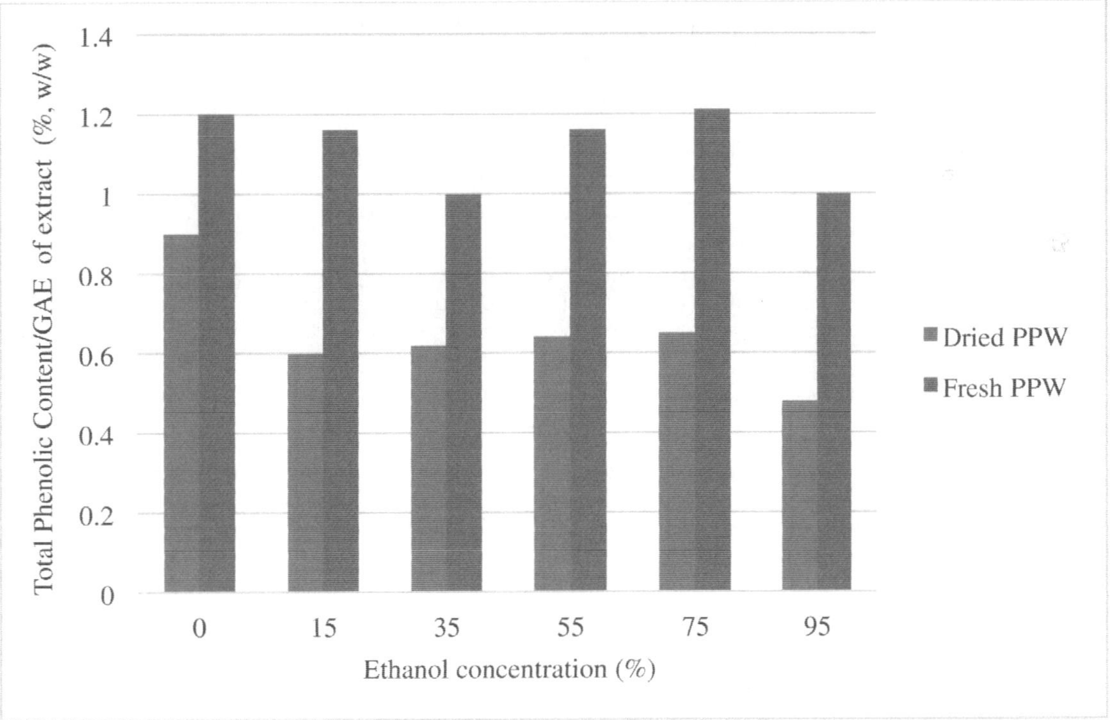

Figure 1. Total Phenolic content/GAE (%, w/w) of PPW extract at various ethanol concentrations in water.

From Figure 1 it shows that the highest total phenolic content of fresh PPW was found in water extract (0% ethanol), whereas for dried PPW was of 75% v/v ethanol extract. The results of total phenolic content in dried PPW almost the same with selected Malaysian pineapple [18].

The different of total phenolic content from fresh and dried PPW was due to heat processing. Fresh PPW was not processed and phenolic compound has yet decomposed. Whereas total phenolic content in water extract is the highest among other. This due to water extract of PPW contain polar compound of which more soluble in water than in aqueous ethanol,

From Figure 1 shows that PPW extract contained 540 – 1260 mg GAE/100 g PPW, nearly about 10 times higher than previous study was of 34.7-54.7 [18] and 22.7-53.8 mg GAE/100 g fruit [11].

4. Conclusion

PPW from Subang, West Java, has succesfully been converted into a potential source of antioxidant. Ethanol/Water of 55% (v/v) provided high yield of antioxidant and avoided microbial contamination.

Acknowledgement

Author thanks to Research Unit for Clean Technology, Indonesian Institute of Sciences, for support this research under Riset Mandiri 2016 and Ms. Rossy Choerun Nissa for her technical help on this research.

References

[1] BPS 2015 *Subang in Figure* BPS Kabupaten Kota Subang hal. 76
[2] Hossain M A and Rahman M M S 2011 *Food Res. Int.* **44** 672-6
[3] Ramalingan C, Srinath R and Islam M M 2012 *Elixir Food Scie.* **45** 7822-26
[4] Ketnawa S, Chaiwut P and Rawdkuen S 2011 *Food Bioprod. Process.* **90** 385-91
[5] Makinde O A, Odeyinka S M and Ayandiran S K 2011 *Livest. Res. Rural Dev.* **23**
[6] Pornpunyapat J, Chotigeat W and Chetpattananondh P 2014 *Adv. Mat. Res.* **875-7** 242–5
[7] Erukainure O L, Ajiboye J A, Okafo O Y, Kosok R S B, Owolab O F O and Adenekan I S O 2012 *J. Food Biochem.* **36** 643–7
[8] Kahl R and Kappus H 1993 *Z. Lebensm. Unters. Forsch.*, **196** 329–38
[9] Ahmad N and Sharma S 2012 *Green Sustain.Chem.* **2** 141-7
[10] Ranjitham A M, Ranjani G S, and Caroling G 2015 *Int.J. Pharm Tech Res.* **8** 750–69
[11] Yuris A and Siow L F 2014 *J. Food Stud.* **3** 40
[12] Kraut J A and Kurtz I 2008 *Clinical J. the American Soc. of Nephrology.* **3** 208–25
[13] Capello C, Fischer U, and Hungerbühler K 2007 *Green Chem.* **9** 927–34
[14] Saraswaty V, Risdian C, Budiwati T A and Tjandrawati M, 2013 *Pros. Teknol. Untuk Mendukung Pembang. Nas.* **1** 196–200
[15] Saraswaty V, Risdian C, Lelono R A A and Mozef T 2015 *Oxid. Antioxid. Med. Sci.* **4** 97
[16] Chew K K, Khoo M Z, Ng S Y, Thoo Y Y, Aida W M W, and Ho C W 2011 *Int. Food Res. J.* **18** 1427–35
[17] Mcdonnell G and Russell A D 1999 *Clinical Microbiol. Rev.* **12** 147–79
[18] Alothman M, Bhat R, and Karim A A 2009 *Food Chem.* **115** 785–8
[19] Yemm E W and Willis A J 1954 *Biochem J.* **57** 508-14

Physical and mechanical properties of five Indonesian bamboos

A H D Abdullah[1*], N Karlina[2], W Rahmatiya[2], S Mudaim[2], Patimah[2] and A R Fajrin[2]
[1]Research Unit for Clean Technology, Indonesian Institute of Sciences (LIPI)
Jalan Cisitu 21/154 D, Bandung 40135, Indonesia
[2] Department of Physics, UIN-Sunan Gunung Djati, Bandung, Indonesia
*Email : dawamdullah@yahoo.com

Abstract. Indonesia is a country with abundant of bamboo. More than 100 species of bamboo are found in Indonesia, and the production of bamboo ranks second after China. Bamboo has great potential to be used as industrial raw materials due to its high mechanical properties. The properties of the bamboo's fiber have been widely reported by several researchers, but the rarely properties of bamboo's culm which is consist of fiber, parenchym, and conducting tissue were reported. This research aimed to identify the culm in five most common bamboos in Indonesia namely Bamboo Temen (*Gigantochloa Atter*), Bamboo Apus (*Gigantochloa Apus Kurz*), Bamboo Kuning (*Bambusa vulgaris schard var. Vitata*), Bamboo Gombong (*Gigantochloa pseudoarundinacea*), and Bamboo Hitam (*Gigantochloa verticillata*). Properties identification were on the density, morphology, tensile strength, and modulus of elasticity. The density of different types of bamboo's culm varied from 0.54 to 0.78 g/cm^3. Morphology of culm showed hollow structure which was consist of vessel and surrounded by fibers. Tensile strength of Bamboo Temen, Bamboo Apus, Bamboo Kuning, Bamboo Gombong, and Bamboo Hitam were of 195, 179, 82, 114, and 118 MPa, respectively whereas corresponding modulus of elasticity were of 16.7; 7.5; 14.3; 16.0; and 10.1 GPa. The best mechanical performance was found in Bamboo Temen.

1. Introduction

Indonesia is a tropical country, rich with the variety of bambu and becoming the biggets exporter of bamboo in the world after China [1]. Indonesia has around 154 types from 1,250 to 1,500 type of world bambu diversity [2]. Bamboo is found in many places, either grow naturally or cultivated deliberately. Bamboo is cultivated since its possess good mechanical strength for structural application.

The strength properties of bamboo increase with decreasing in water content and are closely related to density [3]. The anatomical structure of a transverse section of any culm is determined by the shape, size arrangement and number of vascular bundles which [4]. Due to its vascular shaped like a hollow so the number is more likely to lower the mechanical properties of culm. The culm tissue is mostly parenchyma and the vascular bundles which are composed of vessels and fibres. The total culm compries about 50% parenchyma, 40% fiber, and 10% conducting tissue [4].

The properties of bamboo fiber have been widely reported by several researchers [5-8], but bamboo's culm which is consist of fiber, parenchym, and conducting tissue hardly reported. This research aimed to identify the culm of five most common bamboos in Indonesia, namely Bamboo Temen (*Gigantochloa atter*), Bamboo Apus (*Gigantochloa Apus* Kurz), Bamboo Kuning (*Bambusa vulgaris* Schard var. Vitata), Bamboo Gombong (*Gigantochloa pseudoarundinacea*), and Bamboo Hitam (*Gigantochloa verticillata*). Properties identification were on the density, morphology, tensile strength, and modulus of elasticity.

IOP Conf. Series: Earth and Environmental Science **60** (2017) 012014 doi:10.1088/1755-1315/60/1/012014

2. Experimental

2.1. Material

Five kind of bamboos of which approximately three years old were taken from several places in Bandung, West Java, Indonesia. It was taken about 4 weeks after plant cutting (Figure 1). The skin and inner-most were removed from the culm and resulted the clean culm.

Figure 1. Bamboo Culm

2.2. Density measurement

Each sample of bamboo culm was weighed (mass of bamboo, **m** gr) by using micro-balance (Type Metler-Toledo). Empty pycnometer was weighed, filled fully with distilled water, and reweighed. Sample of bamboo culm were then put into pycnometer and led distilled water comes out through the hole, then weighed. This process was repeated for 3-4 times. Each bamboo's volume was measured as an equal to the volume of water displaced (volume, **v** cm^3). Density calculation was as **m/v**.

2.3. Scanning Electron Microscope

SEM (JEOL JSM-T330A) image was performed to identify morphology of bamboo. Dimension of samples was of 0.5cm x 0.5cm x 0.5cm. Each samples was coated with gold by ion sputtering. The image was obtained by secondary electron at accelerating voltage of 15 kV with 100 and 500 x magnification.

2.4. Tensile Test

Tensile test consist of modulus of elasticity (MOE) and tensile strength test. Specimens was cut with dimensional 25 cm x 2 mm x 3 cm. Tensile test was conducted based on ASTM D3039 reference at 2 mm/min speed, and load cell of 5,000 kgf. Elastic modulus measurement was performed on extensometer. Each sample was measured 2-3 times.

3. Result and discussion

Density of five kind of bamboos is presented in Table 1.

Table 1. Density of bamboos.

Name of bamboo	Density (g/cm^3)
Bamboo Temen	0.76
Bamboo Apus	0.60
Bamboo Kuning	0.62
Bamboo Gombong	0.78
Bamboo Hitam	0.54

Table 1 shows that Bamboo Gombong shows the highest density than other types whereas Bamboo Hitam reveals the lowest density. The density of bamboo can be related with the macro-structure of culm, shape and size of vessel. The vessel are usually in the hollow form, so that more vessels will reduce its density values. Moreover, high density contributes to fiber content in vascular tissue.

Morphology image of bamboos is presented in Figure 2. It reveals that Bamboo Temen shows rare of vascular (vessel) structure indicated by less of hollow structure as dark area on cross-section. The culm looks more solid than four other bamboos. This might be due to the high density of the Bamboo Temen. It is also found that Bamboo Kuning has a hollow structure (vascular) which is different from the other vessel structure. The shape tends to be irregular, unlike other bamboo vessels which possesses a circular form. Irregular shape may produce stress concentrations and affects on its mechanical properties.

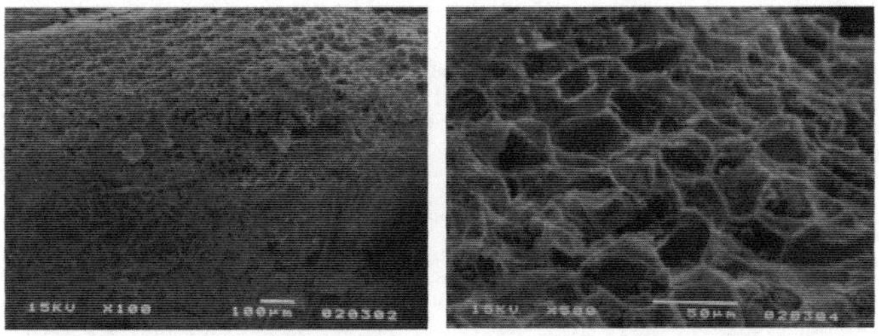

Bamboo Temen

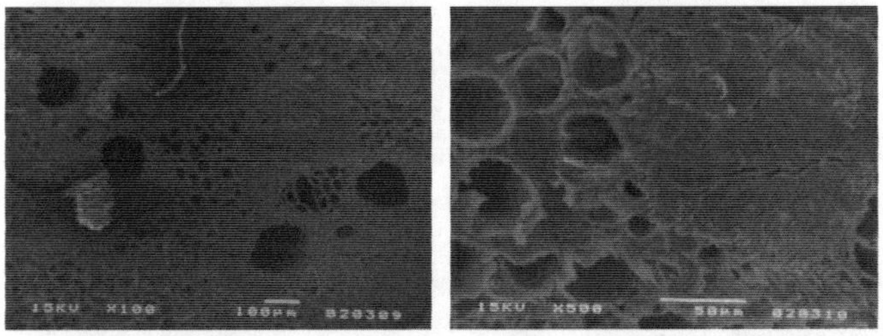

Bamboo Apus

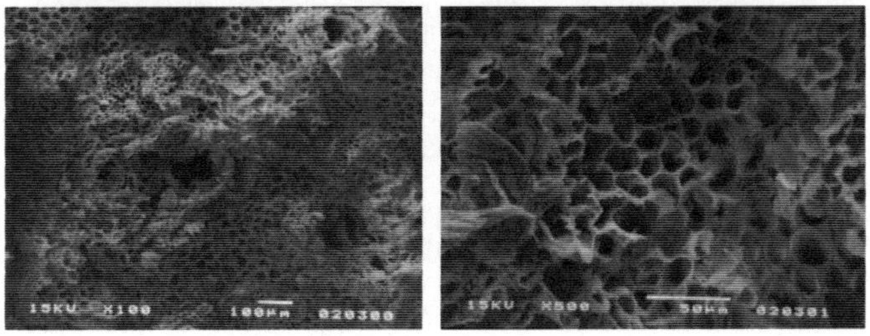

Bamboo Kuning

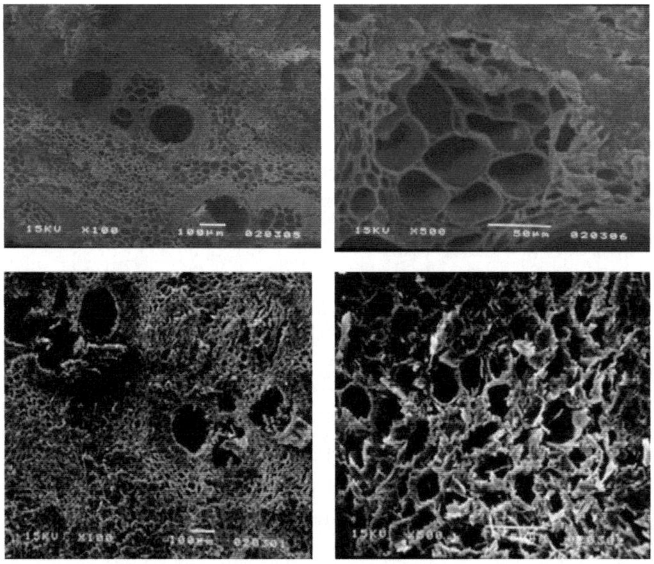

Bamboo Gombong

Bamboo Hitam

100x 500x

Figure 2. Morphology image of bamboos with 100 and 500x magnification.

Tensile strength and modulus of elasticity of bamboos are presented in Figure 3 and 4. Figure 3 and Figure 4 show that Bamboo Temen has the highest strength and elastic modulus among the four others types of bamboo. This is might be due to the structure of Bamboo Temen which is solid and less-vessel as already proved by its density and morphology (Table 1 and Figure 2). Bamboo Kuning has a lowest strength (82 MPa), but the modulus of elasticity values were relatively high (14.35 GPa). As showed in Figure 2, Bamboo Kuning has vascular structure which is tend to be irregular and may caused stress concentrations and finally weaken its strength. Elastic modulus of Bamboo Gombong is quite high which might be due to its high fibers content [9].

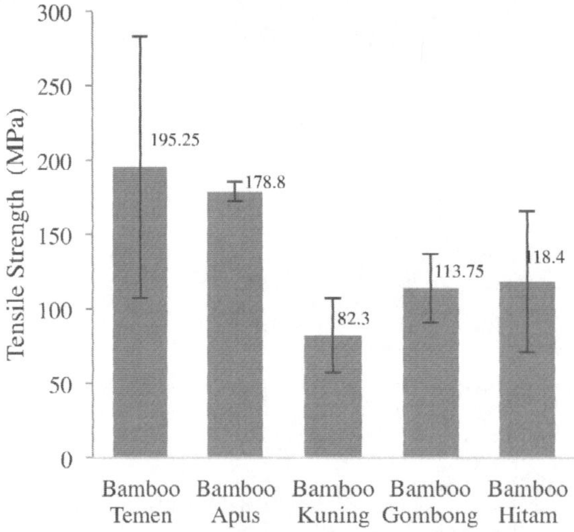

Figure 3. Tensile strength of bamboos.

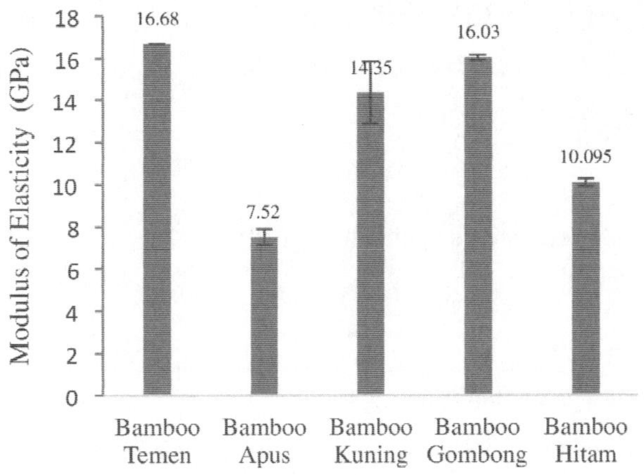

Figure 4. Elastic modulus of bamboos.

4. Conclusion

Based on the results, it can be proved that the macro-structure of the vascular and parenchyma affected the physical and mechanical properties of bamboos. Morphological observation with SEM and verification by density measurements could be performed to analyze these correlations. Bamboo vessels were usually in the form of a hollow and more vessels will reduce its density.

Acknowledgement

The Authors are grateful to the Indonesian Institute of Science through Mandiri Project 2015, Research Unit for Clean Technology LIPI for financial support.

References

[1] H.P.S Abdul Khalil, I.U.H Bhat, M.Jawaid, A. Zaidon, D. Hermawan, Y.S.Hadi 2012 *J. Mat. and Design.* **42** 353-368

[2] Elizabeth A. Widjaja and Karsono 2005 *Biodeversitas.* **Vol. 6 No. 2** 959-99

[3] S Dransfield and E.A. Widjaja 1995 *Plant Resources of South-East Asia (PROSEA) No.7 : Bambus Backhyus Publisher. Leiden*

[4] Liese W 1985 *Institute of Wood and Wood Preservation of the Federal Research Center for Forestry and Forest Product, Federal Republic of Germany.* 196-208

[5] Kabir H, Md. Ghafur A, Ahmed F, Begum F, Md. Qadir R 2014 *Universal J. Mat. Scie.* **2(6)** 119-124

[6] Sreenivasulu S and Reddy A.C 2014 *Int. J. Eng. Res.* **Vol. 3 Issue No: Special 1** 187-194

[7] Shito T, Okubo K, Fujii T 2002 *J. High Perf. Struct. and Compos.* **ISBN 1-85312-904-6** 175-182

[8] Rassiah K and Megat Ahmad MM. H 2013 *Australian J. Basic and Appl. Scie.* **7 (8)**: 247-253

[9] Xiaobo Li 2004 A *Thesis of Graduate Faculty of Lousiana State University Agriculture and Mechanical College* 27-46

Synthesis and characterization of novel TiO_2-ZnO-CoO nanocomposite photocatalyst for photodegradation of methylene blue dye

M Nasir[1]*, D Irnameria[2,3] and M A Zulfikar[2]
[1]Research Unit for Clean Technology, Indonesian Institute of Sciences (LIPI)
Jalan Cisitu 21/154 D, Bandung 40135, Indonesia
[2] *Department of Chemistry, ITB, Bandung*
[2,3] *Department of Health Analysis, Poltekkes Kemenkes Bengkulu*
*Email : mnasir71@yahoo.com

Abstract. TiO_2-ZnO-CoO nanocomposite photocatalyst has been synthezised using sol-gel method. Titanium (IV) ethoxide, zinc acetate ($Zn(CH_3COO)_2.2H_2O$) and cobalt acetate ($Co(C_3COO)_2.4H_2O$) were used as precursors. TiO_2 and ZnO posses good photocatalytic ability in photodegradation of water pollutant. The disadvantage and weakness of photocatalytic reaction of TiO_2 and ZnO mostly happens at ultraviolet range and less at visible light. This research aimed to decrease the disadventage of TiO_2 and ZnO, using cobalt oxide. TiO_2-ZnO-CoO nanocomposite photocatalyst was characterized by using SEM-EDS and XRD. SEM analysis showed homogenous and smooth morphology of TiO_2-ZnO-CoO nanocomposite photocatalyst. EDS analysis revealed the TiO_2-ZnO-CoO nanocomposite photocatalyst had O : Ti : Co : Zn = 32.64 : 37.39 : 11.85 : 18.12. In this study, we expose effect of amount of TiO_2-ZnO-CoO nanocomposite photocatalyst, pH, and irradiation time on methylene blue dye photodegradation. Novel TiO_2-ZnO-CoO nanocomposite photocatalyst was effective in photodegradation of methylene blue dye.

1. Introduction

Fast growing of industry influences the decrease of environment quality. For example; textile industry polluted water and reduced the water quality. Waste water from textile industry contain dyes such as methylene blue. There are many treatments to treat waste water such as membrane process, adsorption, chemical, biological, advanced oxidation, and photodegradation process. Photodegradation of methylene blue by nanometal oxide photo catalyst is one promising process [1].

Many researchers had been interested to study nanoparticle materials since nanoparticle materials can be designed to show a novelty and to increase physical and chemical properties from the materials significantly. Nanoparticle is particle with ranging in size from 1-100 nm. Presently, different metallic nanomaterials are being produced using copper, zinc, titanium, magnesium, gold, alginate, and silver [2]. Nanoparticles such as metals, semiconductors and metal oxides are interesting materials for a wide variety of applications in the field of information, energy, environmental and medical technologies due to their unique or improved properties determined primarily by size, composition, and structure along with their selforganized film structures [3]. Among these metal oxides with semiconductor property, titanium dioxide (TiO_2) and zinc oxide (ZnO) have been extensively studied for their chemical stability dan efficient photocatalytic properties [4]. However, these two metal oxides show disadvantages for application in photocatalytic degradation such as hard for regeneration as well as photocatalytic reaction mostly occurs in ultraviolet light but less in visible light [4]. Aal et.al, 2008 [5] reported the coupling of semiconductors particles with different band-gap widths and carried out photocatalyctic activity experiments using various coupled semiconductor particle systems such TiO_2 – CdS [6], TiO_2 – WO_3 [6], TiO_2 – SnO_2 [7], TiO_2 – MoO_3 [8] and TiO_2 – Fe_2O_3 [9]. However, few researcher focused on the photocatalytic activity of TiO_2 – ZnO coupled oxides [10]. This study aimed

IOP Conf. Series: Earth and Environmental Science **60** (2017) 012015 doi:10.1088/1755-1315/60/1/012015

to synthesize nano photocatalyst to degrade methylene blue dye and to optimize photodegradation process of methylene blue dye.

2. Experimental

2.1. Materials

The chemicals used were titanium (IV) ethoxide (Merck), zinc acetate ($Zn(CH_3COO)_2.2H_2O$) (Merck), cobalt acetate ($Co(CH_3COO)_2.4H_2O$) (Merck), polyvinyl pyrrolidone (PVP) (Merck), methylene blue dye (Merck), ethanol pa (Merck) and demineralization water.

2.2. Synthesis of TiO_2-ZnO-CoO nanocomposite

0,2 M zinc acetate solution was added to 2% PVP solution under vigorous stirring. Cobalt acetate solution of 0.2 M was added drop by drop to the mixture. The mixture was stirred magnetically at room temperature until a homogeneous solution was obtained. Titanium ethoxide in ethanol of 1M was added to the homogeneous solution and stirred for 12 hr then separated using filter to get the precipitate product. The product was dried in oven at 105°C for 6 hr and calcined in furnace at 800°C for 2 hr.

2.3. Characterization of TiO_2-ZnO-CoO nanocomposite

The structural characterization of nanocomposite were investigated by *X-Ray Diffraction* (XRD). The morphology of the product was characterized by *Scanning Electron Microscopy* (SEM/EDS)

2.4. Photodegradation of methylene blue dye using TiO_2-ZnO-CoO nanocomposite

The photodegradation of dye using nanocomposite was evaluated by degradation of a model aqueous solution of methylene blue (MB) after subjection to UV-radiation. The initial concentration of MB aqueous solution was 4 ppm. TiO_2 nanocomposite sample was added to 25 mL dye solution then magnetically stirred in the dark for 30 min. The irradiation source was a UV-C lamp (Philips 15 watt, λ = 253.7 nm). All photocatalytic tests were performed at constant stirring. The parameters for the photodegradation were pH, photodegradation time and nanocomposite mass. After irradiated, mixture were centrifuged. The absorbances of clear filtrate were measured by an Agilent 8453 UV-Vis spectrophotometer at 664 nm, the maximum absorption wavelength of MB.

3. Result and Discussion

3.1. Characterization of TiO2-ZnO-CoO nanocomposite

The XRD patterns of TiO_2-ZnO-CoO nanocomposite are shown in Figure1 - Figure 3. Figure 1 shows that there is titanium dioxide (TiO_2) in the product in rutile form (JCPDS 21-1276). The XRD pattern in 2θ from TiO_2-ZnO-CoO has a similar pattern with the XRD pattern from TiO_2.

Figure 2 shows the XRD pattern from zinc oxide (JCPDS 36-1451) in zincite form. The pattern in 2θ from TiO_2-ZnO-CoO nanocomposite has a similar pattern with that of ZnO. Figure 3 shows the formation of cobalt oxide. The XRD pattern from TiO_2-ZnO-CoO nanocomposite shows similar pattern with that of CoO (JCPDS 42-1300).

IOP Conf. Series: Earth and Environmental Science **60** (2017) 012015 doi:10.1088/1755-1315/60/1/012015

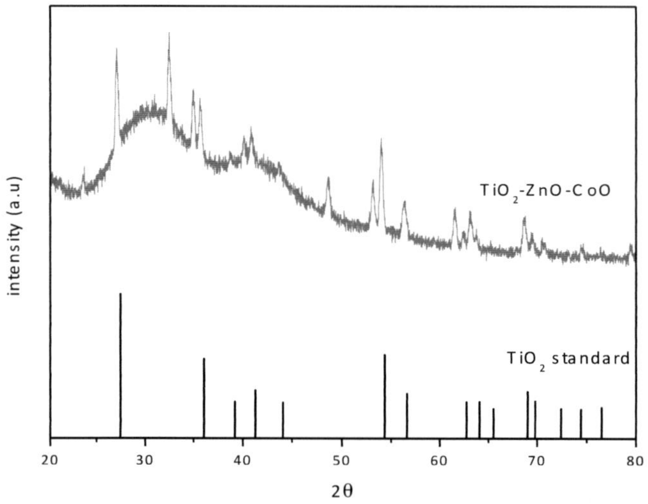

Figure 1. XRD pattern of TiO_2 standard and TiO2-ZnO-CoO nanocomposite.

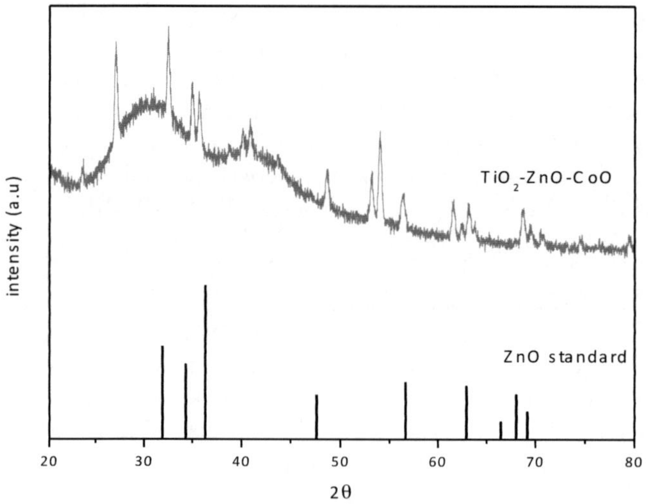

Figure 2. XRD pattern of ZnO standard and TiO_2-ZnO-CoO nanocomposite.

IOP Conf. Series: Earth and Environmental Science **60** (2017) 012015 doi:10.1088/1755-1315/60/1/012015

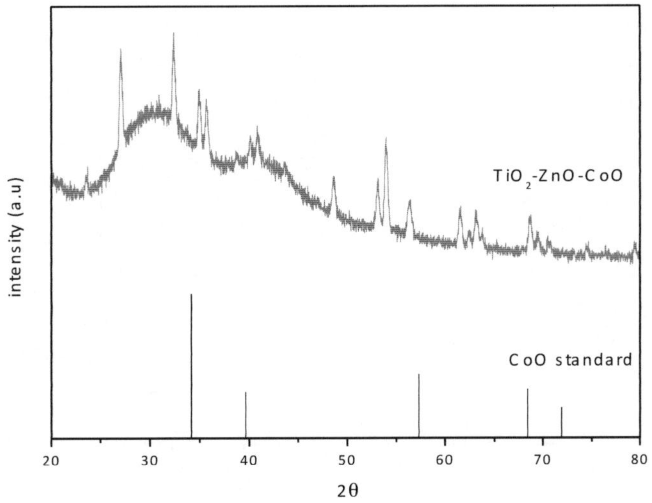

Figure 3. XRD pattern of CoO standard and TiO_2-ZnO-CoO nanocomposite.

SEM image of morphology of the TiO_2-ZnO-CoO nanocomposite is presented in Figure 4. We can observe shape of nanoparticle and its aggregation. Shape of nanoparticle is in spherical form with diameter less 1 μm. Nanoparticle size is relatively homogenous. The EDS analysis shows in Figure 5. The EDS analysis reveals that TiO_2-ZnO-CoO nanocomposite has mass ratio of O : Ti : Co : Zn was 32.64 : 37.39 : 11.85 : 18.12, respectively.

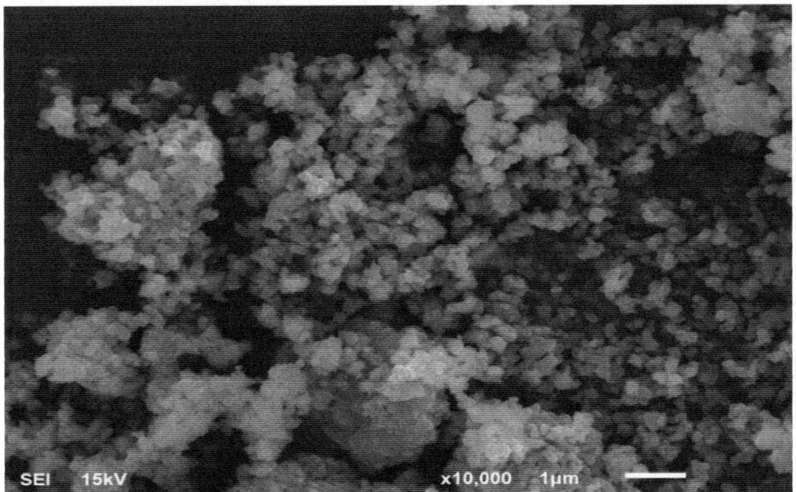

Figure 4. Morphology of TiO_2-ZnO-CoO nanocomposite.

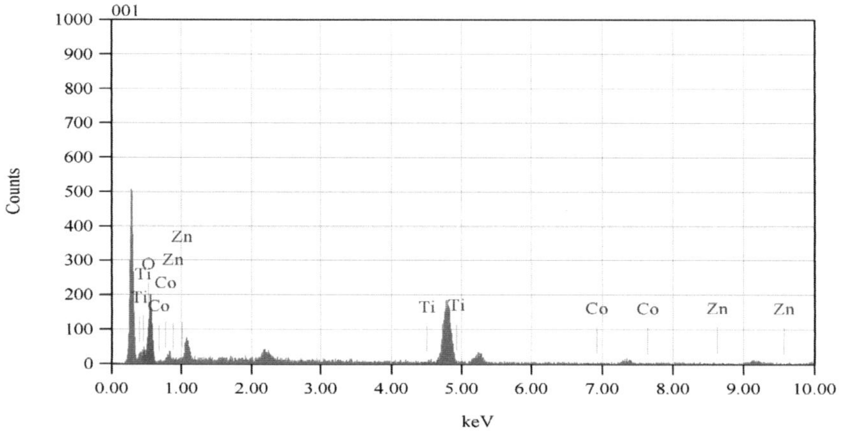

Element	Mass %
O K	32.64
Ti K	37.39
Co K	11.85
Zn K	18.12

Figure 5. EDS of TiO_2-ZnO-CoO nanocomposite.

3.2. Photodegradation of Methylene Blue dye using TiO_2-ZnO-CoO nanocomposite

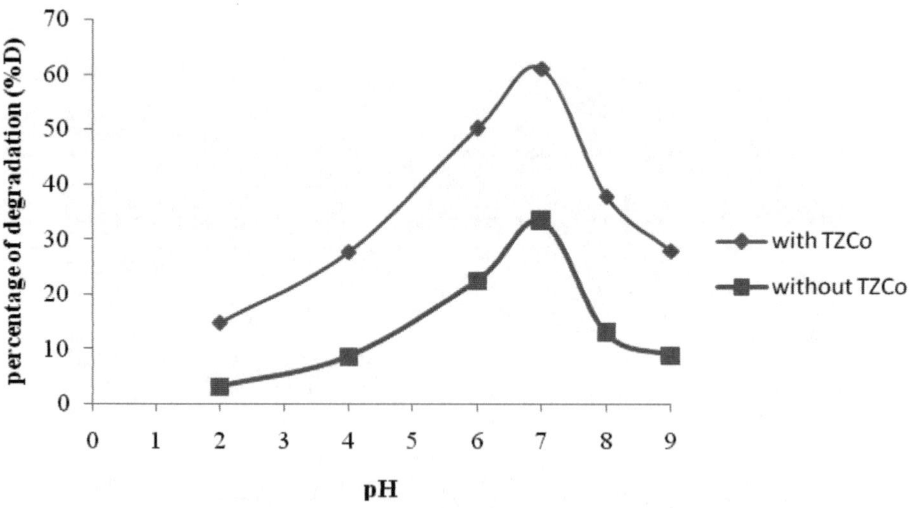

Figure 6. Effect of pH on photodegradation of methylene blue.

Figure 6 shows the effect of pH on photodegradation of methylene blue. Photodegradation was carried out with and without TiO_2-ZnO-CoO (TZCo) as the photocatalyst. Both photodegradation reach the highest value at pH 7 indicated that TiO_2-ZnO-CoO played a role in optimum condition for photodegradation of methylene blue dye at pH 7.

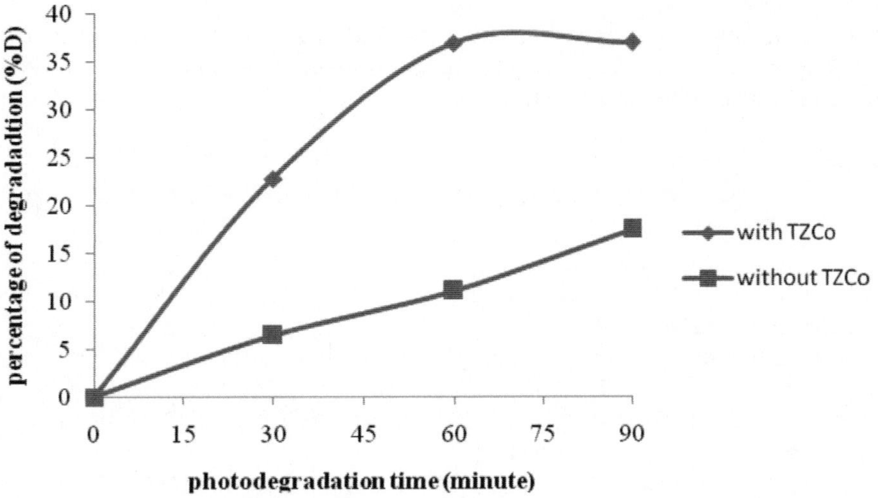

Figure 7. Effect of photodegradation time on photodegradation of methylene blue.

Figure 7 shows that the degree of degradation increased with longer photodegradation time. Sixty minutes was the optimum condition on photodegradation of methylene blue. At 90 min, degradation value similar with of 60 min.

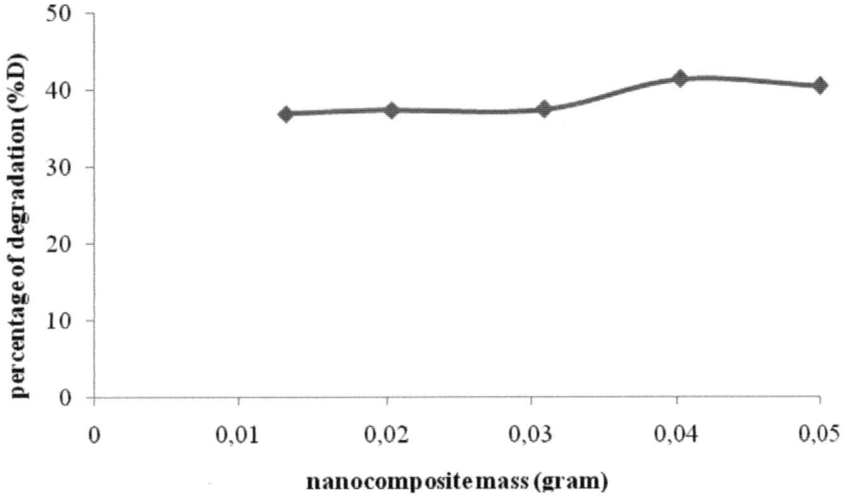

Figure 8. Effect of TiO_2-ZnO-CoO nanocomposite mass on photodegradation of methylene blue.

Figure 8 shows the effect of nanocomposite mass on photodegradation of methylene blue. Addition of nanocomposite mass from 0.01 to 0.04 gr increased the degree of degradation but decreased with the addtion of 0.05 gr of nanocomposite mass. The photodegradation of methylene blue can reach its optimum condition at addition of 0.04 gr of nanocomposite mass.

4. Conclusion

TiO_2-ZnO-CoO nanocomposite has been successfully synthesized for photodegradation of methylene blue dye. The optimum condition for photodegradation of methylene blue was achieved at pH 7 with degree of degradation was of 36.98% for 60 min.

Acknowledgement
This research was financial partly supported by Riset Unggulan LIPI FY 2015-2017.

References

[1] Ramaswamy V, Jagtap NB, Vijayanand S, Bhange DS, Awati PS 2008 *Mat. Res. Bull.* **43** 1152
[2] Hasan S 2015 *Res. J. Recent. Sci.* **4** 1.
[3] Okuyama K, Lenggoro W 2004 *Proceedings of the 2004 International Conference on MEMS, NANO and Smart Systems (ICMENS'04)* 1.
[4] Stoyanova A, Hitkova H, Bachvarova-Nedelcheva A, Iordanova R, Ivanova N, Sredkova M 2013 *J.Chem. Tech. Metallurgy.* **48(2)** 154.
[5] Abdel Aal A, Barakat MA, Mohamed RM 2008 *Appl. Surface Scie.* **254** 4577.
[6] Serpone N, Maruthamuthu P, Pichat P, Pelizzetti E, Hidaka H 1995 *J. Photochem. Photobiol. A.* **85** 247.
[7] Shiyanovskaya I, Hepel M 1998 *J. Electrochem. Soc.* **145** 3981.
[8] Song KY, Park MK, Kwon YT, Lee HW, Chung WJ, Lee WI 2001 *Chem. of Mat.* **13** 2349-55
[9] Pal B, Hata T, Goto K, Nogami G 2001 *J. Mol. Catal. A: Chem.* **169** 147.
[10] Bedja I, Kamat PV 1995 *J. Phys. Chem.* **99** 9182.

Corn stalk as matrix in decomposting toilet for treating urine and feces

N Sintawardani*, **D Nilawati and J T Astuti**
Research Unit for Green Technology, Indonesian Institute of Sciences
Kompleks LIPI, Jalan Cisitu/Sangkuriang, Bandung 40135, Indonesia
* Email : neni001@lipi.go.id

Abstract. Bio-Toilet technology (BT) which is appropriate for the habits of Indonesian people has been studied and developed. BT is a dry toilet technology commonly uses ligno-cellulosic waste materials as matrix to facilitate the growth of natural microbes. In aerobic condition, microbes degrade feces and urine. Mineral as the leftover of feces and urine, such as nitrogen (N), phosphorus (P), and potassium (K) remain in the rest of matrix waste. After certain period. matrix can be harvested and used as soil conditioner. BT uses much less water, mobile, and very useful to be applied in areas where water availability is limited. BT type with different capacities, user amounts and mixing systems has been developed using sawdust for matrix. Since corn stalk is categorized as useless and priceless waste, its application in BT is challenging. Performance of BT with corn stalk as matrix to degrade feces and urine of carnivore imitating the human waste was observed. BT M-15 manual mixing type with paddle was filled with chopped corn stalk as much as 45% of total volume. This BT was designed for 15 person as users per day if 80% reactor volume was filled with ligno-cellulosic matrix. It is assumed that 150 g of feces are discharged once per person/day and 1000 mL of urine 6-8 times per day. Start up process was made in the beginning to initialize the needed microbes in the reactor (matrix). The discharge of feces and urine were increased slowly and gradually the users were increased from 1 to 4 users per day. Performance of BT was indicated by the change in the pile that showed by moisture content, temperature and pH. C/N ratio in matrix decreased significantly from 43 to 17. This result showed that the corn stalk could be used as matrix in BT.

1. Introduction

Recently, the population growth rate tends to decrease in the last four decades, but water requirement and waste increase as well as encumber the environment. One of the environmental factors associated with water and waste is the handling of human sewage from sanitation facilities. Inadequate sanitation facilities and the behaviour of people accustomed to open defecation or direct discharge of waste to water body. Lack of clean water and sanitation facilities, such as sewerage pipeline and other sanitation infrastructures decrease the quality life of the community. Considering the economic condition of developing countries, installation of reasonable, deliberate, matching-local and alternative infrastructure is required. To solve the sanitation problem in the slum or rural area through an inexpensive method, recycling the organic waste back to nature on-site with an environmentally sound technology or process are considered. Dry toilet or de-composting toilet using sawdust as media or bio-toilet (BT) technology is one of an example. The technology which is of Onsite Wastewater Differentiable Treatment System (OWDTS) will give direct positive effects. Water reduction for flushing, unnecessary wastewater canal construction, equipment portability, and possible integration with kitchen waste recycling system are examples of OWDTS. Comprehensive research to develop de-composting toilet system using sawdust as matrix has been carried out [1,2]. Application of such dry toilet using sawdust in Indonesia will overcome the common perception on water utilization for body cleansing [3,4].

Wood waste and sawdust is often regarded as worthless material. But with the increasing price of oil or gas, some households or small-medium scale food industries started to use sawdust as fuel to replace gas or timber. Therefore, finding alternatives material for BT matrix is encouraging. Corn stalk as lignocellulosic material is worthless waste and easy to find. Performance of BT with corn stalk as matrix in BT to degrade feces and urine of carnivore imitating the human waste was observed.

2. Experimental

2.1. Materals

Corn stalk which was collected form traditional market di Bandung, chopped to the size of 1-2 cm, sun dried, and put in the decomposting toilet BT type M15. BT box was made from steel with manual paddle screw installed for mixing. Figure 1 (left side) shows the mechanical and inside part of BT type M15. The cycling part and the box/bowl side are separated by a wall. Cycling is needed to mix the dirt and the matrix after using the BT. Figure 1 (right side) In the right side is the illustration when BT is used.

Figure 1. Left : BT type M15 bio-reactor; Right : Illustration when BT is used.

2.2. Method

BT type 15 was designed to treat feces and urine for 15 users per day if 80% volume of mixing reactor was filled with lignocellulose material. In this research, the reactor was filled with 114 L-of chopped corn stalk, equivalent with its 19,4 kg of fresh chopped corn stalk and represented 45% from the total volume. Using an assumption one person discharges of 150 gram feces/day and 1000 mL total volume of urine/day (6-8 times per day), the BT was gradually filled up. In the first 7 days, a discharge of one person per day has been done to initiate the microbes in the BT reactor (as start up process). The discharge was then increased to 2 persons/day for 5 observation days, 3 persons per day in the next 5 observation days, and end up for 4 persons per day in 3 observation's days.

Since the omnivore animals in the nearby zoo have a diet intake similar to humans, the feces was collected, frozen, and used in the research to imitate the human feces. Human urine from one person was collected, frozen, and used in the research.

Pattern on pH, temperature, dry matter and organic content in corn stalk matrix was measured and observed to determine the performance of reactor.

3. Result and discussion

The input of feces and urine was gradually increased as mentioned in the method. It needed total 23 days from start-up to maximal input (Figure 2).

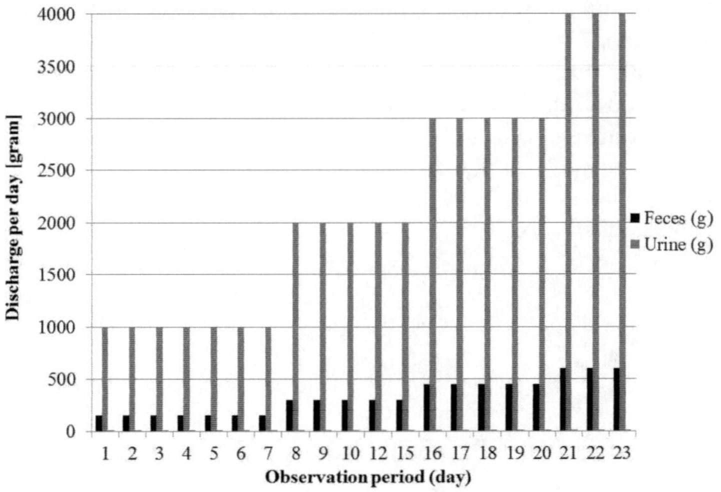

Figure 2. The gradual addition of feces and urine.

During the decomposition, bad odor began slightly detected after days 16. There was no bad odor appeared when 17 users discharged totally 2,550 g feces and 17,000 g urine in 15 days. At this stage odor detected was not too keen to be categorized as a condition rather smelly. This condition lasted for four days, until the observation in day 19. In day 21 the released odor from the reactor increased to be sharp and stinging, but it still got the input of 4 user daily until the day 23 with the total discharged of 6,600 g feces and 44,000 g urine. After 23 days, the input was stopped completely, but the sharp bad smell condition lasted until the day 33.

The bad odor during decomposition process could be generated from the accumulation of un-evaporated liquid from the input materials. In an aerobic decomposting process, moisture content plays a key role together with temperature and aeration. Observation during start-up showed that moisture content in corn stalk increased from 17 to 44, and rapidly to 81% in day 23 (Figure 3).

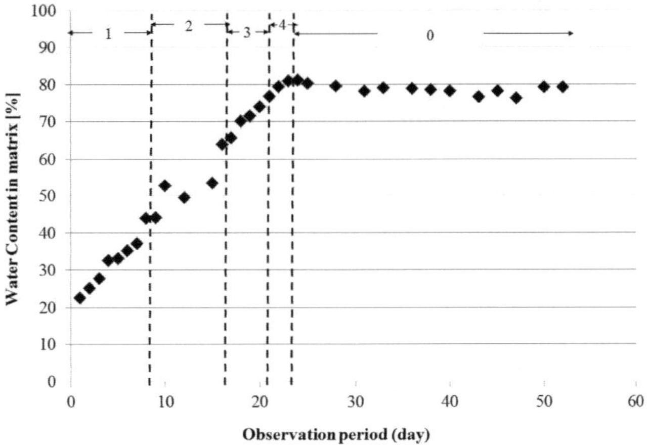

Figure 3. Accumulation of water in the corn stalks matrix in reactor.

Using formula feed for rabbit as model waste in a lab-scale aerobic process, Horisawa et al [4, 5] observed that the degradation rate was nearly constant at moisture content of 30-80%. In our research, after start-up the discharge was increased for 2 users per day until days 15, and the moisture content in corn stalk matrix was stable around 49%. The addition of discharge for 3 users per day and then to 4 users per day linearly increased the moisture content of corn stalk up to 81%. High concentration of water or moisture content in the reactor resulted of the loss of oxygen from the pores of the corn stalk particles and shifted the aerobic condition to anaerobic process. A change of aerobic to anaerobic conditions changed the degradation products. The bad odor could normally produced by anaerobic microorganism in the first step.

The trend of organic matter in the corn stalk during start-up period fluctuated from 97 to 95% dry basis (Figure 4). In the period of day 8 to 23, the decrease of organic matter was linearly from 96 to 93% as the moisture content increased from 20 to 80%. This trend could not convince the degradation of organic material in the discharge or in the corn stalk itself.

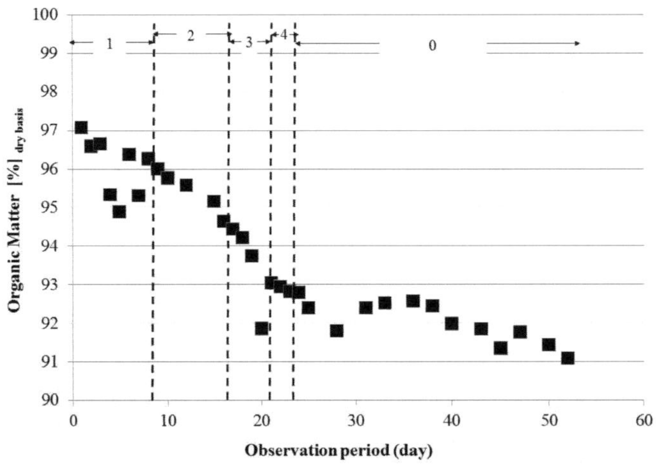

Figure 4. Dry basis organic matter in the matrix during observation's period.

Temperature of corn stalk pile in the reactor showed very interesting trends. By measuring the temperature at the bottom, middle, and top of corn stalk pile in the reactor in 3 different horizontal position (left, middle and right position of the reactor), the changes could be observed in figure 5. With daily ambient temperature during experiment stable in the average of 25°C and 60% relative humidity, the temperature in the reactor (corn stalk pile) increased to 29°C in the first day and the linearly increased to 32°C at the end of start-up period. Addition of a discharge of 2 users per day after start-up increased the temperature in the pile gradually to 41°C in the middle level. The temperature in the middle was not much different with the temperature in the bottom, but much higher in compare with the temperature in the upper level about 4°C. The upper level had more contact with the ambient temperature in compare with the middle or bottom level. The increase of temperature was a desired situation in aerobic decomposting process. A regular discharge could keep the temperature.

As the discharge of feces and urine was increased to 3 users per day, the temperature in the reactor drastically decreased to 32°C in the middle and lower part of the pile, or 30°C on the upper part. The addition to 4 users per day made the temperature in the pile jump to 27°C. The ambient temperature was in average 25°C. Without discharge of feces and urine (after day 23), the temperature in the corn stalk pile increased slowly to 30°C in the middle level. These changes showed the activity of the microbes involved in the process. From organoleptic measurement, the slightly stink smell could be perceived after days 16. It showed a shifting to unfavourable conditions in BT. The comformity of high water content in the pile is shown in Figure 3.

Measurement of the pH in the pile during start-up process showed the increase from 6.2 to 7.3. The addition of 2 users discharge per day decrease the pH in the corn stalk to 6.8 and then slowly back to 7.3 (Figure 6). Further addition of 3 users discharge per day increased the pH in the pile up to 8.5. Further addition of discharge decreased the pH to 8. After discontinuation of the discharge, the pH in the pile slightly fluctuated in the average value of 8. From generated smell it could be presumed that the increase of pH was caused by the ammonia. In anaerobic conditions, the protein from the discharge could be degraded to ammonia.

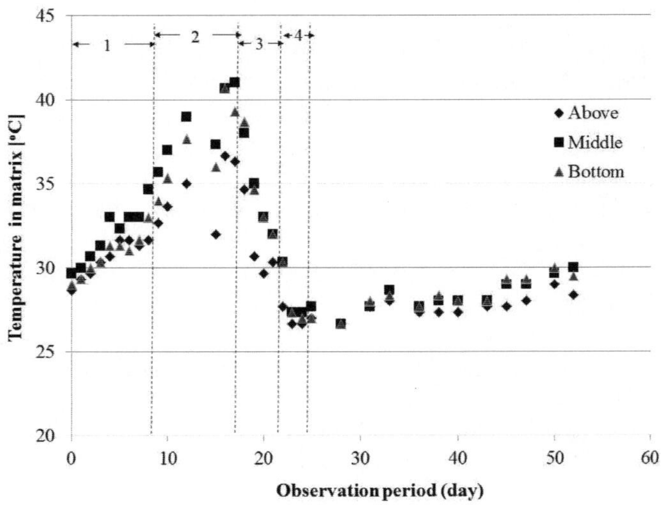

Figure 5. Temperature changes in the corn stalk pile.

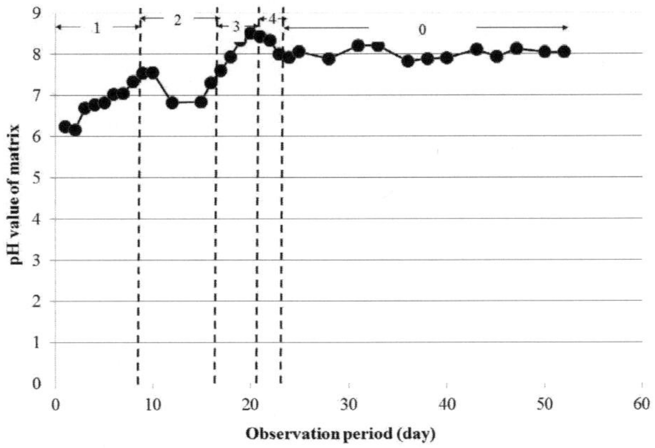

Figure 6. pH value of corn stalk in reactor.

From the day 36 until the last day of observation, the pungent smell inside the reactor was reduced gradually from rather smelly to earthy. After settling for ± 2 months of conditioning, the pungent smell totally disappeared and the material smell resembled soil as well as showed black colour of soil.

During the decomposition process of feces and urine in corn stalk matrix, the C/N ratio decreased gradually, from 43.24 to 24.59 on the day 17 of observation and declined further to 18.74 on the day 23 of observation. The C / N ratio decreased to 16.75 in the final day of observation. The decrease of C/N ratio during the experiment was supposedly due from the process of decomposition by microorganisms in the reactor. The C/N ratio at the end of this research has been approached with the C/N ratio of soil.

4. Conclusion

Corn stalk has been used for matrix in BT. The optimal input to the BT manual type 15 to degrade feces and urine was for 2 users per day, even on the discharge of 3 users per day the performance of BT was slightly reduced. Further addition of discharge to 4 users per day increased the moisture content in the matrix and gave burden to the corn stalk to shift the process to anaerobic condition. The C/N ratio of corn stalk decreased rapidly to soil like condition.

Acknowledgment
Highly appreciations is addressed to Nela Sihole and Mimin Suminar to help in technical aspect, as well as special gratitude to Prof. Minoru Terazawa and Prof. Naoyuki Funamizu for discussions.

References

[1] Terazawa M, Tamai Y, Horisawa S, and Yamashita K 1999 *J. of Wood Scie.* **45** 354-358
[2] Terazawa M 1999 *Proceedings of International Symposium on Bio-Recycling/Composting in Sapporo*, V11-V16
[3] Sintawardani N and Astuti J T 2008 *Proc. : The 6th International Symposium on Southeast Asian Water Environment* **Vol.6** 269-276
[4] Astuti J T, Sintawardani N and Irie M 2008 *J. of Tropical Wood Scie. and Technol.* **6** (1) 26-32
[5] Horisawa S, Sakuma Y, Tamai Y, and Terazawa M 1999 *Internal Report*: Inst. of Wood Tech., Akita Prefectural University, Japan
[6] Horisawa S, Tamai Y, Sakuma Y, Doi S and Terazawa M 1999 Presented Paper at *49th Annual Meeting of the Japan Wood Res. Soc., Tokyo*, Japan.

IOP Conf. Series: Earth and Environmental Science **60** (2017) 012017 doi:10.1088/1755-1315/60/1/012017

Utilization of cotton as carbon nanostructure precursor by pyrolysis method

E R Chaldun[1*], A Hardiansyah[2], H Alliah[3], A N Rohmah[3], M S Mustofa[4], M Ghozali[5] and B S Purwasasmita[6]

[1]Research Unit for Clean Technology, Indonesian Institute of Sciences, Bandung, Indonesia
[2]Department of Metallurgy and Materials Engineering, Faculty of Engineering and Design, Institut Teknologi dan Sains Bandung, Bekasi 17530, Indonesia
[3]Department of Physics, Universitas Islam Negeri Sunan Gunung Djati, Bandung 40614, Indonesia
[4]Center for Science and Technology of Advanced Materials, Indonesia National Nuclear Energy Agency, Kawasan PUSPIPTEK Serpong, South Tangerang, Indonesia
[5]Research Center for Chemistry, Indonesian Institute of Sciences, Bandung, Indonesia
[6]Advanced Material Processing, Department of Engineering Physics, Institut Technologi Bandung, Bandung, Indonesia
*Email : elsyrahimi@gmail.com

Abstract. In this study, cotton-based cellulose was used as precursor to developed carbon nanotube–like structures through modified-pyrolysis method with iron (III) chloride hexahydrate as a catalyst. Reflux process with nitric acid was conducted to purify the resultant of carbon material. The resultant of carbon-based nanostructure were characterized systematically. X-ray diffraction spectra shows the presence of C peaks at 2θ of 26.53°, 42.26°, 44.49°, 54.63° and 77.35° and Fe peaks at 2θ of 44.1° and 64.25°. Scanning electron microscope and transmission electron microscope observation revealed that the resultant of graphitic carbon nanostructures displayed the tube-like structure. Raman spectroscopy results show the presence of D-band and G-band peaks, which confirmed the typical spectrum of carbon-based structures. The D-band peak around 1310-1330 cm^{-1} was assigned to the presence of disorder in graphitic materials whereas the G-band peak around 1580-1590 cm^{-1} was corresponded to the tangential vibrational of the carbon atoms. These condition is a typical spectrum of carbon nanotube-like structures. Eventually, these pyrolysis method could be anticipated as a promising strategy in order to develop the novel carbon nanostructures based on cellulosic material.

1. Introduction

Carbon nanomaterials have received significant attention and development from the scientist and researchers. The excellent propereties make carbon nanomaterial suitable for applications such as optoelectronic, photovoltaic, sensing devices, electron field emission, storage, and production of energy, hydrogen storage, nanocomposites, catalyst support or drug delivery system [1][2][3].

Various techniques have been developed in order to prepare carbon-based nanostructures include electric-arc discharge, laser ablation, or chemical vapor deposition [4]. These techniques are often requires a complex protocol and conducted in high processing temperature which results in expensive and inconvenient procedures. Thus, in order to overcome these challenges, the development of new procedures of which combination with catalyst technology for the easy and low-cost strategies is in needed to develop the carbon-based nanostructures. In this respect, a simple method for preparing graphitic carbon nanostructures is the pyrolisis of carbon precursors in the presence of certain transition metals including Fe, Co, Ni, or Mn that serves as catalyst at moderate temperatures. This method commonly conducted through the carbonization process of as-impregnated polymeric materials. Several investigation have been conducted through pyrolysis techniques using naphthalene/nickel [5] and tetrahydrofuran/nickelocene [6][7] as precursor/catalyst materials. Transition metals are generally used as catalyst owing to their catalytic decomposition of carbon source, ability to form meta-stable carbides and possibility for carbon to diffuse through and over the metals extremely rapidly [8] and the induce formation of graphitic sheets [9].

Recently, the using of natural renewable precursor is gaining interest for selection of carbon precursor. They are inexpensive and widely available substances around the world. Previous investigation have been conducted by using natural resources as a carbon precursor including saccharose [10], bacterial cellulosa [11], zeamays oil [12], and latex [13]. Cotton (*Gossypium sp.*) is a plant-based fiber composed primarily of cellulose which its utilization is still limited as raw material for textiles.

In these study, carbon nanostructure was developed through pyrolysis method by using cotton as carbon source and iron (III) chloride ($FeCl_3$) used as catalyst. The physicochemical properties of the resultant material were characterized systematically.

2. Experimental

2.1. Materials

Cotton fiber was purchased from Polytechnic Textile Bandung, West Java. Iron (III) chloride was purchased from Merck. Deionized water was used throughout the experiment. All chemicals were used without further purification.

2.2. Methods

Preparation of carbon nanostructures was developed through modified-pyrolysis technique. Briefly, 2 g of cotton fiber was immersed in various concentration of iron (III) chloride hexahydrate (0.16 M (A), 0.32 M (B), and 0.48 M (C)). Sample A, B and C were purified in 7 M of nitric acid. The as-prepared materials were then immobilized into the modified-cotton through the sonication-assisted process (Ultrasonic LC 30 H) for 30 min at 60°C. The solvent was evaporated in *vacuo* followed by heat treatment process in quartz reactor under a nitrogen flow with a rate 30 ml/min with three steps of heating at 250 and 500°C for 30 min and 900°C for 2 hr. The carbon samples obtained consisted of a mixture of amorphous carbon and less-crystalline nanostructures. Afterwards, the samples was subjected to purification processes through an effective purification wet-oxidation techniques [16]. In order to investigate the effect of purification process on carbon structure, refluxing with 7 M (A1), 10.78 M (A2), and 14.5 M (A3) of nitric acid (HNO_3) were studied for the optimization with the catalyst concentration of 0.16 M. X-ray diffraction (XRD) analysis was conducted using a Shimadzu XRD 7000 Maxima X to evaluate the crystallographic characteristics by scanning dried powder with Cu K_α (1.5406 Å) radiation. Scanning electron microscope (SEM) JEOL JSM-T330 A was used to observed the structure and morphology of carbon-based material. Transmission electron microscopy (TEM, FEI-Tecnai, Netherland) at an acceleration voltage of 200 kV was used to evaluate the dimensions and size of the resulted carbon material. Raman spectroscopy was conducted using Raman spectroscopy instrument with operating wavelength at 785 nm laser and supported by OPUS software.

3. Result and Discussion

3.1. Formation of carbon structures

Figure 1. shows the diffraction spectra of cotton-based carbon structures with various concentration of catalyst. Diffraction peaks at 26,53°, 26,59° and 26,54° for 0.16, 0.32, and 0.48 M catalyst concentration, respectively, were correspond with the carbon spectra. Table 1 shows particle size of carbon, which was calculated using Scherrer method. According to these method, the as-synthesized carbon were developed in nanometer size with hexagonal crystal system. These results are in accordance with the previous report of diffraction pattern for carbon nanostructures that showed the presence of C atoms at (002) peak at around 26° [16]. The presence of additional peaks at 44,1°, 44,07°, 44,22°, 64,25°, 64,19° and 64,26° might be attributed to the peak of catalyst particles. Carbon nanostructures formation was initiated through the interaction catalyst around the carbon structures. Furthermore, for the growing process of carbon nanostructures, the carbon is incorporated to the end of the growing tube and catalyst particles remains at this end, which is moving away of surface. Afterwards, the catalyst iron nanoparticles diffused the new carbon species that are incorporating to the CNT to radial and axial direction, confirming the development of multi-walled carbon nanotubes (MWCNT) [17].

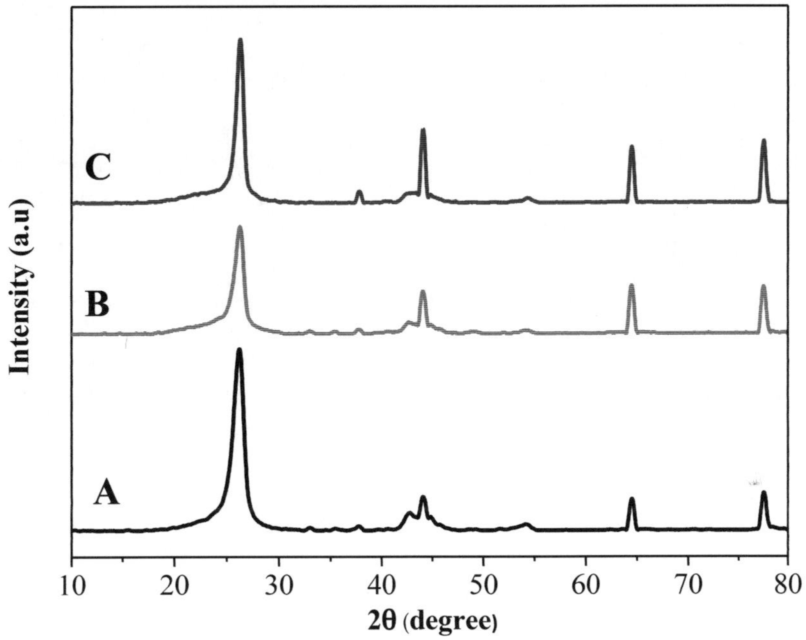

Figure 1. Diffraction spectra of as-prepared carbon structure with catalyst concentration of 0.16 M (A), 0.32 M (B) and 0.48 M (C) and purify with 7 M of nitric acid.

Table 1. Particel size and crystal system of as-prepared carbon structures.

Sample	d (nm)	Element	JCPDS	Crystal System
A	8	C	751621	Hexagonal
B	8	C	751621	Hexagonal
C	10	C	751621	Hexagonal

Figure 2 shows Raman spectra of carbon nanostructure with the presence of D-band and G-band peak, which is in accordance with the well-developed spectrum of carbon nanostructures. The D-band peak was observed around 1350 cm^{-1} which is assigned to the presence of disorder in graphitic materials. Moreover, the G-band peak was observed around 1580 cm^{-1} which is corresponded to the tangential vibrational of the carbon atoms. In parallel with the increasing catalyst concentration, the overall intensity of D-band and G-band peaks was enhanced significantly. These results indicated that iron-based catalyst promoted the growth of carbon nanostructures. Furthermore, Table 2 shows there were no significant different of Raman spectra as increasing catalyst concentration, the ratio of I_D/I_G

was around 1.43-1.49. Theoretically, the quality of the carbon nanostructures, especially carbon nanotubes was depends on the rasio I_D/I_G [18].

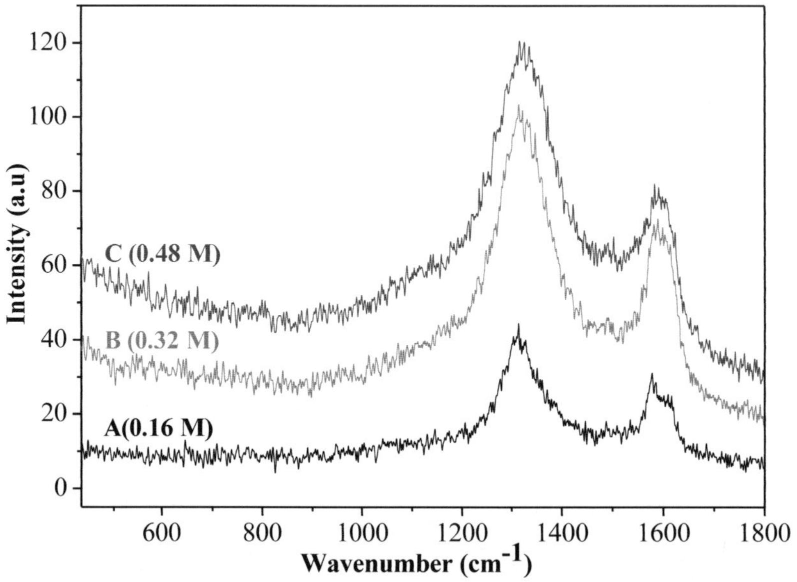

Figure 2. Raman spectra of cotton-based CNT with various catalyst concentration of 0.16 M (A), 0.32 M (B) and 0.48 M (C).

Table 2. Ratio I_D/I_G of cotton-based CNT with catalyst concentration of 0.16 M (A), 0.32 M (B) and 0.48 M (C).

Sample	I_D	I_G	I_D/I_G
A	44.04	30.73	1.43
B	101.2	70.1	1.44
C	117.5	78.,65	1.49

Figure 3. shows the SEM image of CNT samples. All samples showed the morphology of carbon nanotubes with tube-like structures. Sample with the catalyst concentration of 0.16 M pronounced more CNT-like structure. Increasing catalyst concentration led the CNT to be covered by iron particles which is in accordance with XRD and Raman spectra observation.

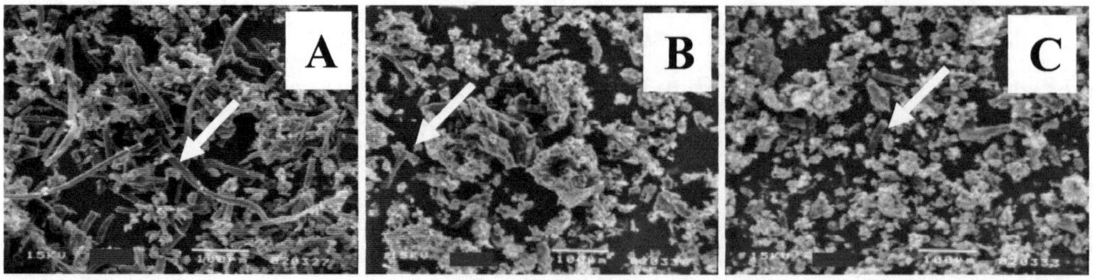

Figure 3. SEM images of cotton-based CNT with catalyst concentration of 0.16 M (A), 0.32 M (B) and 0.48 M (C).

3.2 *Effect of nitric acid to the formation of tube-like structures*

Sample with the catalyst concentration of 0.16 M was used for further purification optimization. Figure 4 shows the typical XRD pattern of C and Fe peak after purification at different concentration of nitric acid whereas Table 3 shows the calculated size. According to the previous report [16], the peak of C (002) was pronounced at $2\theta \approx 26^{o}$ which is in accordance with the XRD pattern of all samples prepared. Generally, there was no obvious changes in the crystal structure of carbon. Moreover, the XRD spectra of sample A2 and A3 peak exhibites less-crystaline in compared with sample A1. This might be due to the increasing concentration of nitric acid tend to disrupted more chemical bonding in carbon structures, thus CNT become less crystalline [19].

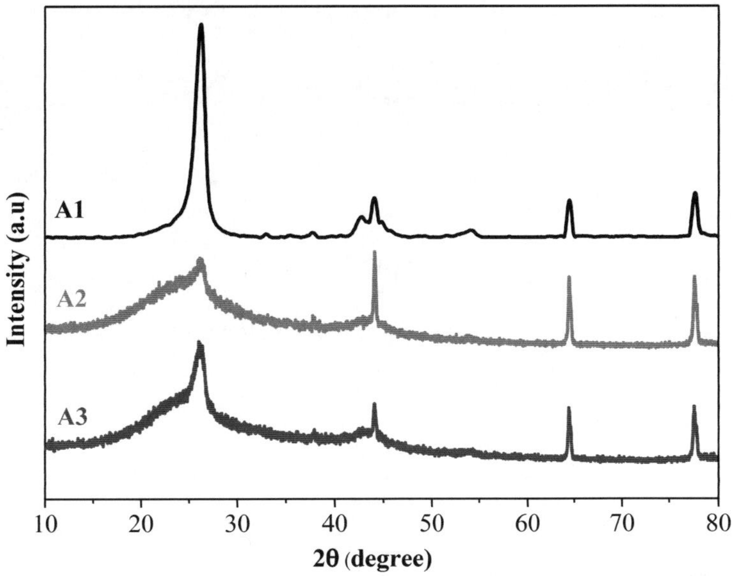

Figure 4. Diffraction pattern of cotton-based CNT with 0.16 M of catalyst purified by nitric acid with concentration of 7 (A1), 10.78 (A2), and 14.56 (A3) M.

Table 3. Particel size (d, nm) and crystal system of as-prepared carbon structures.

Sample	d (nm)	Element	JCPDS	Crystal system
A1	8	C	751621	Hexagonal
A2	28	C	751621	Hexagonal
A3	3	C	751621	Hexagonal

Raman spectra of CNT samples using different concentration of nitric acid are presented in Figure 5. Raman analysis results show the peak of D-band and G-band, which typically spectrum of the CNT. The D-band peak around 1350 cm^{-1} was assigned to the presence of disorder in graphitic materials whereas the G-band peak around 1580 cm^{-1} was corresponded to the tangential vibrational of the carbon atoms. Table 4 shows the ratio of I_D/I_G in the range about 1.22-1.43.

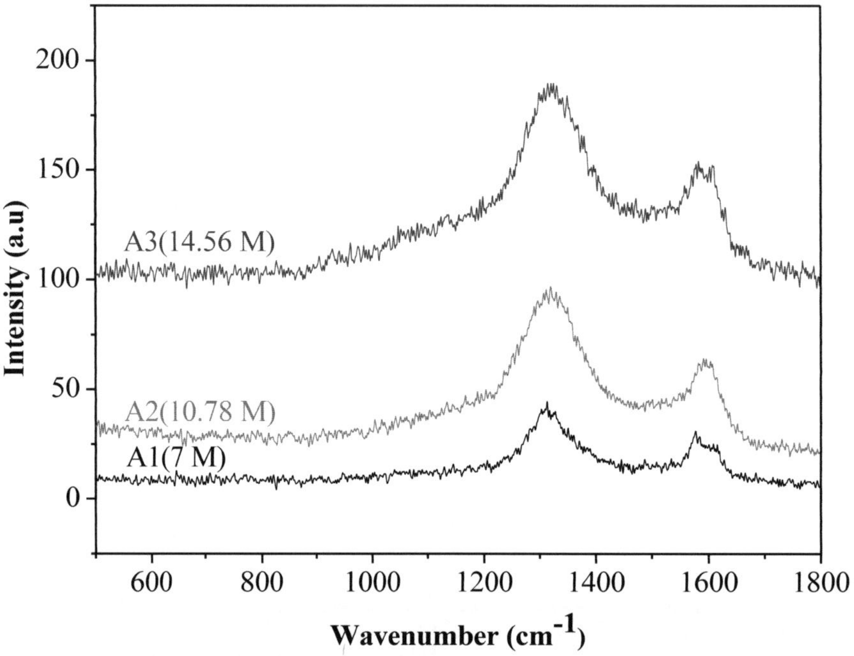

Figure 5. Raman spectra of as-prepared carbon structures with various nitric acid concentration of 7 (A1), 10.78 (A2) and 14.56 (A3) M.

Table 4. Ratio I_D/I_G of as-prepared carbon structures.

Sample	I_D	I_G	I_D/I_G
A1	44.04	30.73	1.43
A2	91.77	64.23	1.42
A3	186.48	151.66	1.22

Figure 6 shows SEM images of cotton-based CNT with various nitric acid concentration of 7 M (A1), 10.78 M (A2) and 14.56 (A3) M. Generally, the samples showed the morphology of carbon nanotubes-like structures. Through purification process, most of the metal catalyst was removed. A portion of remaining catalyst particle surrounded on carbon at sample A2 and A3 might be due to the high concentration of catalyst particles. Interestingly, increasing nitric acidconcentration has affected to length of carbon nanotubes-like structures. Sample A2 and A3 were slightly shorter than sample A1. This is might be due to the high concentration of nitric acid used for the purification process of which distructed chemical bonding in CNT and resulted in less-crystalline structure [19]. This results are in accordance with aforementioned characterization.

Figure 6. SEM images of as-prepared carbon structures with 0.16 M catalyst with various nitric acid concentration of 7 (A1), 10.78 (A2) and 14.56 (A3) M.

Figure 7 shows the TEM images of sample A1, cotton-based CNT produced by 0.16 M of catalyst and purified by 7 M of nitric acid. Figure 7a shows the tube-like structure of the as prepared carbon-based materials. The sheets of graphite are orderly arranged in concentric cylinders i.e., the tubular structure of nanotubes. Furthermore, Figure 7b shows the formation of multiwalled-like type, and most graphene layers grow perpendicularly to the growth axis of the tubes. The multiwalled carbon nanotube (MWCNT)-like structures was consists of 23 layers and each layer within 2 μm.

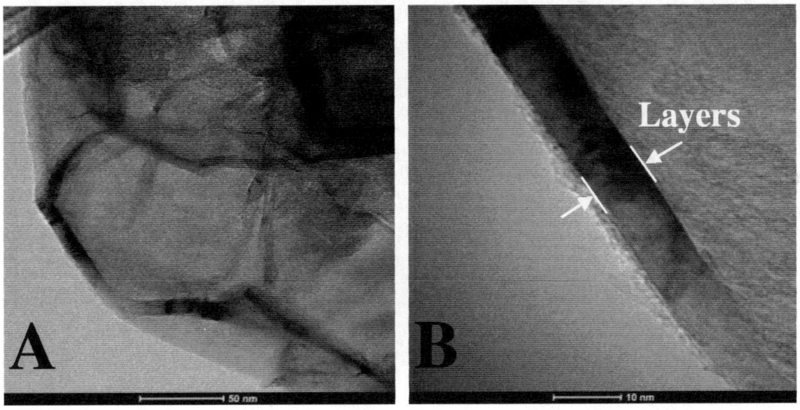

Figure 7. TEM images of as-prepared carbon structures of sample A1 (0.16 M of catalyst and purify with 7 M nitric acid) (A) and its magnified image (B).

4. Conclusion

In summary, multiwalled carbon nanotubes (MWCNT) structures were obtained through modified-pyrolysis process by using cotton as a cellulose-based precursor and iron (III) hexahydrate as catalyst. The formation of MWCNT was optimize at 0.16 M catalystconcentration as in good agreement with XRD, Raman, SEM, and TEM characterizations. The results has obtained I_D/I_G ratio of about 1,43 at 7 M of nitric acid for purification. Eventually, these results show that the utilization of cotton fiber *via* modified-pyrolysis techniques demonstrate a novel strategies to developed carbon-based nanostructure.

Acknowledgement

The authors gratefully acknowledge the support for this research from Indonesian Institute of Sciences No 1130/F/2015; Advanced Material Processing, Institut of Technology Bandung; Center for Science and Technology of Advanced Materials, Indonesia National Nuclear Energy Agency; and Department of Physics, Sunan Gunung Djati State Islamic University.

References

[1] Sevilla M and Fuertes A B 2010 *Chem. Phys. Lett.* **490 (1–3)** 63–68
[2] Jariwala D, Sangwan V K, Lauhon L J, Marks T J and Hersam M C 2013 *Chem. Soc. Rev.* **42** 2824–2860
[3] Zhang Q, Huang J Q, Qian W Z, Zhang Y Y and Wei F 2013 *Small.* **9(8)** 1237–1265
[4] P. Mahanandia, P. N. Vishwakarma, K. K. Nanda, V. Prasad, K. Barai, A. K. Mondal, S. Sarangi, G. K. Dey and S. V. Subramanyam 2008 *Solid State Commun.* **145(3)** 143–148
[5] Wang C.-G., Shao X.-H. and Xue R.-S. 2015 *J. Nanomater.* 1–5
[6] Mahanandia P. , Vishwakarma P N , Nanda K K, Prasad V., S. V. Subramanyam S V, S. K. Dev S K and P. V. Satyam P V 2006 *Mater. Res. Bull.* **41(12)** 2311–2317
[7] P. Mahanandia, V. P. Arya, K. K. Nanda, P. V. Bhotla and S. V. Subramanyam 2009 *Mater. Sci. Eng. B.* **164(3)** 140–150
[8] Mageswari S, Shabudeen P S S, Kanakachalam N, and Karthikeyan S.2014 *J. Environ. nanotechnol.* **3(2)** 62–68
[9] Lee C J, Park J, and Yu J A 2002 *Chem. Phys. Lett. Elsevier.* **360** 250–255
[10] Sevilla M. and Fuertes A B 2009 *Mater. Chem. Phys.* **113(1)** 208–214
[11] Chaldun E R, Karina M, and Purwasasmita B S 2013 *Adv. Mater. Res.* **789** 232–236
[12] Angulakshmi V S, Karthikeyan S, and Shabudeen P S S 2015 *RASAYAN J.Chem.* **8(1)** 1–7
[13] Viswanathan G, Bhowmik S, and Sharon M 2014, *Int. J. Mater. Mech. Manuf.* **2(1)** 25-28
[14] Paulo J, Morais S, Freitas M D, Souza M D, Dias L, Magalhães D, and Ribeiro A 2013 *Carbohydr. Polym.* **91(1)** 229–235
[15] Lu P and Hsieh Y, “Author’s personal copy Preparation and properties of cellulose nanocrystals : Rods , spheres , and network.”
[16] Cao A, Xu C, Liang J, Wu D, and Wei B 2001*Chem. Phys. Lett. Elsevier.* **344** 13–17
[17] Antu W, Alonso G, and Espinosa F, 2006 *Diam. Relat. Mater. Elsevier.* **15** 1329–1335,
[18] Dresselhaus M S, G. Dresselhaus G, Saito R, and Jorio A 2005 *Phys. Reports Elsevier.* **409** 47–99
[19] Hou P, C. Liu C, and Cheng H, “Author’s personal copy Purification of carbon nanotubes.”

Incorporation of citrus essential oils into bacterial cellulose-based edible films and assessment of their physical properties

L Indrarti*and Indriyati
Research Unit for Clean Technology, Indonesian Institute of Sciences (LIPI)
Jalan Cisitu 21/154 D, Bandung 40135, Indonesia
*Email : luciaindrarti 1@gmail.com

Abstract. The use of edible films in food protection and preservation has recently gained more interest since they offer several advantages over synthetic packaging materials. Biocellulose (BC) offers great opportunity as edible film due to their unique physical and mechanical properties. In this study, biocellulose films were prepared by solution casting with addition of 30% carboxymethyl cellulose (CMC) and 30% glycerol as the homogenizer and plasticizer, respectively. Furthermore, various citrus essential oils (EOs) including lemon, lime, and sweet orange were added at 50% w/w of BC dried weight. The solutions were then cast on the tray and allowed to dry in the air convection oven at 40°C overnight. The films were characterized for water solubility, tensile strength (TS), elongation at break (EB), water vapour transmission rate (WVTR), and color. Those characteristics may influence consumer acceptability of the packaged products. Results revealed that addition of lemon and sweet orange EOs into BC-based edible film decreased water solubility and TS, but improved EB, as these oils acted as plasticizers in the film. However, different trend was observed for BC-based film incorporated with lime oil, which had higher solubility and TS, but lower EB and WVTR compared with that of control film. Addition of citrus EOs into BC-based films did not have much effect on color properties as stated in L*, a*, and b* values.

1. Introduction

Packaging has an important role in food protection and preservation. Research on edible films and coatings on food system has been conducted in order to improve food quality, add value to the edible polymer materials, and reduce synthetic packaging materials. An edible film or coating could be defined as a packaging made from edible materials [1]. Edible films are distinguished from edible coating by their manufacturing method and application in food. Edible films are dried forming thin material structures and are used to wrap the product or make a pouch or bags, while edible coatings are usually applied as a liquid to the outer surface of food product and allowed to dry to form thin layer structure.

Polysaccharides such as starch, cellulose derivatives, chitosan, and carrageenan, are most often used for edible films among others because of their film-forming properties. As one form of celluloses, biocellulose (BC) is produced by bacteria *Gluconacetobacter xylinus* (formerly known *as Acetobacter xylinum*) that secretes large quantities of cellulose as microfibrils and these microfibrils merge to form a large ribbon of cellulose in a medium containing carbon sources. Due to BC structure that consist only glucose monomer, it exhibits great properties such as unique nanostructure, high water holding capacity, high mechanical strength and high crystallinity [2-5]. BC has been identified as one of the suitable materials for edible film [6-7].

There are several requirements to be considered for edible films application in food area, such as appropriate water barrier properties, good mechanical strength, low cost of raw materials, and simple technology production [8]. Our previous study revealed that by simple production technology and with low cost of raw material, edible film from BC can be produced. Addition of glycerol to BC film gave higher efficiency in plasticizing BC-based films which was shown in increased elongation at break andin tensile strength [7]. Nowadays, one of the major challenges for edible film technologies is the design of antibacterial edible film using minimum amounts of active compounds, such as phenolic or essential oil (EO) compounds. Edible films can be used as a carrier of EO. EO interacts with polymer and plasticizer and forms part of chemical structure of the film [9]. EOs are "essential" in common sense that they carry a distinctive scent of the source plant and are generally extracted by steam distillation and other processes like maceration, cold pressing, or solvent extraction. Some studies have demonstrated the use of EO in edible film to inhibit the growth of microorganisms [10-11]. However, the use of essential oils also affects the physicochemical properties of edible films, particularly BC-based edible films, and these characteristics are relatively unknown. Therefore, it is important to evaluate the effect of EO additon on BC film characteristics. Thus, the aim of this work is focused on analyzing the effect of EO incorporation on the physicochemical behavior of BC-based edible films. The characteristics including solubility, mechanical properties, water vapour transmission rate (WVTR), and color at three different kinds of citrus essential oils was discussed. Those characteristics may influence consumer acceptability of the packaged products.

2. Experimental

2.1. Preparation of BC-based edible films

BC gels were purchased from local industry in Cianjur, West Java province, Indonesia. Washing, purification, and preparation of BC slurry were conducted as previously report [7]. BC slurry, CMC (Himedia) and glycerol (Merck) were mixed under magnetic stirring condition (named as control film). Lime, lemon and sweet orange essential oils purchased from Lansida were added at concentration of 50 %v/w over dried BC. The solutions were degassed under a vacuum bell jar to remove the bubbles prior to cast onto the tray. The trays were placed overnight at 40 °C in gear oven, and then cooled to room temperature before peeling the films off the trays.

2.2. Characterization of BC-based edible films

Water solubility of the films was determined according to the method of Ojagh et al. [12]. Pieces of film (2 cm x 2 cm) were cut from each film and dried at 110 °C in an oven. Samples were weighted to the nearest 0.0001 g and noted as initial dry weight (w_o). Samples were then placed into beaker with 20 mL aquadest and heated at 60°C for 3 hr. The remaining pieces of film after soaking were filtered through nylon filter, followed by oven drying at 110 °C until its weight get constant and noted as final dry weight (w_f). Samples were measured in 2 replicates and the percentage of total soluble matter (% solubility) was calculated as follow.

$$\%solubility = \frac{(wo - wf)}{wo} x100\% \qquad (1)$$

Mechanical properties were characterized by Orientec UCT-5T universal testing with 100 kgf load cell according to ISO 527- 1993E standard method. Dumbbell-shaped specimens were obtained from each film according to ISO 527-2 type 5A. The measurement was conducted at 23°C. At least five specimens of each sample were measured and computerized calculated to obtain the average value.

Water vapor transmission rate (WVTR) was measured using gravimetric modified cup method based on ASTM E96 [13-14]. The thickness of each sample was determined in five random positions. The edible films were firmly fixed onto the opening of glass cups containing of silica gel and then, these cups were placed in a dessicator with distilled water. These cups were weighed using an analytical balance (±0.0001 g) at interval of 24 hr during 5 days. The samples were measured in 3 replicates. WVTR was calculated from the slope of the regression analysis of weight gain versus time divided by the exposed film area.

Color values of films were measured using color difference meter (Murakami Color Research Lab) and expressed in L*, a* and b* color space. L* represents the lightness, a* represents chromaticity on a green (-) to red (+) axis, and b* represents chromaticity on a blue (-) to yellow (+). For measurement, films were placed on a white standard plate (L*=93.8, a*=-0.7 and b*= 1.4). Whiteness index was estimated using the following equation [15].

$$WI = 100 - ((100 - L)^2 + a^2 + b^2)^{1/2} \quad (2)$$

The color was expressed as difference of color (ΔE^*), accordingly to the equation [16].

$$\Delta E^* = ((\Delta L)^2 + (\Delta a)^2 + (\Delta b)^2)^{1/2} \quad (3)$$

where ΔL, Δa, and Δb correspond to the variation between the color parameter of film and that of white standard used as background. Film opacity was calculated with reflectance measurements of each film (n=2) with black and white backing plates, using the following equation [17].

$$Opacity = \frac{Y_{blackbaking}}{Y_{whitebacking}} x100\% \quad (4)$$

where Y is the color values of the film with the black or white backing plates.

3. Result and discussion

3.1. Film appearance

The BC-based edible films formed using variations of citrus essential oils were visually homogeneous, no crack or brittle areas, and no visible pores or bubbles. They were easy to handle and easy to peel from the casting plates.

3.2. Water solubility

Film solubility is an important factor for edible applications. Film solubility at temperature above room temperature is advantageous if the films are consumed with a pre-heated product prior to consumption and it may also be an important factor to determine biodegradability of film when used as packaging wrap. The water solubility of BC-based edible films with and without EOs is shown in Table 1. It would be expected that hydrophilic compounds should increase a film's solubility, whereas hydrophobic compounds should decrease it [18]. In this study, incorporating lemon and sweet orange EOs into BC-based film decreased water solubility of edible films, which is in accordance with the expectation. Lower solubility of film contained sweet orange EO compared to film contained lemon EO may be explained by an increase in interaction between the hydroxyl group of BC chains and sweet orange EO components, leading to a decrease in availability of hydroxyl group, and thus reducing polysaccharide-water interactions. On the other hand, incorporating lime EO into BC-based edible film increased water solubility, probably due to less lime EO components that interact with hydroxyl group of BC chain, leading to an increase in availability of hydroxyl group, and thus enhancing BC-water interactions.

Table 1. Water solubility of BC-based edible films with and without essential oils.

Formulation	Solubility (%)
BC+CMC+glycerol (control)	26.21±0.56
BC+CMC+glycerol+lemon EO	25.79±0.48
BC+CMC+glycerol+lime EO	28.24±2.61
BC+CMC+glycerol+sweet orange EO	19.71±1.96

Cellulose is a hydrogen bond cross-linked polymer. Despite being hydrophilic, cellulose is insoluble in common solvents, because it has rigid long chains and strong intra and inters hydrogen bonds [19]. Usually, BC-based film is insoluble in water. However, in this study, water solubility of BC-based films was in the range of 20-28%. It can be explained that water went inside the cellulose's structures and broke apart the intermolecular attractions between CMC and BC. Therefore, the low percentage of solubility of BC-based edible films elucidates only CMC was dissolved in water while the remaining solid was fragmented BC fibers.

3.3. Mechanical properties of BC-based edible films

Mechanical properties of films were characterized by measuring the tensile strength (TS) and elongation at break (EB), which are key indicators of a films strength and flexibility. TS exhibits the maximum stress during tensile testing before failure, while EB means film strechability prior to breakage.

The effect of citrus essential oils and control on the mechanical properties of BC film samples is shown in Table 2. Compared with the control film, incorporation of lemon and sweet orange EOs into BC-based films showed a decreased in TS and increased in EB of the BC-based films compared to the control due to the effect of plasticized behavior of these oils. This is common result when EOs were introduced into the films and also in agreement with the literatures [15,20]. On the other hand, different trend was observed for BC-based film enriched with lime oil, which had higher TS and lower EB than that of control film, presumably because lime oil EO might consist mostly of terpene-like compound rather than lipids. It might interfere the interaction between polymer chains and reduce flexibility of BC-based film [13].

Table 2. Mechanical properties of BC-based edible films with and without essential oils.

Formulation	TS (MPa)	EB (%)
BC+CMC+glycerol (control)	106.6±6.6	22.9±1.4
BC+CMC+glycerol+lemon EO	101.8±4.2	27.4±1.1
BC+CMC+glycerol+lime EO	133.1±22.5	20.6±3.6
BC+CMC+glycerol+sweet orange EO	92.8±22.5	26.6±2.7

3.4. Water vapor transmission rate (WVTR)

WVTR is an important property for films used as edible food coatings, because most natural biopolymers are very sustain to water. WVTR of BC films containing citrus essential oils is shown in Table 3.

Table 3. Water vapor transmission rate (WVTR) of BC-based edible films with and without essential oils.

Formulation	WVTR (gr/cm^2/day)
BC+CMC+glycerol (control)	1.69±0.19
BC+CMC+glycerol+lemon EO	1.61±0.17
BC+CMC+glycerol+lime EO	1.41±0.08
BC+CMC+glycerol+sweet orange EO	1.78±0.11

Incorporating lemon and lime EOs caused a slight decreased of WVTR compared to that of control film. In contrast, addition of sweet orange EO slightly increased the WVTR of the film. Water vapor transfer generally occurs through the hydrophilic portion of the film. Essential oil's chemical nature also plays an important role in the barrier properties of edible films. The composition of the citric fruits is generally composed of 90% terpenes, 5% oxygenated compounds, and less than 1% non-volatile compounds such as waxes and pigments [21]. Different amounts of monoterpenes could influence the hydrophobicity. In other words, monoterpenes is the governing factor in determining the hydrophobicity [13]. The presence of a hydrophobic disperse phase, even at small ratio, limits water vapour transfer since it introduces the interfere in hydrophilic phase and increases the tortuosity factor of mass transfer [15].

3.5. Color

Color attributes are of prime importance because they directly influence consumer acceptability [22]. Addition of EOs in BC-based films may change the native color of BC-based films. However, in this study, there was no significant change in L* value for all films, whereas incorporation of citrus EOs on BC slightly influenced a* and b* value as shown in Table 4. As a result, ΔE values changed upon of citrus EOs on BC-based films. The interaction of citrus EOs did not significantly influence the white index the BC film. Furthermore, the opacity of films increased with addition of citrus EOs. Sweet orange EO gave the highest opaque percentage, followed by lime EO. This property is important in various film applications, particularly if the film is used as a surface food coating or for improving product appearance.

Table 4. Color parameters of BC-based edible films with and without essential oils.

Film sampel	L*	a*	b*	Color difference (ΔE)	White index (WI)	Opacity (%)
Control	92.7	0.4	10.1	8.8	87.53	46.39
Lemon	92.4	0.3	8.6	7.4	88.52	48.38
Lime	92.4	0.2	10.6	9.3	86.96	53.68
Sweet orange	92.4	-0.4	10.4	9.2	87.13	74.89

4. Conclusion

Some physicochemical characterizations of BC-based films blended with citrus essential oils (EO) have been carried out. Addition of various citrus oils into BC-based films showed an effect on solubility, mechanical properties, WVTR and opacity of BC films. The film that contained sweet orange EO presented the most opaque color compared to others. Tensile analyses showed that addition of sweet orange and lemon oils could enhance films' flexibility. WVTR decreased when lemon and lime EOs were incorporated. However, whiteness index were not significantly affected by addition of citrus EOs-BC films.

Acknowledgement

Authors would like to acknowledge Indonesian Institute of Sciences under Riset Unggulan 2016 for financial support.

References

[1] Pascall M A and Lin S-J 2013 *J. Food Process. Technol.* **4** e116.
[2] Chen P, Cho S Y and Jin H J 2010 *Macromol. Res.* **18** 309–20.
[3] Saibuatong O A and Phisalaphong M 2010 *Carbohyd. Polym.* **79(2)** 455–60.
[4] Castro C, Zuluaga R, Putaux J L, Caro G, Mondragon I and Gañán P 2011 *Carbohyd. Polym.* **84(1)** 96–102.
[5] Keshk S M A S 2014 *J. Bioproces. Biotech.* **4** 150.
[6] Tang X, Kumar P, Alavi S and Sandeep K 2012 *Crit. Rev. Food Sci.* **52** 426–42.
[7] Indrarti L, Indriyati, Syampurwadi A and Pujiastuti S 2016 *AIP Conference Proceedings* **1711**, 050007.
[8] Debeaufort F, Quezada-Gallo J A, Voilley A 1998 *Crit. Rev. Food Sci.* **38** 299-313.
[9] Ponce A G, Roura S I, del Valle S E and Moreira M R 2008 *Postharvest Biol. Technol.* **49** 294-300.

IOP Conf. Series: Earth and Environmental Science **60** (2017) 012018 doi:10.1088/1755-1315/60/1/012018

[10] Rojas-Graü M A, Avena-Bustillos R J, Friedman M, Henika P R, Martin-Belloso O and Mchugh T H 2006 *J. Agric. Food Chem.* **54** 9262-7.

[11] Du W X, Olsen C W, Avena-Bustillos R J, McHugh T H, Levin C E and Friedman M 2009 *J. Food Sci.* **74(7)** M372-8.

[12] Ojagh S M, Rezaei M, Razavi S H and Hosseini S M 2010 *Food Chem.* **122 (1)** 161-6.

[13] Yanwong S and Threepopnatku P 2015 *IOP Conf. Ser. Mater. Sci. Eng.* **87** 012064.

[14] Sobral P J A, Menegalli F C, Hubinger M D and Roques M A 2001 *Food Hydrocoll.* **15** 423-32.

[15] Atares L, Bonilla J and Chiralt A 2010 *J. Food Eng.* **100** 678-87.

[16] Garcia F T and Sobral P J d A 2005 *LWT – Food Sci. Technol.* **38** 289-96.

[17] Anonymous, HunterLab method 2008 *Application note: Opacity* **8**, 4.

[18] Kavoosi G, Dadfar S M M and Mohammadi Purfard A 2013 *J. Food Sci.* **78** E244-50.

[19] Pandey M, Abeer M M and Amin M C I 2014 *Int. J. Pharm. Pharm. Sci.* **6** 89-93.

[20] Ekthamasut K and Akesowan A 2001 *AU J. Technol.* **5** #3.

[21] Colecio-Juárez M C, Rubio-Núnez R E, Botello-Álvarez J E, Martinez-González G M, Navarrete-Bolanos J L and Jiménez-Islas H 2012 *Chil. J. Agric. Res.* **72(2)** 276-80.

[22] Kunte L A, Gennadios A, Cuppett S L, Hanna M A and Weller C L 1997 *Cereal Chem.* **74** 115-8.

IOP Conf. Series: Earth and Environmental Science 60(2017) 012019 doi:10.1088/1755-1315/60/1/01201

Co-cracking of real MSW into bio-oil over natural kaolin

I M Gandidi[1*], M D Susila[1] and N A Pambudi[2]
[1]Department of Mechanical Engineering, Faculty of Engineering, Lampung University, Jalan S. Brojonegoro No. 1, Bandarlampung, Indonesia
[2]Mechanical Engineering Educations, Universitas Negeri SebelasMaret, Jalan Ir. Sutami 36 A Surakarta 57126, Indonesia
*Email : indra.mamad@eng.unila.ac.id

Abstract. Municipal solid waste (MSW) is a potential material that can be converted into bio-oil through thermal degradation process or pyrolysis. The efficiency and productivity of pyrolysis can be increased with the use of natural catalyst like kaolin. The addition of catalyst also reduces the overall cost of conversion process. In this study conversion of MSW into Bio Fuel using Pyrolysis in the presence of of natural kaolin as catalyst has been investigated for 60 min at 400°C temperature. During the process 0.5 w/w catalyst to MSW ratio was maintained. Gas chromatography-mass spectrometry (GC-MS) was used to analyse the chemical composition of bio fuel. It is found that bio-oil production increases substantially with the use of catalyst. It is observed that the production of bio-oil is 23.6 % with the use of catalyst in process, which was only 15.2 % without the use of catalyst. The hydrocarbon range distribution of oil produced through pyrolysis reveals that gasoline and diesel fuel (C_5-C_{20}) are its main constituents. The functional group detected in bio-oil by GC-MS analysis is similar to that of diesel-48 in which paraffin and olefin are major mass species.

1. Introduction

Open dumping of municipal solid waste (MSW) has created a lot of problems in cities of developing countries. No suitable place can be found for waste dumping as limited free area is available in urban area and for places outside the city high transportation, maintenance and labour cost is needed [1]. Also, MSW landfills provide habitat for insects, pest and bacteria that can spread different diseases. Furthermore, the piles of garbage in landfills produce leachate that contaminates ground and surface water [2]. The bio decomposition of waste releases methane gas into atmosphere, which is twenty times more dangerous than pollution caused by CO_2 emissions [3]. The landfills are also considered as one of the cause of floods, global warming and green house effects [4,5]. In contrast to these problems MSW has a great amount of chemical energy stored in it. This energy is in the form of chemical bonds between molecules of carbon, hydrogen and oxygen [6]. When these bonds break, organic material is converted into gas, liquid and solid which is known as biofuel [7]. The calorific value which can be generated from municipal waste is around 20.57 MJ/kg [8].Therefore, MSW has a high potential for producing bio-energy. Several methods have been proposed to produce bio-energy from MSW, one of them is pyrolysis method.

Pyrolysis is a thermal degradation process in which MSW is converted into solid, liquid and gas at elevated temperature in the absence of oxygen [9]. Normally operating temperature for this process is 400-600°C [10-13]. In this process, long hydrocarbon chains break into smaller. The products obtained from the pyrolysis can be divided into non-condensable gas, liquid and solid fractions. From the liquid fraction paraffin, olefins, naphthalene and aromatics can be recovered. The hydrocarbon range of these species lie in the gasoline (C_5-C_{12}), diesel (C_{12}-C_{20}) and heavy oil (>C_{20}) ranges [14].

The focus of current research in this field is to improve production of bio-oil from MSW because it can be stored, transported and upgraded easily. The pyrolysis process has less thermal efficiency. It is because of the fact that cracking reaction is highly endothermic due to which very high working temperature is required. Also, the products formed by thermal pyrolysis have low molecular weights and wide carbon number range. The addition of catalyst in pyrolysis process can overcome these problems. Catalyst reduces the activation energy and improve the decomposition reaction and product yield [15-19]. However most of the research publication presented deal with the application of commercial or synthetic catalyst such as Y-Zeolite, ZSM-5, Y-zeolite, β-zeolite, MoO_3, Ni-Mo-catalyst, HZSM-5, $Al(OH)_3$ etc [15-19]. In the pyrolysis process. The use of these catalyst increase the overall production cost. It is due to the fact that impurities and contaminants can cause faster deactivation of catalyst. Therefore, the use of natural catalyst instead of commercial catalyst is a better option. These natural catalysts are found abundantly in nature due to which they are very economical. Many researcher have investigated thermal degradation process in the presence of natural catalyst [20-23]. But most of them focus only on single component pyrolysis or specified MSW such as pure and mixed plastic. Whereas in actual scenario MSW contains huge variety of different kind of waste material and it is impossible to separate them because of huge quantity. Also, more information regarding the characteristics of bio-oil obtained from mixed MSW is needed before it can be used as engine fuel.

Therefore, this research focuses on the pyrolysis process to produce bio-oil from un-separated MSW. The MSW contained different kind of plastics, biomass, textile, paper and rubber wastes. The experiment was performed for non-catalytic and natural catalytic pyrolysis processes. Then the product characteristics such as its composition, yield, chemical and physical properties are studied. In addition, a comparison was done between bio-oil and conventional oil to check the possibility of bio-oil as a replacement of conventional oil.

2. Experimental

2.1. Waste and catalyst for pyrolysis

The MSW used in study contained different kinds of waste including vegetable, garden and household waste. It also contained LDPE, HDPE, paper/cardboard, PS, rubber and textile waste. Initially all samples were dried using sunlight. The drying process took 3-5 days. After drying, each component was separated to determine the percentage composition of each type of waste. Figure 1 shows the initial physical form of MSW and percentage composition of each type of waste. After drying process MSW was then chopped into smaller pieces of 5-10 cm. This was done to make MSW size proportional with reactor capacity. Kaolin as natural catalyst for this experiment was obtained from local source in Lampung province of Indonesia. The catalyst was then thermally activated at 500°C for 2 hr to improve the active site and crystalline behavior of catalyst. XRD and BET techniques were used to determine the chemical composition, structure and surface of Kaolin.

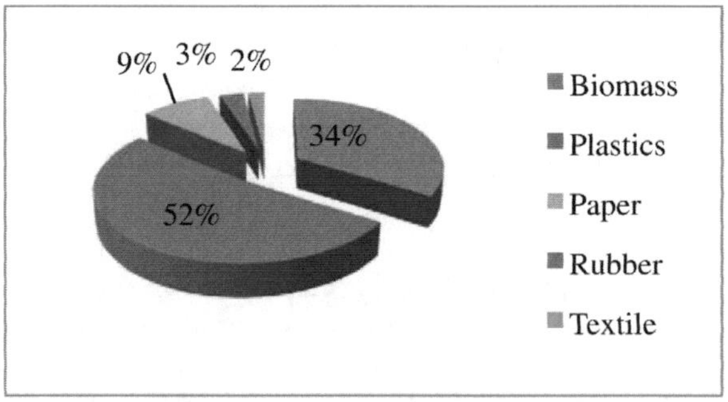

Figure 1. Composition of MSW.

2.2. Pyrolysis experiment

The schematic diagram of whole experimental setup is shown in Figure 2. Laboratory scale fixed bed reactor was used in pyrolysis of MSW sample. The reactor was set vertically up and the N_2gas is passed through it for 3 min. The N_2 gas introduced in the reactor from top and after circulating in whole body it passed out from the top of reactor. The purpose of introducing N_2 was to remove the air from reactor to create inert environment for pyrolysis reaction. In these experiments, 500g MSW sample for non-catalytic pyrolysis and total 750 g MSW and catalyst for natural catalytic pyrolysis were fed into the reactor. 250:500 ratio of catalyst to MSW was maintained. Reactor was heated at heating rate of 12°C until the reactor temperature attained the set point of 400°C. Total reaction time was 60 min. The gas formed during pyrolysis left the reactor from the top and entered into the vacuum gas cleaner where first condensation took place. After passing through Vacuum gas cleaner gas entered into a shell and tube condenser where second condensation took place at 20°C. Second condensation separated liquid from non-condensable gas. The gaseous products were burned off to prevent emissions from hydrocarbon gases. The liquid product from condenser was mixture of bio-oil and water which were later separated by precipitation and filtration process. The char produced during burning of gases remained in the reactor and was collected after the completion of experiment.

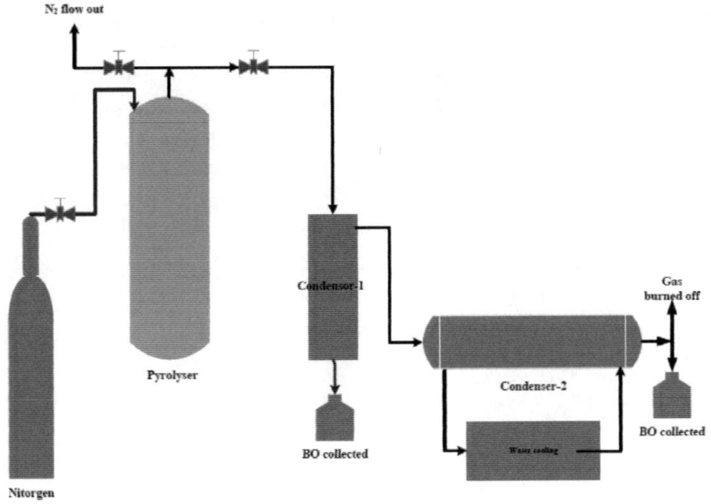

Figure 2. Apparatus.

2.3. Bio-oil analysis

Chemical composition of bio-oil obtained from filtration process was studied to determine the impact of catalyst. Gas chromatography-mass spectrometry (GC-MS, QP2010S Shimadzu) was used to determine mass fraction in bio-oil. The column DB-1 (Agilent J 100% dimethylpolysiloxane) was capillary column having 30m length, 0.25 mm diameter and 25 μm film thickness. Helium was used as the gas carrier. Initially temperature was set at 60°C for 5 min followed by heating rate of 5°C/min up to the temperature of 280°C and held for 51 min. To estimate the heating value carbon-hydrogen weight ratio [24] approach was used, according to which paraffin has greatest heating value followed by iso-paraffin, olefin, naphthalene, and aromatic (PIONA). Table 1 shows the ASTM standards which were used to analyse the physical and chemical properties.

Table 1. Properties of bio-oil.

Propertis	Units	Standard	Diesel 48	Pertamina Dex	Gasoline	Non-Catalyst	Catalyst
Cetane Octane Number		-	48 (cetane)	51 (cetane)	88 (octane)	-	-
Density@15ºC	g/cm^3	ASTM D-1298	0.815-0.87	0.82-0.86	0.7-0.78	0.8133	0.8038
Kinematic Viskosity@40º C	cSt	ASTM D-445	2.0-5.0	2.0-4.5	0.4-0.8	2.662	2.0345
Flash Point	ºC	ASTM D-93	min 60	min 60	min 40	min 5	min 5
Water Content	%Vol	ASTM D-1744	max 0.05	max 0.05	-	1.5	3
Oil Content	%Vol	ASTM D-1744	-	-	-	95	86
Solid Content	%Vol	ASTM D-482	max 0.35	max 0.05	-	3.5	11

3. Result and discussion

Figure 3 shows the comparison between the mass yield of products obtained by non-catalyst and natural catalyst (kaolin) pyrolysis methods. It is clear that catalytic pyrolysis produced higher bio-oil product than the non- catalytic pyrolysis. However, pyrolysis process produces significant amount of solid residue. It is due to the presence of ash and impurities in MSW. It is found that kaolin addition remarkably increased the decomposition reaction and the production of bio-oil up to the 55.2%. Similar result has been reported of which an increase results using kaolin as a catalyst [25, 26]. Kaolin also reduced the solid, showed that kaolin has excellent productivity behaviour in pyrolysis process. This behaviour is explained on the basis of its active site and acidity which facilitate the cracking reaction. However, water content and significant amount of syngas in pyrolysis product suggest the need of further drying of MSW. It also discovered that the condensation temperature of 20°C was not favourable as it reduced the liquid fraction in product.

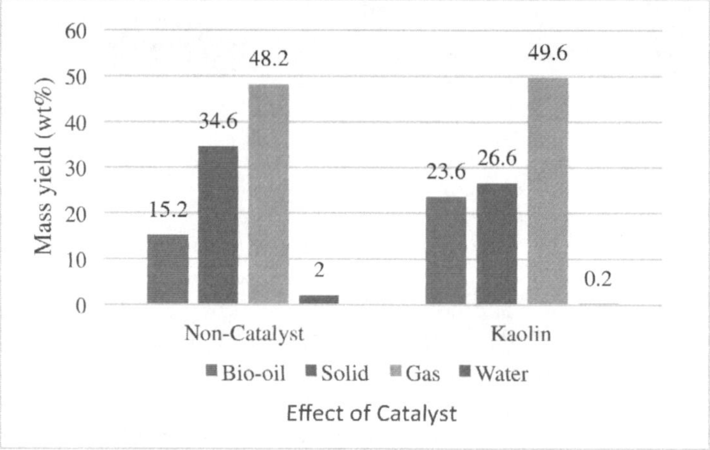

Figure 3. Mass yield of pyrolysis products.

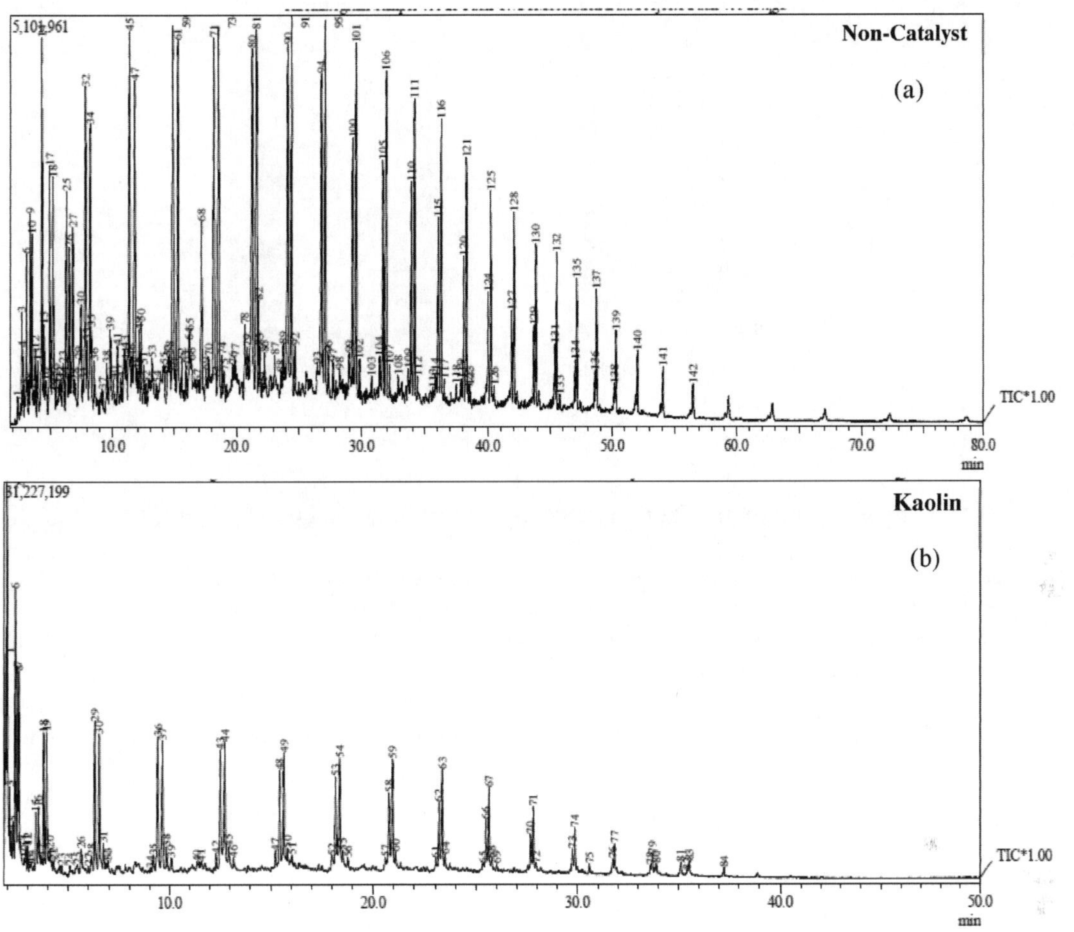

Figure 4. Chromatogram of the bio-oil mass fraction.

On the basis of petroleum mass fraction, bio-oil obtained from Pyrolysis of MSW can be classified into three categorize i.e. gasoline (C_5-C_{12}), diesel (C_{12}-C_{20}) and heavy oil (>C_{20}). The number of different carbon ranges found in bio-oil are represented by peaks obtained by GC-MS which are shown in Figure 4. Figure 5 shows the percentages of carbon ranges in different kind of fuels. It can be seen that both gasoline and diesel are the main organic component in bio-oil. Also, the addition of kaolin increases the amount of gasoline and reduces diesel content as catalyst degradation enhances with the addition of kaolin. Due to the high percentage of both gasoline and diesel content in bio-oil, it has physical properties between gasoline and diesel fuel (Table 1). This implies that bio-oil from pyrolysis has a great potential to be used as fuel instead of conventional fuels.

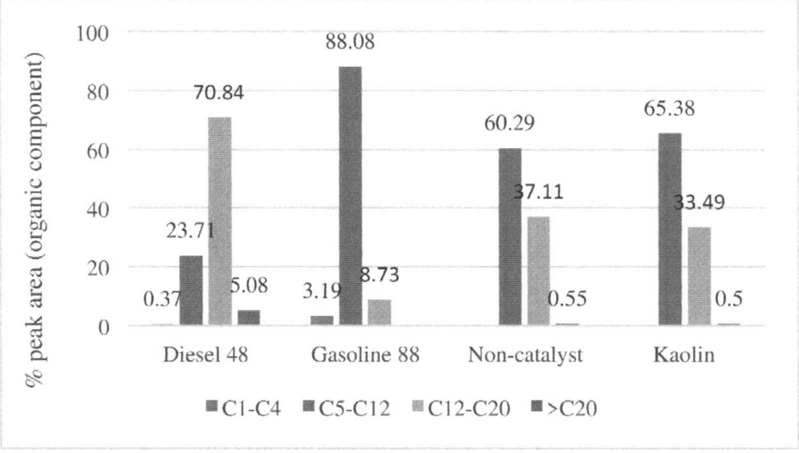

Figure 5. Carbon distribution range on conventional and pyrolysis fuel.

Percentages of mass species found in bio-oil and conventional fuels are shown in Figure 6 and Figure 7, respectively. Diesel-48 contains high amount of paraffin, gasoline-88 contains aromatic and paraffin whereas pyrolysis oil contains mainly paraffin and olefin (C_5-C_{20}) and small concentration of aromatics. Similar result was reported by Panda et al [27] and difference result by use of silica-alumina catalyst by Wang et al[19]. This shows that the addition of kaolin does not impact the percentage of different mass species in final product and it only slightly improves the product yield.

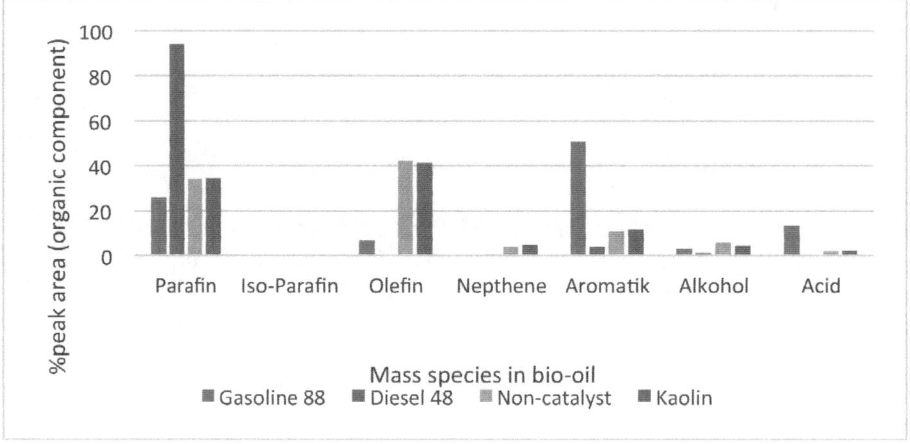

Figure 6. A comparison of mass fraction between pyrolysis oil and conventional fuel.

This results indicates that kaolin has no influence on selectivity behavior and mass species content in bio-oil. Both bio-oil have heating values between gasoline and diesel fuel. Presence of oxygenated functional group and aromatics matter reduces the calorific value of bio-oil. It is simply because of the fact that Carbon oxygen bonds do not release energy during combustion.

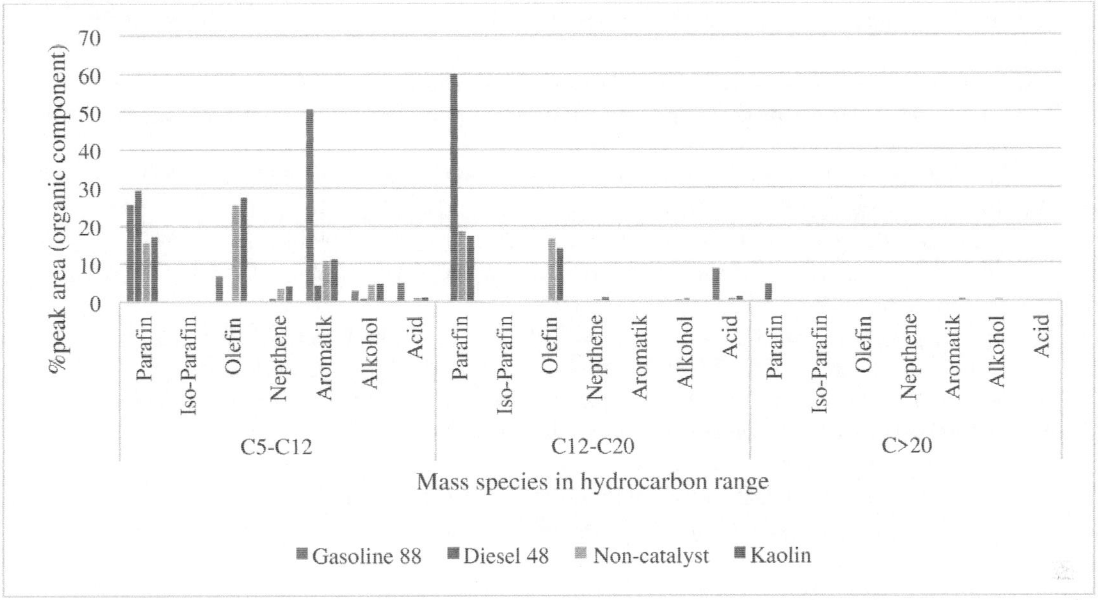

Figure 7. Percentage of organic compound in mass species hydrocarbon range.

4. Conclusion

The study concluded that the use of kaolin as natural catalyst improved the product yield but did not affect the percentage of species found in bio-oil. The physical properties of bio-oil laid between diesel and gasoline because of which it can be used instead of conventional fuels. Low calorific value of oil and production of water and high level of non-condensable gas during pyrolysis process was found and addressed to make bio-oil as an alternative of commercial fuels.

Acknowledgment

This research is partly supported by National Competitive Research Program Grant of Directorate of Higher Education, Ministry of Research, Technology, and Higher Education of the Republic of Indonesia. The authors would like to thank University of Lampung and Gadjah Mada University for their support.

References

[1]. Arinal H,Gandidi I M, and Harmen B 2010 *Final report of strategic research grant*. 12-144.
[2]. Alexandra L C 2012 *World Bank*. 2-4.
[3]. Haradhan M 2011 *Int. J. Economic and Political Integration.* **2 (1)** 3–10.
[4]. Unila LP 2010 *Final Report of Waste Management for Rockefeller foundation*
[5]. Hari B D and Sameer A 2013 *Habitat International.* **38** 100-105.
[6]. Ebru A and Ahmet D 2009 *Int. Advanced Technologies Symposium.* 1-4.
[7]. McKendry P 2002 *Bioresource Technol.* **83** 37–46.
[8]. Ari D P, Kilbergen W, Gultom and Aryadi S 2007 *Final Report of Feasibility Study.* 8-57.
[9]. Hossain A K and Davies P A 2010 *Renew and Sustainable Energy Rev.* **21** 165–189.
[10]. Liu Y Q, Wang J, Dong J and Thay J H 2012 *International Conference of Industrial and Hazardous Waste Management.* **1-7.**
[11]. Seung S K, Jinsoon K, Jong K J, Young K P and Chan J P 2013 *Renewable Energy.* **54** 241-247.
[12]. Velghe I, Carleer R, Yperman J and Schreurs S 2011 *J. Anal. App. Pyrolysis.* **92** 366–375.
[13]. Ayhan D 2004 *J. Anal. Appl. Pyrolysis.* **72**97–102.

[14]. Aguado J, Serrano D P, Guillermo S M and Madrid S 2007*J. Anal. Appl. Pyrolysis.***79 (1)** 415-423.
[15]. Chika M, Jude A O and Paul T W 2015*Energy Research Institute.* 1-29.
[16]. Aida Isma M I, Salmiaton A and Nur Dinie K B 2015*Int.J. Environmental Science and Development.* **6** 606-609.
[17]. Funda A, Miskolczi N and Borsodi N 2013*Bioresource Technology.* **133** 443–454.
[18]. Norbert M., Ates F and Borsodi N 2013*Bioresource Technology.* **144** 370–379.
[19]. Wang J L and Wang L L 2011 *Energy Sources.* **33 (21)** 1940-1948.
[20]. Mochamad S, Harwin S, Tinton N, Putri N, Shou C, Zainal A and Kunio Y 2014*Energy Procedia.***47** 180 – 188.
[21]. Wenger Jorn 2015 *J. Materials Science and Engineering.***5 (3-4)** 183-185.
[22]. Kyaw K T and Hmwe C S S 2015 *Int.J.Advances in Engineering & Technology.* **8** 794-802.
[23]. Abbas A S and Mohamed F A 2015 *Diyala Journal of Engineering Sciences.* 819-834.
[24]. Riazi M R 2005 Characterization and properties of petroleum. 30-84.
[25]. Panda AK and Singh R K 2013*Advances in Energy Engineering.* **1** 74-84.
[26]. Pattaraprakorn W and Bhasaputra P 2014*GMSARN International Journal.* **8** 67-72.
[27]. Panda A K and Singh R K 2014*Int. J. Environment and Waste Management.* **13** 104-114.

IOP Conf. Series: Earth and Environmental Science **60** (2017) 012020 doi:10.1088/1755-1315/60/1/012020

Removal of nickel ion from electroplating wastewater using double chamber electrodeposition cell (DCEC) reactor partitioned with water hyacinth (*Eichhornia crassipes*) leaves

Djaenudin, Widyarani*, H R Hariyadi, D R Wulan and S Cahyaningsih
Research Unit for Clean Technology, Indonesian Institute of Sciences (LIPI)
Jalan Cisitu 21/154 D, Bandung 40135, Indonesia
*E-mail: widyarani@lipi.go.id

Abstract. Nickel is a heavy metal present in many types of industrial wastewater, and its contamination to the water bodies should be prevented. The objective of this research was to study the performance of Double Chamber Electrodeposition Cell (DCEC) for nickel ion removal. Water hyacinth (*Eichhornia crassipes*) leaves were used to separate the two chambers. The experiment was performed with synthetic electroplating wastewater in a batch system for 72 h. Changes of pH, electric current, and nickel ion concentration in the catholyte were monitored. An experiment with Single Chamber Electrodeposition Cell (SCEC) was also performed as comparison. After 72 h operation of DCEC, nickel ion concentration in the catholyte decreased from 2200 g.m^{-3} to 0.4 g.m^{-3}, equivalent to 99.98% removal. DCEC reactor performed better than the SCEC reactor that only achieved 59% removal. The results show that an almost-complete removal of nickel ion can be achieved with DCEC. Water hyacinth leaves can be used as low-cost alternatives for industrial membranes.

1. Introduction

Nickel is a heavy metal with broad industrial applications; more than half of global nickel production is used in stainless steel while electroplating accounts for 11% of total nickel processing [1]. The wastewater generated from nickel processing still contains a substantial amount of nickel that should be removed before the final discharge to the environment. Exposure to nickel may cause health problems including skin irritation, asthma, and conjunctivitis. High exposure of nickel e.g. in workplace has been linked to cancer [2,3].

Nickel ion can be removed from nickel-containing wastewater by several techniques: chemical precipitation, ion floatation, ion exchange, adsorption, membrane filtration, and electrochemical treatments [4,5]. Chemical precipitation is currently the most used treatment because the operation is simple and requires low capital cost. The drawback is the high volume generation of nickel-containing sludge that may still create environmental problem if not handled properly [4]. Electrochemical treatments enable recovery of nickel in the form of pure nickel, nickel compounds, or concentrated stream that can be reused in industrial processes [6–8]. The drawbacks of these treatments are the high energy consumption and possible fouling in the processes that use membranes [4].

Electrochemical treatments can be performed by electrodeposition, electrodialysis, electrodeionization, or combination of these processes [9]. Electrodeposition occurs based on the following reactions:

Cathodic reaction : $Ni^{2+}_{(aq)} + 2\ e^- \rightleftharpoons Ni_{(s)}$

Anodic reaction : $H_2O \rightleftharpoons \frac{1}{2}\ O_{2(g)} + 2\ H^+_{(aq)} + 2\ e^-$

Competing reactions sometime occur in the cathode, for instance hydrogen gas ($H_{2(g)}$) formation and water hydrolysis. These reactions may influence nickel ion reduction on the cathode [10]. Membranes are sometimes used to separate the cathode and the anode, and optimize the reaction in each electrode. The use of membranes requires high capital cost and maintenance, therefore an alternative is envisaged.

Water hyacinth leaves were used in this experiment as the partition between the anode and the cathode chamber due to the similarities with industrial membrane's properties. The cell wall of water hyacinth leaves contains 20-25% cellulose, 35-40% hemicellulose, and 6-8% lignin [11]. Cellulose isolated from water hyacinth has been used to prepare cellulose diacetate, a membrane precursor, with comparable properties to commercial cellulose diacetate [12].

A previous study compared the performance of Single Chamber Electrodeposition Cell (SCEC) and Double Chamber Electrodeposition Cell (DCEC) partitioned with water hyacinth leaves [7]. During the 4 hr operation, DCEC had a higher current efficiency and a lower specific energy compared to SCEC. However, only 27% of nickel ion was removed from the initial concentration of 2402 $g.m^{-3}$ to the final concentration of 1756 $g.m^{-3}$. The objective of this research was to study the nickel ion removal from DCEC partitioned with water hyacinth leaves during prolonged operation. An experiment with SCEC was also performed for comparison.

2. Experimental

2.1. Materials

Synthetic wastewater was prepared using $NiSO_4.6H_2O$ dispersed in distilled water to the concentration of 2200 $g\text{-}Ni.m^{-3}$. All chemicals used were of analytical grade.

Water hyacinth leaves were obtained from Cikapundung River, Bandung, West Java, Indonesia. Prior to experiments, the leaves were rinsed with distilled water to remove dirt.

2.2. Experimental setup

The experiment was performed in a previously described setup [7]. A Double Chamber Electrodeposition Cell (DCEC) consisted of two conjoined cylindrical compartments partitioned with water hyacinth leaf, each having 500 ml working volume. The synthetic wastewater was placed in the compartment housing a copper plate cathode. The second compartment housed a graphite plate anode and 0.2 M H_2SO_4 was used as the anolyte. The experiment was performed in duplicate at 26.0 ± 0.3°C, 5V DC input for 72 h. Nickel ion concentration, pH, conductivity, and electric current were measured over time.

In Single Chamber Electrodeposition Cell (SCEC), a cylindrical compartment (500 ml working volume) housed both the cathode and the anode. The experiment was performed at similar conditions as the experiment with DCEC.

2.3. Analysis

Nickel ion (Ni^{2+}) concentration in the solution was determined using Atomic Absorption Spectrophotometry (AAS). The absorbance was measured with Shimadzu AA 6501S (Japan) at 232 nm [13].

The pore size of water hyacinth leaves was measured using BET surface area analysis. Ion exchange capacity was measured using potentiometric titration [14].

2.4. Calculations

Reactor performance was measured by nickel ion removal (%), which was calculated based on initial nickel ion concentration (C_0, $g.m^{-3}$) and nickel ion concentration at time t (C_t):

$$\text{Removal}=\frac{C_0-C_t}{C_0}\text{x}100\% \quad (1)$$

Nickel ion removal may follow first order kinetic with removal rate constant k [15]:

$$\frac{dC_t}{dt}=-kC_t \quad (2)$$

Specific energy of removal ($kWh.g\text{-}Ni^{-1}$) is the amount of energy required to remove *m* g of nickel ion:

$$\text{Specific energy of removal} = \frac{V \cdot I \cdot t}{m} \quad (3)$$

V = applied voltage (V); I = measured electric current (A); t = time (h)

3. Result and discussion

3.1 Nickel ion removal

Nickel ion concentrations were measured over time, and the results are presented in Figure 1.

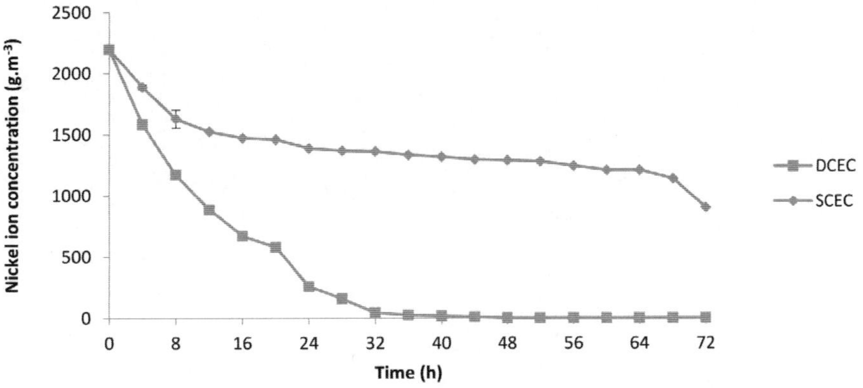

Figure 1. Nickel ion concentrations in Double Chamber Electrodeposition Cell (DCEC) and Single Chamber Electrodeposition Cell (SCEC).

Figure 1 shows that the nickel ion concentration in DCEC decreased rapidly between $t = 0$ (2200 $g.m^{-3}$) and $t = 32$ h (44 $g.m^{-3}$), and the final concentration at $t = 72$ h was 0.4 $g.m^{-3}$. This result showed the decrease of nickel ion concentration as 99.98% removal and therefore had met the Indonesian standard for liquid waste of 1 $g.m^{-3}$ for metal plating industry [16]. DCEC performed better removal than SCEC, which still contained 900 $g.m^{-3}$ nickel ion (59% removal) after 72 h.

Nickel ion removal in DCEC followed first order kinetic with removal rate constant $k = 0.12\ h^{-1}$ ($R^2 = 0.956$). On the other hand, nickel ion removal in SCEC only followed first order kinetic during the first 16 h with $k = 0.03\ h^{-1}$ ($R^2 = 0.907$). After $t = 16$ h, nickel ion was removed from SCEC at the linear rate of 7 $(g.m^{-3}).h^{-1}$.

During metal electrodeposition on the cathode, the involved mechanisms are diffusion, convection, and migration of metal ions and electrons toward the cathode surface, followed by electron adsorption to form adatom, and diffusion of the adatom to form crystal [10]. Without partition between the anode and the cathode chambers, hydrogen ions (H^+) that were formed in the anode were transferred onto cathode and formed $H_{2(g)}$ that reduced the mass transfer toward the cathode surface. The transfer of H^+ still occurs when sulfonated membranes are used as partition, since these membranes are capable of proton transfer [17]. The main functional groups in water hyacinth leaves are C=O, CH_3, and –COOH [12], and our results suggest that water hyacinth leaves were able to prevent H^+ transfer from anode to cathode. Furthermore, a high nickel ion concentration in the cathode chamber also optimized the mass transfer.

Water hyacinth leaves used in the experiments had surface pore sizes of 2-50 nm, ion exchanger capacity of 0.64 $meq.g^{-1}$, and conductivity of 0.13 $mS.cm^{-1}$. Conductivity and current in DCEC were higher than SCEC, and relatively constant throughout the experiment. The differences in conductivity were only 2% or less (Figure 2a), confirming that water hyacinth leaves had a good conducting property as expected.

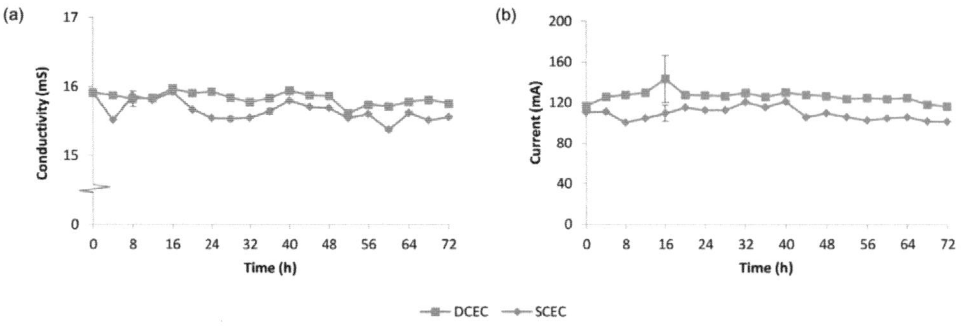

Figure 2. Conductivity (a) and current (b) in DCEC and SCEC.

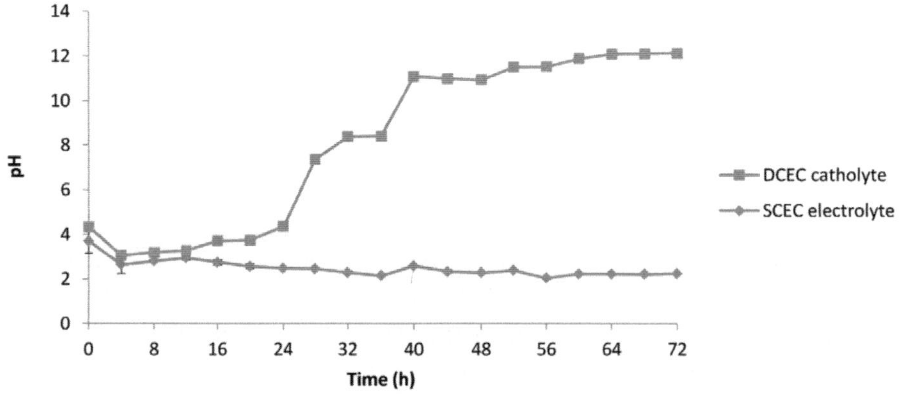

Figure 3. pH changes in DCEC and SCEC.

Compared to conductivity, the differences in current were more evident. This suggests that compared to SCEC, the reactions occurred more efficiently in both chambers of DCEC, therefore enabled a higher electron transfer rate.

Figure 3 shows that between t = 0 and t = 24 h, the pH in the cathode chamber of DCEC was around 4. The pH started to increase after t = 24 h and reached 12 at the end of the reaction. At t = 24 h, 88% nickel ion removal was already achieved and the competing reaction that hydrolyze water into $H_{2(g)}$ and hydroxide ion (OH^-) started to occur. The increase in OH^- concentration occurred faster as nickel ion concentration decreased, and consequently the pH increased. Since the optimum pH for nickel ion precipitation is 10–11 [18], the increase of pH particularly between t = 24 h and t = 48 h increased the deposition of nickel ion and consequently its removal from wastewater.

In SCEC, the pH decreased from 4 at the beginning of the reaction until 2.2 after 72 h. At low pH, more H^+ are available for reduction reaction at the cathode, competing with nickel ion reduction and therefore lowering the efficiency of nickel ion removal at the cathode.

After 72 h, the difference between initial and final concentration showed that 1.1 g of nickel ion was removed from synthetic wastewater in DCEC. However, based on the initial and final weight of the cathode, it was observed that only 0.7 g nickel ion was deposited on the cathode or equivalent to 67% of the removed nickel ion. This discrepancy suggests that around one third of the nickel ion was adsorbed by water hyacinth leaf. The ability of water hyacinth leaves to adsorb heavy metals has been documented previously [19–21]. Therefore, water hyacinth leaves increased nickel ion removal through two mechanisms: preventing H^+ transfer and nickel ion adsorption.

3.2 Energy consumption

The hourly energy consumption in DCEC was relatively constant at 0.0005 kWh. As the energy increase was in a higher order compared to the increase of removed nickel ion, the overall specific energy of removal almost linearly proportional to energy consumption. For 72 h of operation, the total specific energy of removal was 0.041 kWh.g-Ni^{-1}.

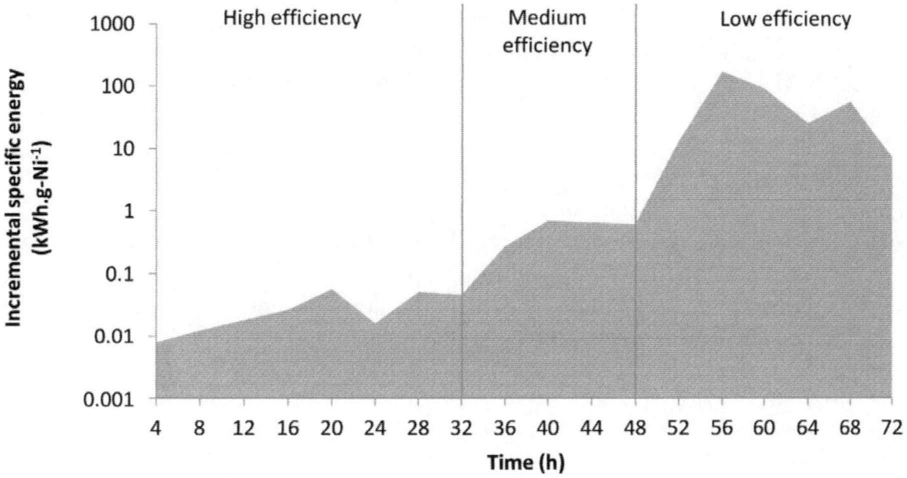

Figure 4. Incremental specific energy of removal in DCEC.

Figure 4 shows the required energy per g of removed nickel ion for every 4 h period. Between $t = 0$ and $t = 32$ h when nickel ion concentration was still higher than 40 g.m^{-3} (Figure 1), the specific energy of removal was less than 0.1 kWh.g-Ni^{-1}. To reduce nickel ion concentration from 44 g.m^{-3} ($t =$ 32 h) to 1.5 g.m^{-3} ($t = 48$ h), 0.3-0.7 kWh.g-Ni^{-1} (0.5 kWh.g-Ni^{-1} on average) was required. The least efficient removal occurred after $t = 48$ h to reduce nickel ion concentration from 1.5 to 0.4 g.m^{-3}. On average 30 kWh.kg-Ni^{-1} was required, more than 1000 times the energy requirement before $t = 32$ h. It is suggested that if a complete removal is not expected, 48 h operation can be applied. However, if the aim is to meet the effluent standard of 1 g.m^{-3} [16], it might be more energy-efficient to combine DCEC with a polishing step e.g. adsorption or membrane filtration [5,22], that can remove nickel ion at low concentration.

4. Conclusion

The results show that nickel ion can almost be completely removed from artificial wastewater using a DCEC reactor partitioned with water hyacinth leaves. The prevention of hydrogen gas attachment on the cathode, high pH in the cathode chamber, and adsorption of nickel ion by water hyacinth leaves contribute to this high removal. This also shows that water hyacinth leaves can be low-cost alternatives for industrial membranes.

Acknowledgment
This research was funded by Indonesian Institute of Sciences (LIPI) through the Deputy of Engineering Sciences Thematic Program.

References

[1] Kuck P H 2016 2013 Minerals Yearbook: Nickel *USGS Miner. Inf.*
[2] Cempel M and Nikel G 2006 *Pol. J. Environ. Stud.* **15** 375–82
[3] Denkhaus E and Salnikow K 2002 *Crit. Rev. Oncol. Hematol.* **42** 35–56

[4] Barakat M A 2011 New trends in removing heavy metals from industrial wastewater *Arab. J. Chem.* **4** 361–77
[5] Coman V, Robotin B and Ilea P 2013 *Resour. Conserv. Recycl.* **73** 229–38
[6] Benvenuti T, Siqueira Rodrigues M A, Bernardes A M and Zoppas-Ferreira J *J. Clean. Prod.* doi:10.1016/j.jclepro.2016.05.139
[7] Djaenudin, Wulan D R and Hariyadi H R 2015 *Procedia Chem.* **16** 155–63
[8] Robotin B, Ispas A, Coman V, Bund A and Ilea P 2013 *Waste Manag.* **33** 2381–9
[9] Peng C, Jin R, Li G, Li F and Gu Q 2014 *Sep. Purif. Technol.* **136** 42–9
[10] Oztekin Y and Yazicigil Z 2006 *Desalination.* **190** 79–88
[11] Abdelhamid A M and Gabr A A 1991 *Arch. Für Tierernaehrung* **41** 745–56
[12] Istirokhatun T, Rokhati N, Rachmawaty R, Meriyani M, Priyanto S and Susanto H 2015 *Procedia Environ. Sci.* **23** 274–81
[13] APHA 1998 *Standard Methods for Examination of Water & Wastewater* (Washington: American Public Health Association)
[14] Son W-K, Kim S H and Kim T I 2000 *J. Polym. Sci. Part B Polym. Phys.* **38** 3181–8
[15] Doan H D, Weli A and Wu J 2009 *Chem. Eng. J.* **151** 51–8
[16] Minister of Environment, Republic of Indonesia Peraturan Menteri Lingkungan Hidup Republik Indonesia No. 5 Tahun 2014 tentang Baku Mutu Air Limbah
[17] Neelakandan S, Kanagaraj P, Nagendran A, Rana D, Matsuura T and Muthumeenal A 2015 *Renew. Energy* **78** 306–13
[18] McAnally S, Benefield L and Reed R B 1984 *Sep. Sci. Technol.* **19** 191–217
[19] Komy Z R, Abdelraheem W H and Ismail N M 2013 *J. King Saud Univ. - Sci.* **25** 47–56
[20] Mohanty K, Jha M, Meikap B C and Biswas M N 2006 *Chem. Eng. J.* **117** 71–7
[21] Schneider I A H, Rubio J, Misra M and Smith R W 1995 *Miner. Eng.* **8** 979–88
[22] Fu F and Wang Q 2011 *J. Environ. Manage.* **92** 407–18

Pre-concentration of Pb, Cd, and Ni in river water using back extraction method

T A Koesmawati[1,3*], M Moelyo[2], A Rizqiani[3] and S Tanuwidjaja[3]

[1]Research Unit for Clean Technology, Indonesian Institute of Sciences (LIPI) Jalan Cisitu 21/154 D, Bandung 40135, Indonesia

[2]Research and Development Center for Water Resources, Jalan Ir. H. Juanda, Bandung, Indonesia

[3]Bakti Asih Higher Education for Analyst, Jalan Padasuka Atas No. 233, Bandung, Indonesia

*E-mail : tiny001@lipi.go.id ; tiny.agustini@gmail.com

Abstract. Dissolved heavy metals such as Pb, Cd and Ni in natural water can be the source of harmful contaminant for human health and aquatic life in aquatic systems. Particular treatment is needed for low concentration of metals in water samples using Flame Atomic Absorption Spectrophotometer (FAAS) measurement. The sample was collected from Cikapundung River in Siliwangi and Dayeuhkolot, Bandung. Back-extraction procedure was used as an alternative method for pre-concentration in trace metal analysis. Ammonium Pyrrolidine Dithiocarbamate (APDC) and Diethylammonium Diethyldithiocarbamate (DDDC) in Methyl Isobuthyl Ketone (MIBK) were used as chelating agent. In this study, the application of back-extraction procedure using APDC/DDDC chelating agent in MIBK resulted the detection limit of 2.2 μg/L for Pb; 1.6 μg/L for Cd; and 1.0 μg/L for Ni. The result showed that sensitivity was increased 4 times for Pb, 6 times for Ni, and none for Cd. The highest concentration of Pb in Cikapundung River was found 8.0 μg/L. All Ni concentration in water samples were under limit of detection. Method validation was conducted in this study and found that the value of correlation coefficient (r) for Pb, Cd and Ni was 0.9995, 0.9960 and 0.9994, respectively. The recovery value for Pb, Cd and Ni were 112.0, 81.5 and 85.9%, respectively in Cikapundung River, and 119.0, 83.5 and 85.9 %, respectively, in groundwater samples. The value of Relative Standard Deviation (RSD) for Pb, Cd and Ni in Cikapundung River were 2.01, 2.15 and 6.40%, respectively. On the other hand, the value of RSD of Pb, Cd and Ni in groundwater were 2.70, 3.30 and 7.69%, respectively. In conclusion, back-extraction method can be applied as pre-concentration for low concentration of Pb and Ni in river water samples.

1. Introduction

Water analysis is very important to quantify the heavy metals water pollution. Analytical method should be developed to get an accurate and precision result. Sample preparation is usually needed in chemical analysis to convert analytes to a more suitable detectable form, to separate them from the sample matrix or to concentrate species for trace analysis. This step is the most difficult part in chemical analysis, due to the risks of analyte losses, sample contamination and incomplete sample decomposition as well as analyte extraction. In addition, sample preparation is often the most time-consuming step in analytical procedures. It is contrasting with the impressive development of modern instrumental techniques that significantly improved detectability and reduced the analysis time. In spite of this development, sample treatment is still needed to improve its selectivity_and to minimize matrix effects. Sample preparation techniques such as hot plate and microwave digestion widely used for this purpose and as the established standard methods for years [2], ISO standard 11466 [3], and preparation for material standard in The Community Bureau of Reference (BCR) [4,5] for trace element dissolution from a large number of matrices. The digestion techniques, however, require the use of concentrated mineral acids, high temperatures and, in the case of microwave applications, high pressures.

Another pre-concentration techniques for metals include solid-phase extraction (SPE) has received much attention in recent years for the analysis of trace concentrations in samples. This technique reduces consumption and exposure to solvent, disposal costs, and extraction time [6,7]. Octadecyl-bonded silica (ODBS) membrane disks have been utilized for the extraction and analysis of many different organic and environmental matrices [8,9]. Moreover, these membrane disks modified by suitable ligands are also successfully used for the separation and sensitive determination of metal ions [10,11].

Metal exchange back-extraction was developed by Lo et al. [12], using Hg to overcome the slow kinetics of acid back-extraction. This method required a single 20-min extraction into chloroform to pre-concentrate (> 90%) of Cd, Co, Cu, Fe, Ni, Pb and Zn. Improvements to this method were recently made by Sachsenberg et al. [13] of which a 90-s single extraction into di-isobutyl ketone (DIBK), with Pd back-extraction is required to quantitatively determine Pb, Cu, Ni and Cd in seawater. This technique produced high pre-concentration factors from relatively small sample volumes, but was limited by the range of metals determined. Back extraction method is an alternative method which was used for water samples. APDC/DDCC as chelating agent in Freon-TF at pH 5 was used in back extraction method. Dithizone was used as chelating agent for Pb and Cd extraction in xylene solvent [11]. The detection limit of Pb and Cd were 0.4 and 0.008 μg/L, respectively. GFAAS is used in back extraction method to determine soluble Cd, Cu, Ni, Pb, and Zn in water sample. The chelating agent was mixture of APDC/DDDC in MIBK solvent and reached the detection limit of 0.001 μg/L [14].

FAAS is as still widely used in Indonesian laboratories to measure metal concentration in water. However, pre-concentration is needed to improve the detection limit. In this study, heavy metals such as Pb, Cd and Ni from river water were measured using back extraction method. Chelating agents used in this study were APDC-DDDC in MIBK solvent at pH 5.

2. Experimental

A total of 12 samples were collected from 6 different areas along Cikapundung River, Bandung City, West Java, as shown in Table 1. The collection method was based on Indonesian National Standard (SNI 06-2412-1991) [15]. Temperature and pH were measured at the sampling sites. Pb, Cd, and Ni analysis was done at an analytical laboratory, Research and Development Center for Water Resources, Bandung.

The extraction was carried out based on Moelyo, et.al. [14]. Prior to analysis, all samples was stored in refrigerator. The sample was filtered using Whatman filter paper 0.45 μm. For every 500 mL volume of filtrate 0.5 mL of concentrated HNO_3 was added. Prior to extraction, the sample was added with 1% ammonium acetate buffer to get pH 4-5. The sample was put in separation funnel, then added with 1 mL of chelate mixture solution of 1% APDC/DDDC and 20 mL of MIBK. The mixture was mixed for 2 min and allowed into two layers. One mL of concentrated HNO_3 was added to the organic layer. The solution was mixed slowly for 20 min, then 9 mL of aqua-millipore was added and mixed for another 30 sec, allowed the mixture to separate, water layer was transferred to a bottle, and measured by FAAS (Hitachi Zeeman 2300 with auto-sampler systems). The FAAS working condition is showed in Table 2.

Validation was carried out by measuring linearity, precision, accuracy, LOD and LOQ. To obtain linearity equation, the instrument respond was plotted in different standard concentration of each analyte. Precision was measured six times for sample concentration of each analyte. The accuracy was measured by adding standard solution to the sample solution and the accuracy of the method was calculated as % recovery. LOD and LOQ was determined by measuring the reagent blank.

Table 1. Sample location and measured parameter on site.

No	Location	pH	Temp	Local Time	Weather	Geography Position	
						South	East
1	Cikapundung River-Gandok	6,80	25,1ºC	10.10	Clear	06º 51' 47,52"	107º 37' 13,5"
2	Ground water-Gandok	6,19	23,8ºC	10.20	Clear	06º 52' 51,24"	107º 36' 56,34"
3	Cikapundung River -Siliwangi	6,86	23,9ºC	10.50	Clear	06º 53' 4,8"	107º 36' 27,24"
4	Ground water -Siliwangi	7,07	24,6ºC	10.55	Clear	06º 53' 3,48"	107º 36' 28,44"
5	Cikapundung River -Wastukencana	7,13	24,2ºC	11.33	Clear	06º 54' 19,56"	107º 36' 27,48"
6	Ground water -Wastukencana	7,05	24,0ºC	11.48	Clear	06º 54' 19,32"	107º 36' 32,4"
7	Cikapundung River -Dalem kaum	7,09	24,0ºC	12.25	Clear	06º 55' 23,76"	107º 36' 31,32"
8	Ground water -Dalem kaum	7,09	24,0ºC	12.35	Clear	06º 55' 23,76"	107º 36' 31,32"
9	Cikapundung River -Soekarno Hatta	7,22	25,7ºC	13.02	Clear	06º 56' 57"	107º 37' 18,78"
10	Ground water -Soekarno Hatta	6,97	24,0ºC	13.58	Clear	06º 56' 55,26"	107º 37' 17,94"
11	Cikapundung River -Dayeuh kolot	7,03	26,9ºC	14.23	Clear	06º 59' 3"	107º 37' 48,3"
12	Ground water -Dayeuh kolot	7,10	26,1ºC	14.34	Cloudy	06º 59' 2,7"	107º 37' 47,28"

Precision method was carried out by measuring the water samples 6 times. The standard deviation (s) and relative standard deviation (RSD) were calculated and the result was compared to the coefficient of variation (CV) Horwitz of $2^{1-0.5 \log C}$. Accuracy of method was carried out by adding standard solution to water sample. The percentage of recovery was calculated and compared with the acceptable recovery value based on AOAC [16].

Table 2. FAAS condition.

	Pb	Cd	Ni
Lamp current (mA)	7.5	7.5	10
Wavelength (nm)	217.0	228.8	232.0
Slit (nm)	1.3	1.3	0.2
Burner head	Standard	Standard	Standard
Burner height (mm)	7.5	5.0	7.5
Fuel gas support	Air-acetylene	Air-acetylene	Air-acetylene
Oxidant gas pressure (Kpa)	160	160	160
Fuel gas flow rate (L/min)	2.0	1.8	1.8

3. Result and discussion

The range of all metal concentration for standard calibration curve was between 10 to 100 μg/L. Table 3 showed the correlation coefficient for Pb, Cd and Ni and their linearity equations. The correlation coefficient for all metals were found higher than 0.99.

Table 3. Calibration curve for each metal.
(Standard range 0.01-0.10 mg/L).

Metal	Linearity equation	Correlation coefficient (r)
Pb	y = 1.327x	0.9995
Cd	y = 9.060x	0.9960
Ni	y = 2.014x	0.9994

LOD value using back extraction method for Pb, Cd and Ni was found as 2.2, 1.6 and 1.0 μg/L, respectively whereas LOQ was found 7.3, 5.2 and 3.2 μg/L for Pb, Cd and Ni, respectively (Table 4). Those values were 4 and 6 times lower than acid sample preparation. LOD was calculated as three times of standard deviation of the reagent blank. The LOQ was calculated as 10 times standard deviation of the reagent blank measurement. Back-extraction method had successfully improve the detection limit for Pb and Ni metals to 0.002 and 0.001 mg/L, respectively, while for Cd, there is no significant difference.

Relative standard deviation (RSD) was calculated from standard deviation divided by average concentration multiplied by 100%. RSD value was between 2.0-7.7 % for all metals. Recovery was calculated from final concentration minus sample concentration divided by standard concentration added multiply by 100%. The recovery value was found between 81-119%. This value was acceptable by AOAC for low concentration (65-120%) [16]. Accuracy and precision results as showed in Table 5.

Table 4. LOD and LOQ value using back-extraction method for Pb, Cd and Ni metals.

Metal	LOD (μg/L)	LOQ (μg/L)
Pb	2.2	7.3
Cd	1.6	5.2
Ni	1.0	3.2

Table 5. Accuracy and Precision data for Pb, Cd and Ni metals.

Metal	% Recovery	% RSD	0.67 x CV Horwitz
River water sample			
Pb	112	2,01	18,89
Cd	81,5	2,15	19,82
Ni	85,9	6,40	19,66
Ground water sample			
Pb	119	2,70	18,74
Cd	83,5	3,30	19,75
Ni	91,1	7,69	19,49

Table 6. Heavy metal concentration.

No	Location	Metal concentration		
		Pb (μg/L)	Cd (μg/L)	Ni (μg/L)
1	Cikapundung River-Gandok	ND	ND	ND
2	Cikapundung River -Siliwangi	ND	ND	ND
3	Cikapundung River -Wastukencana	ND	ND	ND
4	Cikapundung River -Dalemkaum	ND	ND	ND
5	Cikapundung River -Soekarno Hatta	ND	ND	ND
6	Cikapundung River -Dayeuh kolot	**8.0**	ND	ND
7	Ground Water-Gandok	ND	ND	ND
8	Ground Water -Siliwangi	ND	ND	ND
9	Ground Water -Wastukencana	ND	ND	ND
10	Ground Water -Dalemkaum	**13.0**	ND	ND
11	Ground Water -Soekarno Hatta	ND	ND	ND
12	Ground Water -Dayeuh kolot	ND	ND	ND

ND = not detected

Table 6 shows, Pb concentration in Cikapundung River from Dayeuhkolot area was found 8 μg/L, below the permissible level of water quality class 1 of 30 μ g/L based on the Indonesian Government Regulation (Table 7). As Dayeuhkolot area is surrounded by textile industries, Pb concentration was found higher than other area which is mostly housing area. On the other hand, Pb concentration for Ground Water from Dalemkaum area was below the acceptable level of 13 and 15 μg/L, respectively, according to the Indonesian Government Regulation and the USEPA. As Dalemkaum is a Central Business District (CBD) for Bandung, water quality could be polluted from transportation and electroplating industry nearby the Bandung CBD.

Table 7. Allowable metal concentration in water samples (μg/L).

Metal	Internasional Standard*	Indonesian Standard **			
		Class 1	Class 2	Class 3	Class 4
Arsen	10	1.0	2.0	2.0	5.0
Lead	15	30	30	30	1000
Cadmium	10	10	10	10	10
Nikel	100	-	-	-	-
Cobalt	100	-	-	-	-
Mangan	50	100	-	-	-
Chrom	100	50	50	50	10
Iron	300	300	-	-	-

Source : * USEPA [16]
** Indonesian Government Regulation [17].

4. Conclusion

The back-extraction method can be used as an alternative method for determination of Pb, and Ni in water sample. Back-extraction method was found more sensitive compared to both USEPA and ISO

11466 methods for determination of Pb and Ni metal in water sample by FAAS. The LOD for Pb and Ni was lower 4 and 6 times, respectively. However, the extraction method using APDC/DDDC was not demonstrated for the determination of cadmium in water sample. Water sample taken from Cikapundung was classified as class 1 and confirm to use for drinking water or any other use with the similar requirements.

Acknowledgement
The authors thank to Analytical Laboratory, Research and Development Centre for Water Resources, Bandung for their laboratory facilities.

References

[1] Fulazzaky M A 2010 *Environ. Monit. Assess*, **168**: 669-684.
[2] USEPA Method 3050 1986 U.S. Environmental Protection Agency Acid digestion of sediment, sludge and soils USEPA (Ed.), Test Methods for Evaluating Soil Waste SW-846, USEPA, Cincinnati, OH, U.S.A.
[3] ISO 11466 1995 Determination of elements in aqua regia and nitric acid digests by flame atomic absorption spectrometry
[4] Quevauviller Ph, Lachica M, Boratora E, Gomez A, Rauret G, Ure A, Muntau H 1998 *Fresenius J. Anal. Chem.* **360** p. 505
[5] Quevauviller Ph, Rauret A, Muntau H, Ure A M, Rubio R, López Sánchez J F, Fiedler H D, Griepink B 1994 *Fresenius J. Anal. Chem.* **349**, pp. 808–814
[6] Hagen D F, Markell C G, Schmitt G A, Blevins D D 1990 *Anal. Chim. Acta.* **236**, 157.
[7] Yamini Y, Khorasani A 1994 *J. High Resolut. Chromatogr.* **17**, 634.
[8] Messer D C, Taylor L T 1995 *J. Chromatogr. Sci.* **33,** 290.
[9] Shamsipur M, Ghiasvand A R, Yamini Y 1999 *Anal. Chem.* **71**, 4892.
[10] Yamini Y, Shamsipur M 1996 *Talanta,* **43**, 2117.
[11] Carasek, Eduardo et al. 2002 [Online] *www.sciencedirect.com* [downloaded 07 Juli 2014].
[12] Lo J M, Yu J C, Hutchison F I and Wal C M 1982 *Anal. Chem.* **54**: 2536-2539.
[13] Sachsenberg S, Klenke Th, Krumbein W E and Zeeck E 1992 *Fresenius J. Anal. Chem.* **342**: 163-166.
[14] Moelyo, Moelyadi, dkk.. 1995 Determinasi Logam Berat Kelumit dalam Sumber Air dengan Metode Ekstraksi Ulang dan SSA-Tungku Karbon, Local Research Report, **p. 23**
[15] Standar Nasional Indonesia (SNI) 1991 [Online]. *http://www.pip2bdiy.org* [downloaded at 07 June 2014].
[16] AOAC 2002 Guidelines for Single Laboratory Validation of Chemical Methods for Dietary Supplements and Botanicals
[17] USEPA 1996 National Primary and Secondary Drinking Water Regulations
[18] Peraturan Pemerintah Republik Indonesia Nomor 82 *2001* [Online]. *http://www.minerba.esdm.go.id [*downloaded at 11 December 2014].

The presence of bromuconazole fungicide pollutant in organic waste anaerobic fermentation

H R Hariyadi
Research Unit for Clean Technology, Indonesian Institute of Sciences (LIPI)
Jalan Cisitu 21/154 D, Bandung 40135, Indonesia
Email : hari.rom.haryadi@lipi.go.id

Abstract. The presence of bromuconazole fungicide pollutant in organic waste anaerobic fermentation was carried out as well as the influence phenol and benzoate, and biodegradation of bromuconazole. Bromuconazole is a fungicide effective against *Ascomycetes*, *Basidiomycetes* and fungi imperfecti in cereals, grapes, top fruits and vegetables. It is also effective against *Alternaria* and *Fusarium sp.* The remaining fungicide in leaves might contaminates landfill. One month of organic waste added with bromuconazole was anaerobically incubated in 500 mL bottles at 30°C without shaking in dark room. High-Performance Liquid Chromatography (HPLC) with UV detector and a 100 RP 185µm Lichrosphere column was used to determine bromuconazole concentration. Methane content was determined by Gas Chromatography (GC) method equipped with a flame ionization detector and a metal column packed with 5% neopentyl glycol sebacate and 1% H_3PO_4 on Chromosorb W-AW (mesh 80-100). After incubation for 225 days, bromuconazole of 200 mg/L inhibited the production of methane (99.5 mM) significantly, but did not inhibit the production of volatile fatty acids. The addition of 100 mg/L phenol or 146 mg/L benzoate increased the production of methane, 143 mM and 135.2 mM, respectively compared with control (121.8 mM). In anaerobic conditions, the presence of toxic pollutants such as fungicide bromuconazole in landfills sites may cause further problems with the accumulation of volatile fatty acids in leachate. Further study to determine the threshold, the presence of bromconazole in low concentration (less than 200 mg/L) on the methane production is recommended.

1. Introduction

Azole compounds, particularly triazole, has been used as the active ingredient in the manufacture of fungicides and pharmaceutical preparations since 1970^s. Although the production and its use has increased, but it is still lack of information about the nature of these compounds in environment either as well as landfill site and process effluent. The problem may arises when these potentially toxic compounds pollute the environment and threaten both the aquatic organism as well as human health. Activities of microorganisms might be disturbed by the presence of such compounds, and decrease the fermentation process performance. Therefore, the experiment to observe the presence of the compound in anaerobic process of the refuse is important. There are about 16 different fungicide triazole derivatives and one of them is bromuconazole (Figure 1). This compound has been developed by the Rhone Poulenc Agrochemie UK since 1991. Bromuconazole ($C_{13}H_{12}BrCl_2N_3O$, MW = 377) or 1[(2RS,4RS,2SR,4SR)-4-bromo-2-(2,4-dichlorophenyl) tetra- hydrofurfuryl]-1H-1,2,4-triazole is a heterocyclic-aliphatic compound belongs to the groups of systemic fungicide and it is active against *Ascomycetes*, *Basidiomycetes*, fungi imperfecti in cerealia and fruits as well as active against fungus *Alternaria sp* and *Fusarium sp.* Toxicity by 50% (LD_{50}) for rats and mice, respectively, 365 mg/kg and 1151 mg/kg. Towards fish such as Blugill and trout, toxicity of bromuconazole (LC_{50}) is 3.1 mg/L and 1.7 mg/L, respectively [1]. Bromuconazole is one of the 79 substances of the third stage Part A of the review program covered by Commission

Regulation (EC) No 1490/2002 [2]. Similar products such as Amitrole (3-amino-1 h -1,2 4-triazole) reported to be transformed into 3-(3-amino-1 h-triazole-1-yl) alanin by *E. coli* [3]. Benzotriazole transformed to 1-deoxyribosida by *Aerobacter aerogeneses* ATCC9621 supplementation with 5 '-thymidylic acid and *Aspergilus niger* (Van Teigh) was able to reducing Triadimefon became Triadimol [4].

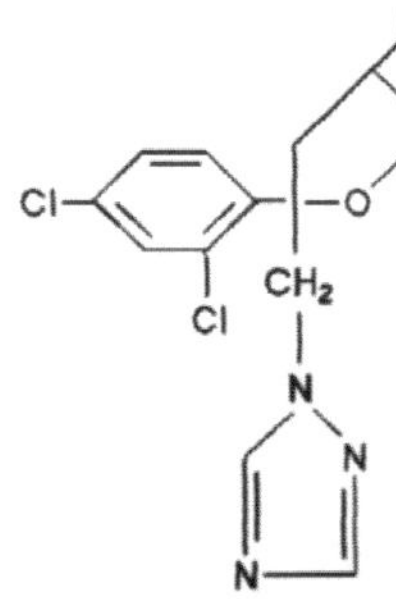

Figure 1. Chemical structure of bromuconazole.

As pollutant, bromuconazole might be degraded biologically in aerobic and anaerobic condition. In anaerobic fermentation of organic compounds, the resulted xenobiotic compounds such as pesticide will accumulate if not degraded into methane [5]. The production of methane, as a source of energy, is one of the advantages of anaerobic fermentation process. Anaerobic degradation occurs due to the role of either one type or mixture of microbes. For example, the benzoate was degraded by *Moraxella sp* or *Pseudomonas* to diethyl ester [6] and degradation of phenols by a community of microbes produced methane [7][8]. Xenobiotic compound can be degraded if the appropriate enzyme produced by microbes during the process of evolution. Degradation is influenced by the ability of the enzyme to accept a substrate with a structure similar (but not identical) to that of natural and less complex compound that also presence in media. The degradation is also influenced by the ability of substrate to stimulate synthesis of the necessary enzyme [9].

2. Experimental

2.1. Anaerobic experiment

This study was conducted in Waste Technology Laboratory Strathclyde University Glasgow. The presence of anaerobic bromuconazole was carried out using one month partially degraded waste substrate taken from local domestic waste disposal as source of microbes. As many as 25 gr of refuse was filled into 500 mL volume of bottle and added with water as much as 250 mL. Bottle sealed was then sealed with the rubber sub a seal No. 45. Prior to incubation at 30°C without stirring in a dark room, the bottle purged with oxygen-free nitrogen gas assuming the anaerobic conditions was obtained. The experiment was conducted using three replicates. Bromuconazole is an aromatic compound with solubility in water is 50 mg/L, so for preparing a solution with a high concentration, organic solvent such as acetone needs to be used. Bromuconazole with final concentration of 100 mg/L of which dissolved with acetone was added each on day 0 and 31 (treatment 1). On the different bottle (treatment 2) on day 31 was added with 100 mg/L (final concentration) phenol. And bottle as treatment 3 was added with 146 mg/L benzoate (final concentration). Bottles with phenol and benzoate was then added with bromconazole 200 mg/L (final concentration) on day 103. Bottle without bromuconazole, benzoate, or phenol was used as treatment 4 or control. Sample was taken on a regular basis to determine the concentration of methane in the headspace bottle, volatile fatty acids, phenols, and benzoate.

2.2. Chemical analysis: bromuconazole, phenols and benzoate

Determination the concentration of bromuconazole, phenol and benzoate was carried out by High-performance Liquid Chromatography (HPLC) method. Bromuconazole standard was received from the Rhone-Poulence Agrochemie whereas HPLC solvents such as acetonitrile and other chemicals purchased from E-merk. The equipment consists of a pump Binary LC250 (Perkin Elmer); Lichrosphere 100 RP18 column um 5:150 x 4.6 mm (Altech); UV detector LC 90J (Perkin Elmer). An interface of Nelson and Nelson 900 Series system integrators software PC version 5.1 was used to process the data. The determination was carried out isocratically with mobile phase acetonitrile/water (65/35% v/v) on the conditions of flowrate, 1 mL/min; wavelength, UV 230 nm; response, 0.5; sensitivity, 1 AUFS; maximum pressure, 3500 psi; duration 15 min; injection volume 20 μl. Bromuconazole concentration in the sample was determined based on the area and the concentration of the standard of bromuconazole dissolved in mobile phase on the same analysis conditions. Concentration of phenols and benzoate in the sample was determined by the same method.

2.3. Methane and volatile fatty acid

The concentration of methane in the headspace bottle was determined using Gas Chromatography methods, GC8700 (Perkin Elmer) equipped with Flame Ionisation Detector (FID). The temperature of the injector, detector, and oven each set on 200, 210 and 80°C, respectively. Pure methane (uL 10 x 5) and the triplicates sample separately injected through a gas syringe (Hamilton). Concentration of methane was calculated based on the standard temperature (°K) and local pressure (mm). Oxygen-free nitrogen gas used as carrier with speed 20 lbf/sq-inc), column (Phase Sep) made of metal of 2 m x 2 mm. Type of packing column used was 5% neopentyl glycol sebacate and 1% H_3PO_4 on Chromosorb w-aw (80 – 100 mesh). For the concentration of volatile fatty acids (VFA), the temperature of the injector and detector were set at 200 and 300°C. The oven was set at 100 °C for 2 min and then up 30 °C per min until the preserved 150°C for 3.6 seconds. VFA standard (0-10 mm) consists of acetate, propionate, butyrate, hexanoate and valerate. Both standard and acidified samples with concentrated formic acid (0.7:0.3) as much as 1 μL were injected separately through the liquid syringe (Hamilton). Through comparison with a standard that has been analyzed at the same day, the VFA concentration in the sample was determined.

3. Result and discussion

By HPLC, bromuconazole was separated into two peaks with a proportion of 51 : 49% at 5.07 and 6.58 min, respectively. HPLC separated 2 sets of diastereomers and but not the 2 sets of enantiomers. These two peaks, corresponded to diastereomer consists of 1,2,4- and 1,3,4-bromuconazole and diastereomer which consists of 1,2,3- and 1,2,5-bromconazole, respectively. Figure 2 shows the production of methane from fermentation of refuse without carbon source as control. Total methane concentration in the headspace increased quickly after the first 2 months of incubation whereas the volatile fatty acid concentration reached maximum point (18 mM) on day 31 and then declined. After 100 days of incubation, volatile fatty acid was not detected. On day 173, the total methane concentration in the headspace reached 121.78 mM. In this experiment, benzoate and phenol has been selected as the additional carbon source or induced substrate due to the basic structure of the compound similar with one (phenyl ring) of the rings that formed bromuconazole. It was expected that with the additional carbon source depletion, microbes degraded bromuconazole through its phenyl ring.

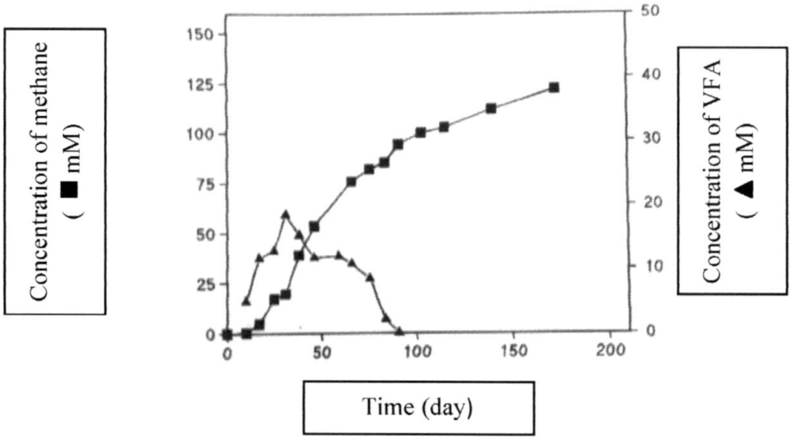

Figure 2. Production of methane and volatile fatty acid (VFA) from refuse fermentation.

Figure 3 and Figure 4 show the ability of the microbes to degrade phenol or benzoate as a source of carbon. Phenol or benzoate added separately on day 31 when the main carbon source of refuse has been depleted as a result of degradation by microbes and was indicated by maximum concentration of volatile fatty acids produced (Figure 2).

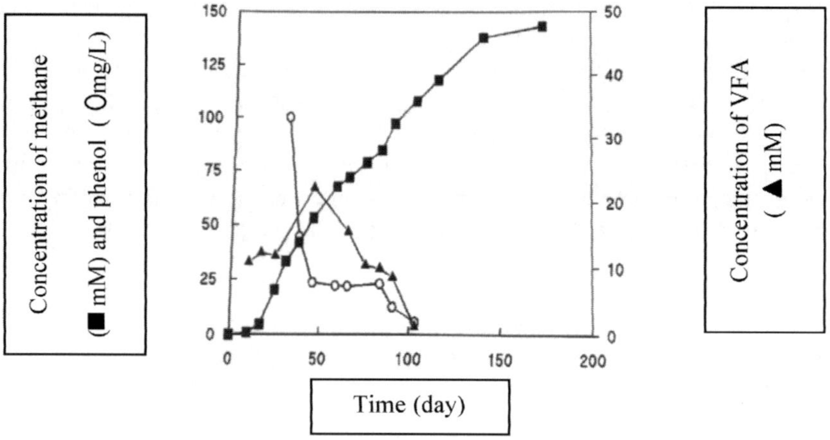

Figure 3. Effect of phenol addition on methane production at day 31.

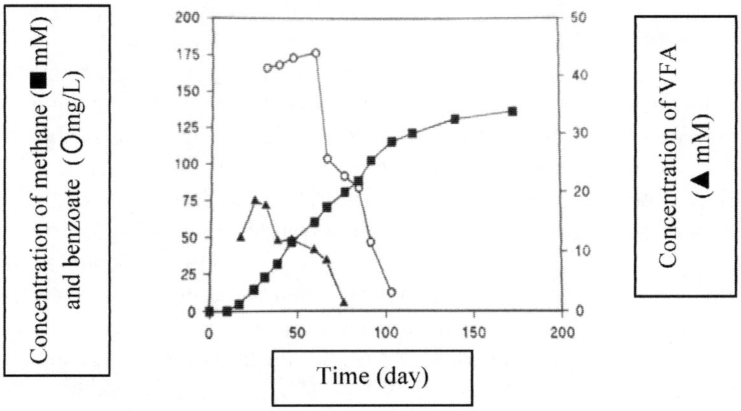

Figure 4. Effect of benzoate addition on methane production at day 31.

Figure 3 shows that by the addition of phenol on the substrate, volatile fatty acid production continued and reached a maximum of 22.30 mM on day 46 and then declined to 1.20 mM on day 103. At day 46, phenol concentration decreased rapidly to 23.46 mg/L whereas at day 103, phenol concentration was very low (5.64 mg/L). Meanwhile, methane production increased rapidly since day 17 and reached 142.98 mM/l at day 173, higher than control was of 121.78 mM . The effect of benzoate addition on methane production also showed similar pattern (Figure 4). At day 173, total accumulation of methane (135.15 mM) in the headspace of the bottle with the addition of benzoate was also higher than control, but lower than that of the addition of phenol (142.98 mM). On the other hand, the concentration of benzoate, of which was not declined initially, started to decrease quickly on day 59 and was very low (12.11 mg/L) on day 103. Effect of bromuconazole addition is in Figure 5.

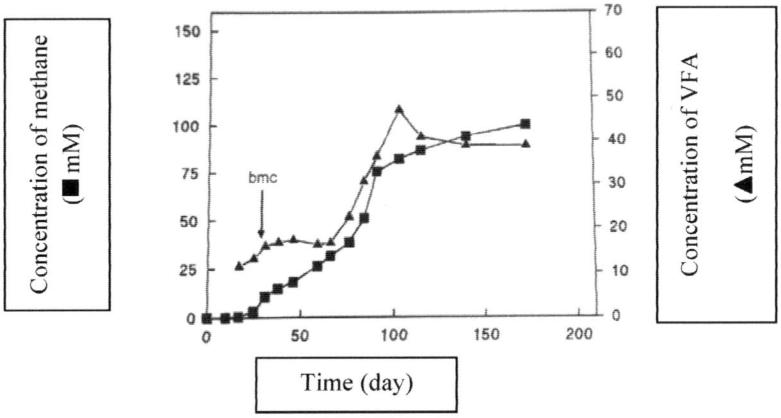

Figure 5. Effect of bromuconazole addition on methane and volatile fatty acid production at day 0 and 31.

In this experiment, bromuconazole added on day 0 and 31 so that the total concentration in the bottles (final concentrations) was 200 mg/L. The methane production after bromuconazole addition on day 0 was lower than the control (Figure 2) but did significantly influence against the volatile fatty acid concentration up to day 31. Since day 31, where as another 100 mg/L of bromuconazole were added, methane was produced but volatile fatty acid production was inhibited until day 66. The concentration of volatile fatty acid continued to accumulate until 38.77 mM on the day 173, higher than control. Although methane was produced continuously but on the day 173 its concentration was 99.58 mM, lower than control and other treatments with phenol and benzoate. It suggested that bromuconazole had greater influence on methanogen activities rather than fermentative bacteria producing volatile fatty acid. The influence of bromuconazole inhibition against the production of methane is shown in Figure 6. Methane production stopped as so bromuconazole (200 mg/L final concentration) was added on the day 103, whereas concentration of phenol or benzoate was very low (each 5.64 mg/L and 12.11 mg/L). Inhibition of methane production continued until the end of the experiment at day 173. Methane was still generated on bottles that was not added with bromuconazole on day 103. This indicated that the contamination of bromuconazole highly influential on the production of methane and the concentration of 200 mg/L might be too toxic to methanogen bacteria. Further study to determine the threshold, the presence of bromconazole in low concentration (less than 200 mg/L) on the methane production from refuse and also its toxicity based on bacterial growth inhibition by 50% (IC_{50}) is recommended.

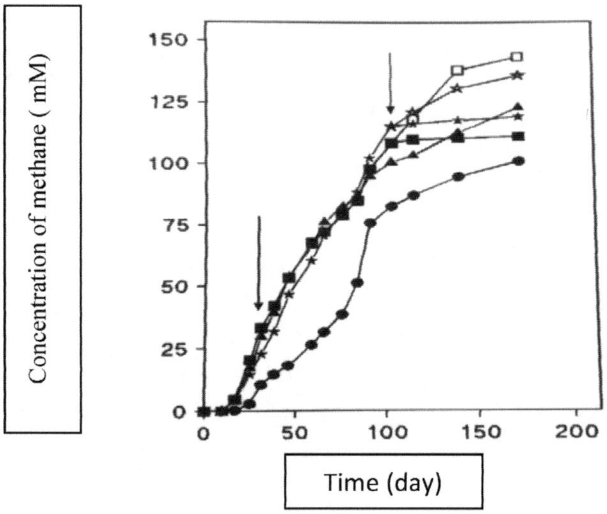

Figure 6. Methane production from refuse supplemented with:
● bromuconazole 100 mg/L at day 0 and 31.
■ Phenol 100 mg/L at day 31 and bromuconazole 200 mg/L at day 103; ★Benzoate 146 mg/L at day 31 and bromuconazole 200 mg/L at day 10□☆ Phenol or benzoate at day 31 and no added bromuconazole at day 103; ▲Control: without phenol, benzoate or bromuconazole

4. Conclusion

It was concluded that phenol or benzoate as a source of additional carbon improved methane production. By improving methane production, phenol or benzoate indicated no inhibition on methanogen bacterial activity. Therefore the presence of phenol or benzoate in refuse were in tolerant by fermentative bacteria as well as methanogen bacteria to degrade refuse. In contrast, bromuconazole indicated its toxicity against methanogen bacterial activity. The addition of phenol and benzoate improved adaptation of the anaerobic microbes to bromuconazole. The presence of bromuconazole in anaerobic process caused the refuse imperfectly fermented and so the production of methane was inhibited and volatile fatty acids accumulated. Accumulation of the volatile fatty acids may cause further pollution in aquatic environment around the landfills site. This phenomenon indicated the persistence of bromuconazole and this compound might not be degraded in the process.

Acknowledgement

The Author is grateful to Irene Watson-Craik Ph.D for her supervision and Rhone Poulenc Agrochemie UK for the supply of bromuconazole and analytical method.

References

[1] Worthing C R 1991 *Pesticide Manual 9th ed.* The British Crop Protection Council.
[2] European Food Safety Authority 2010 *J. EFSA.* **8(8)** 1704–1788.
[3] Kieslich K 1976 Microbial transformation of nonsteroid cyclic compounds, John Wiley and sons.
[4] Deas A H B and Clifford D R 1982 *Pest. Biochem. and Physio.* **17** 120–133.
[5] Fewson C A 1981 *FEMS Symp.* **12** 141–179.
[6] Evans W C and Fuchs G 1988 *Ann. Review of Microb.* **42** 289–317.
[7] Balba M T and Evans W C 1977 *Biochemical Soc. Trans.* **5** 302–304.
[8] Fery J G and Wolfe R S 1976 *Arch. Microb.* **107** 33–40.
[9] Hutzinger O and Veerkamp W 1981. *FEMS Symp.* **12** 3–45.

IOP Conf. Series: Earth and Environmental Science **60** (2017) 012023 doi:10.1088/1755-1315/60/1/012023

Flow rate analysis of wastewater inside reactor tanks on tofu wastewater treatment plant

Mamat[1*], N Sintawardani[1], J T Astuti[1], D Nilawati[1], D R Wulan[1], Muchlis[1], L Sriwuryandari[1], T Sembiring[1] and N W Jern[2]
[1]Research Unit for Clean Technology, Indonesian Institute of Sciences (LIPI)
Jalan Cisitu 21/154 D, Bandung 40135, Indonesia
[2]NEWRI, Nanyang Technological University (NTU) - Singapore
[*]E-mail: mamat@lipi.go.id, sik_lik@ymail.com

Abstract. The research aimed to analyse the flow rate of the wastewater inside reactor tanks which were placed a number of bamboo cutting. The resistance of wastewater flow inside reactor tanks might not be occurred and produce biogas fuel optimally. Wastewater from eleven tofu factories was treated by multi-stages anaerobic process to reduce its organic pollutant and produce biogas. Biogas plant has six reactor tanks of which its capacity for waste water and gas dome was 18 m^3 and 4.5 m^3, respectively. Wastewater was pumped from collecting ponds to reactors by either serial or parallel way. Maximum pump capacity, head, and electrical motor power was $5m^3/h$, 50m, and 0.75HP, consecutively. Maximum pressure of biogas inside the reactor tanks was 55 mbar higher than atmosphere pressure. A number of 1,400 pieces of cutting bamboo at 50-60 mm diameter and 100 mm length were used as bacteria growth media inside each reactor tank, covering around 14,287 m^2 bamboo area, and cross section area of inner reactor was 4,9 m^2. In each reactor, a 6 inches PVC pipe was installed vertically as channel. When channels inside reactor were opened, flow rate of wastewater was $6x10^{-1}$ $L.sec^{-1}$. Contrary, when channels were closed on the upper part, wastewater flow inside the first reactor affected and increased gas dome. Initially, wastewater flowed into each reactor by a gravity mode with head difference between the second and third reactor was $15x10^{-2}$m. However, head loss at the second reactor was equal to the third reactor by 8,422 x 10^{-4}m. As result, wastewater flow at the second and third reactors were stagnant. To overcome the problem pump in each reactor should be installed in serial mode. In order to reach the output from the first reactor and the others would be equal, and biogas space was not filled by wastewater, therefore biogas production will be optimum.

1. Introduction

A multi-stages biogas plant which was built in Giriharja, Kebonjati village, Sumedang district could treat 18-20 m^3. day^{-1} wastewater from eleven local tofu factories. This wastewater contains 7000-8000 gr $COD.L^{-1}$ that exceeds the effluent standard for soybean-based industry. By wastewater treatment, the tofu factories could comply an environmental regulation, and the water safely discharged into Cileuweung River. Tofu production resulted two types of wastewater namely the dilute and concentrated wastewater, both of wastewater was then collected into the ponds. Dilute wastewater was then fed to reactor to be processed anaerobically to produce biogas. The biogas plant has 6 reactor tanks, every three reactor tanks were set up serial mode with capacity of each reactor tank is 18 m^3 of wastewater and 4.5 m^3 of biogas. Biogas production was designed until 120 $m^3.day^{-1}$ of gas as an LPG substitution for 91 households.

IOP Conf. Series: Earth and Environmental Science 60(2017) 012023 doi:10.1088/1755-1315/60/1/01202

Wastewater collected in the pond was pumped into equalizing tank and drained by gravity to all reactors. Bamboo as bacteria growth media were placed horizontally on each reactor, between buffer zone, which may cause a resistance of wastewater flow rate. Due to bamboo placed between barriers, it caused a resistance to wastewater flow and raised the wastewater level inside reactor tanks. Therefore, space of biogas was occupied by wastewater. Consequently, biogas was not produced. Therefore, flow rate input was not equal to its output, resulted wastewater level inside the first, second, fourth, and fifth tanks were higher than the set up levels. This configuration was set as shown in Figure 1.

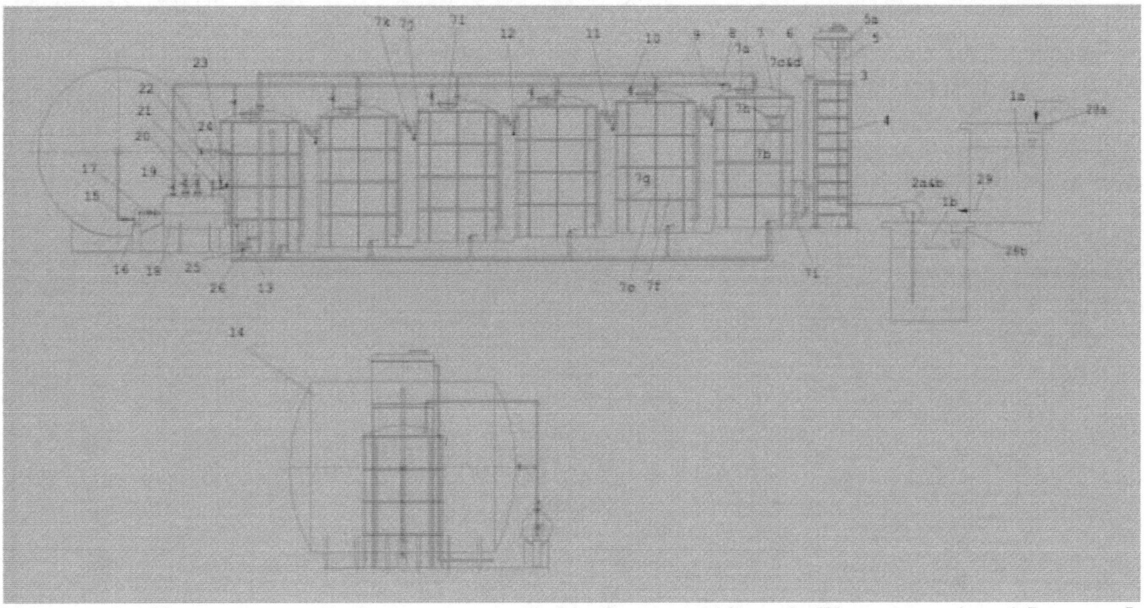

Notation: 1a. Upper storage, 1b. Lower storage, 2a&b. Pump machines, 3. Wastewater pipe, 4.Support, 5. Equalizing tank, 5a. Filter, 6. Inlet pipe, 7. Reactor tank, 7a. Manhole, 7b. Bamboo cutting, 7c&d. Reinforce, 7e. Reinforce ring, 7f. Bolt nut, 7g. Vertical reinforce, 7h.Nylon net, 7i.Anchore, 8. Gas pipe, 9. Back wash pipe, 10. Back wash valve, 11.Outlet pipe, 12. Valve, 13. Gas flow meter, 13. Gas scrubber, 14. Gas holder, 15. Gas valve, 16. Gas compressor, 17.Check valve, 18. Distribution tank, 19. Back wash valve, 20. Gas valve, 21.Pressure gauge, 22. Relief valve, 23. Gas pipe, 24. Sight glass, 25. Concrete base, 26. Drain pipe.

Figure 1. Arrangement of reactor tanks.

At the beginning, wastewater flowed from the first reactor to another by gravity mode with a head difference of each reactor was 0.15 m. An analysis based on observations, theory of fluid mechanic and discussion, it could be known that reactor head loss was lower than head losses at each reactor, therefore flow stagnation was occurred at the second and third reactor for the first unit, and at the fifth and the sixth for the second unit. When a remained of biogas space was occupied by wastewater, the condition could decrease biogas production.

This research aimed to solve problem caused by a resistance of wastewater flow rate occurred inside existing reactor tanks filled with bamboo, therefore the ideal flow rate of wastewater inside reactor tank could be created to produce biogas optimally.

2. Experimental

2.1. Materials

Reactor tanks were built by 10 mm-thickness composite materials with 2500 mm in diameter, 4000 mm height, and 650 mm dome height. The reactor tanks have a space of 80% wastewater and 20% of biogas. Tofu wastewater was pumped from the collection pond into an equalizing tank and distributed to all reactors. Capacity of pump is 5 $m^3.hr^{-1}$ and its head is 50 m operated by an electrical power of

IOP Conf. Series: Earth and Environmental Science **60** (2017) 012023 doi:10.1088/1755-1315/60/1/012023

motor at 0.75 HP [1]. Equipment that applied at the experiment were water flow meter, stop watch, ball valve and water pump.

2.2. *Experimental setup*

Each reactor tank was equipped with bottom and top buffers, where perforated-corrugated composites at bottom and nylon nets at top barriers with 2000 mm distance were set up and in between were filled by 1400 pieces bamboo cutting as bacteria growth media to produce methane gas. Each bamboo has 50-80 mm in diameter and 100 mm length. The 2500 mm diameter and 3 mm thickness bottom barrier [2] has 200 holes and each hole has a 50 mm in diameter. The nylon net as a top barrier which was submerged at 1200 mm below water level has a (30x30) mm^2 hole.

Residence time of wastewater existing inside each reactor is about eight hr before the wastewater moved to the next reactors. Initial pH of wastewater was 4.5-5.0 and wastewater pH at the effluent must be around neutral.

Inlet and outlet of reactors were equipped a water flow meter, therefore fed and drained wastewater quantity could be recorded.

2.3. *Calculations*

To analyse the flow rate of wastewater inside the reactor tanks, the steady flow equation for unit mass flow rate under uniform was applied, continuous flow condition based on Figure 2.

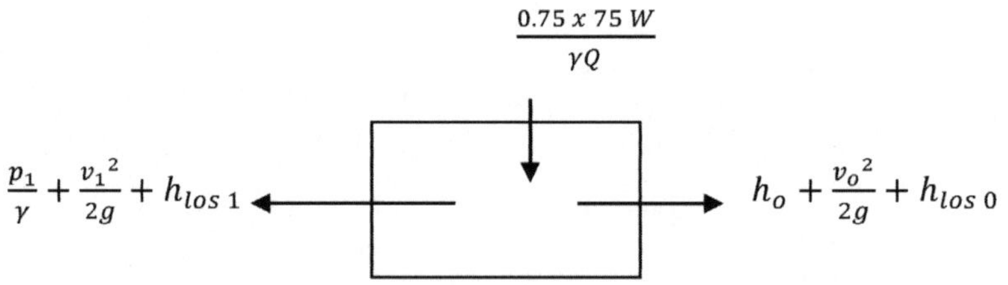

Figure 2. Energy balance.

Energy needed to raise wastewater from collection ponds to pump machine is $\dot{m}\left(\frac{v_o^2}{2g} + h_o + h_{los\ 0}\right)$, while energy needed to press wastewater flowed to reactor tank is $\dot{m}\left(\frac{v_1^2}{2g} + \frac{p_1}{\gamma} + h_{loa\ 1}\right)$ and power of pump obtained is $\dot{m}\left(\frac{0.75x75W}{\gamma\dot{Q}}\right)$. Where inlet velocity of wastewater to pump is $v_o(m/sec)$, outlet velocity of wastewater from pump is $v_1(m/sec)$, head wastewater level in collection pond is $h_o(m)$, static pressure of wastewater went out from pump is $p_1(kg/m^2)$, head losses of inlet is $h_{los\ 0}\ (m)$, head losses of outlet is $h_{los\ 1}(m)$, power of pump is $W(kgm/sec)$, flow rate of wastewater is $\dot{Q}(m^3/sec)$, density of wastewater is $\gamma(kg/m^3)$ and gravity is $g(m/sec^2)$. Thereby an equation of energy balance could be obtained as follows:

$\frac{56.25W}{\gamma\dot{Q}} = \frac{p_1}{\gamma} + \frac{v_o^2+v_1^2}{2g} + h_o + h_{loso} + h_{los1}$, diameter of inlet pipe was equal to outlet pipe and $v_o = v_1$, therefore outlet velocity of the pump could be obtained by an equation (1).

$$v_1 = \sqrt{2g\left\{\left(\frac{56.25wW}{\gamma\dot{Q}}\right) - \left[\frac{p_1+\gamma(h_0+h_{loso}+h_{los1})}{\gamma}\right]\right\}} \qquad (1)$$

Actual head of the pump could be obtained by an equation of $\frac{v_1^2}{2g}\ equal\ h_p$.

Head losses occurred at wastewater flow to reactor depends on some factors, such as: friction of flow on wall face, bend flow, expansion and contraction flow. Particulate number contained by wastewater had a significant effect on a friction at pipe wall and vessel. The condition could be assumed as a face roughness of drain wall, and friction value affected by wall roughness of $k\ (mm)$, length of pipeline is l (m), diameter of pipeline is D (m) and Reynolds number dimensionless. Friction value of wastewater flow could be obtained by an equation as follows:

$$f = 0.001375\left[1 + \left(200\frac{k}{D} + \frac{10^6}{R}\right)^{\frac{1}{3}}\right] \quad (2)$$

Head losses caused by some factors for examples; contraction factor, expansion factor, friction factor, bend factor, and discharge factor. Contraction and expansion factor obtained from graphic of area ratio versus contraction and expansion factor. Head losses factor of bend, tee joint, foot valve, ball valve and check valve, therefore could be obtained by [3]. Head losses at pump inlet caused by energy needed to raise wastewater from collection ponds to pump machine, friction of pipe wall, bend and contraction caused a diameter difference between inlet and outlet of pump. Therefore head losses of pump inlet could be obtained by an equation as follows [4]:

$$h_{los} = \left(f\frac{l}{d} + k_{fv} + k_v + +k_{cv} + k_T + k_k + k_c + k_e\right)\frac{{v_1}^2}{2g} \quad (3)$$

where head losses factor of foot valve is k_{fv}, had losses factor of ball valve is k_v, head losses factor of check valve is k_{cv},head losses factor of tee joint is k_T, head losses factor of knee is k_k, head losses factor of contraction is k_c, and head losses factor of expansion is k_e.

Inlet to the first reactor tank formed velocity of wastewater pumped which was might higher than head losses of bamboo cutting, head of wastewater inside reactor tank and pressure of biogas. Head losses at the first reactor tank, which was lower than head of pump and could be formulated by an equation as follows:

$$h_r + p_g + h_{loss} < h_p \quad (4)$$

Wastewater inside the first reactor had to flow into the second reactor and wastewater inside the second reactor had to flow into the third reactor then went out from the third reactor tank to the collection ponds. Head differences between the first reactors was equalled to the second and the third that is $h_r(m)$. Based on Figure 2, outlet of the first reactor be analysed using equation as follows:

$$\frac{p_1}{\gamma} = \frac{p_2}{\gamma} + \frac{{v_2}^2}{2g} + h_{los2} + h_2 \quad (5)$$

Wastewater flowed from the first reactor to the second and the third one, when each head difference of reactor tanks must more than head losses of bamboo cutting, expansion, contraction and friction. Therefore flow rate of reactor output could be obtained by an equation as follows [5]:

$$\dot{Q}_r = c_d A v \quad (6)$$

Flow rate of reactor tank output is $\dot{Q}_r$ (m^3/sec), discharge factor is c_d, discharge area of reactor tank is A (m^2) and wastewater velocity is v (m/sec).

3. Result and discussion

3.1 Flow rate observation

The results show that wastewater density was 1,050 kg.m^{-3}, inlet pipe of pump was equipped a foot valve losses factor was 1.55, at pipeline from collection ponds equipped a 10 standard 90° elbows, and losses factor of each elbow was 0.9, therefore losses factor obtained was 9, the pipeline equipped by two ball valves and losses factor of each ball valve was 10, therefore losses factor of ball valves obtained was 20, pipeline length from pump machine to the reactor was 26m and roughness of pipeline was 15x10^{-5} mm , wastewater velocity was 0.708 m.sec^{-1}, pipeline diameter was 5x10^{-2} m, kinematic viscosity was 0.804x10^{-6} cs , Reynolds number was 17.7x10^{3} , therefore Re >2100 . It means that turbulent flow will occur.

Wastewater contained particulate of soybeans, therefore it was assumed that those roughness was 45x10^{-3} mm. Thereby friction factor could be obtained by an equation (2) that was 10.67x10^{-5}. Pump capacity of 1.39 L.sec^{-1} and wastewater contained particulate of soy beans roughness of particulate was 45x10^{-3} mm. Based on an equation (3), head losses occurred at inlet pipeline to reactor could obtained be 1.6528 m. Wastewater velocity inside reactor obtained was 2.8367x10^{-4} m.sec^{-1}, inlet diameter of pipe was 100mm, so velocity of wastewater flowed inside pipe 17.7 x 10^{-2} m.sec^{-1}. Cross section of bamboo was 94.2x10^{-4} m^2 and the clearance between bamboo were tightly, therefore velocity of wastewater flowed passed among bamboo could be $1.15x10^{-3}$ $m.sec^{-1}$, velocity of wastewater passed screen obtained was $55.33x10^{-2}$ $m.sec^{-1}$, velocity of wastewater flowed inside reactor obtained was $4.43x10^{-2}$ $m.sec^{-1}$ and velocity of wastewater flowed passed buffer hole obtained was $35x10^{-4}$ $m.sec^{-1}$. Thereby Reynolds number obtained was $56.135x10^{4} > 2100$, so wastewater flowed to pass bamboo would be a turbulent mode, while friction factor at bamboo obtained was $46x10^{-4}$. Therefore head losses occurred at the reactor obtained was $84.22x10^{-2}$ m.

Biogas pressure inside reactor tank was set up at 55 mbar or 550 mmH_2O, therefore total head losses obtained was 11.295 m, it was lower than head pump. Therefore, wastewater could flow inside reactors. Discharge factor obtained was $98x10^{-2}$ and contraction factor obtained was 0.8, wastewater velocity went out from buffer obtained was $2.8x10^{-4}$ $m.sec^{-1}$ and wastewater velocity inside outlet pipe obtained was 0.175 $m.sec^{-1}$. Based on equation (6), wastewater output of the first reactor obtained was $1.08x10^{-3}$ $L.sec^{-1}$. In the experiment a number of 1400 pieces of bamboo cutting were placed inside all reactor tanks, wastewater in the ponds pumped into equalizing tank and flowed by gravity way to reactor tanks. Each reactor equipped by a float stick to indicate wastewater increase inside reactor tank. When the channel cup was opened wastewater from equation tank to the first and second reactor tank flowed in normal way. On the contrary, when the channel cup was closed wastewater flow was stagnant, therefore a float stick raise at 42.5 cm for only in 27 min, therefore wastewater occupied gas space in the first reactor other reactors as well. Experiment data were presented in Table 1.

Table 1. Flow rate observation.

Trial order	Time Observation (min)	Δh_{Stick} (cm)
1	5	30.0
2	5	38.0
3	25	41.5
4	12	42.0

These data show that wastewater flow from the first to other reactors was resistance, therefore wastewater occupied biogas space.

3.2 Energy consumption

To prevent the accident and to obtain an optimum of biogas production, a channel was installed to reduce resistant of wastewater flow rate at 0.6 $L.sec^{-1}$, therefore wastewater level inside the reactor was concur with the level of output drain. Therefore, based on analysis amount of bamboo cutting placed inside each reactor should not more than 1400 pieces and channelled.

Input and output of drained pipe was of 50 mm in diameter, mm length, number of bamboo was of 1400 pieces, 50-60 mm bamboo diameter, 100 mm bamboo length, 1.8 m^3 volumes of equalizing tank, difference of height between equalizing and reactor tank is 600mm, input drain was installed at 400 mm and output drain at 3,600 mm from bottom of reactors. Cross section of equalizing tank is 1.2 m^2 and cross section of reactor tank is 4.9 m^2. The velocity of wastewater flow inside pipe obtained was $1.085 m.sec^{-1}$, ambient temperature of 30°C, kinematic viscosity of $804x10^{-3}$ cs or $8.04x10^{-7}$ $m^2.sec^{-1}$, Reynolds number was 6,747. It was higher than 2,300, therefore the flow of wastewater inside pipe was a turbulent flow. Based on the equation (1) the addition of bamboo cutting placed in the reactor increased the flow rate resistant. The time consumption by wastewater flowed from the first to the last reactor was 351 to 1,961sec. Therefore, flow rate of wastewater went out from drain-out pipe of the last reactor could be 150 to 1,925 L. hr^{-1}. The value of roughness (k) of materials commonly used to manufacture pipe is 0 to10^{-4} mm, based on the equations (2) and (3) friction factor obtained was $865x10^{-5}$.

Number of bamboo inside reactor was 1400 pieces and flow rate of wastewater went out from reactor was $69x10^{-3}$ $L.sec^{-1}$. It was small wastewater output in comparison with the demand for fermentation process. The heads between reactor tanks were 0.15 m, gas space among reactors were connected each other, therefore gas pressure at each reactor tank was balanced. Head losses of reactors were higher than head of reactor tanks, therefore wastewater inside the reactor difficult to flow to the other reactors. To increase wastewater flowrate, therefore inside reactor tanks were completed by perforated channels built from PVC pipe schedule AW with 150 mm inside diameter installed between the both buffer plates.

Wastewater flow rate obtained was $6x10^{-1}$ $L.sec^{-1}$ and most of wastewater flowed through the perforated channel tube. Wastewater flow rate demand for the aerobic process at the reactor was $625x10^{-3}$ $L.sec^{-1}$. Therefore, flow rate of wastewater went into reactor still less than $25x10^{-2}$ $L.sec^{-1}$. The channel tube was 100 holes-perforated and having 13mm in diameter. So the objective of installing the channel tube is to reduce resistant of wastewater flow and to change saturated wastewater by fresh one by setting up the time. When residence time is 10hr, wastewater flow rate achieved was 21,600 L for 10 hr, it was higher than flow rate needed at about 17,280 L.

When the channel tube was closed, it was no wastewater went out from reactor, however, after 30 min wastewater went out from reactor at flowrate of $69x10^{-3}$ $L.sec^{-1}$, therefore it was impossible that output of reactor at opened channel was lower than the closed one. This is because barrier plate was opened, the case was caused by wastewater pressure at bamboo cutting and barrier plate. The condition was not ideal, because food supply for bacteria was not sufficient.

When top barrier plate was changed with nylon net and the channel was closed, wastewater level inside the first reactor tank raised 42 cm on 22 min, therefore wastewater filled the gas space. The condition was also occurred at the second reactor. Number of bamboo cutting inside reactor was 1,400 pieces and head losses achieved was $8,422x10^{-4}$ m. It was higher than head of wastewater inside reactor $15x10^{-2}$ m, therefore wastewater difficult to flow out from reactor, it could be shown in Table 1. The condition inside reactor, the wastewater filled the biogas space, therefore biogas production from the reactor could not be achieved optimally.

To overcome the case, therefore each reactor equipped with two pump machine different types, i.e.: (1) A centrifugal pump to supply wastewater to the first and the fourth reactor tanks which has capacity of 1.11 $L.sec^{-1}$ and head of 20 m, (2) A magnetic pump at capacity of 1 $L.sec^{-1}$, 6 m head and 100 watt power to supply wastewater to the other tanks. The head of both pumps were higher than head losses, therefore wastewater could flow well as capacity needed from the first to other reactors. Thereby, wastewater inside reactor did not occupy the biogas space, therefore biogas production could be optimum.

Existing bamboo cutting inside reactors caused resistance and affected flow rate of wastewater inside reactors, the more bamboo cutting inside reactors the higher resistance occurred and the less wastewater discharged from reactors. Number of bamboo cutting inside every reactor were the same and hope that wastewater flow inside reactor tanks could achieve a balance between input and output. New arrangement of the reactor tanks is shown in Figure 3.

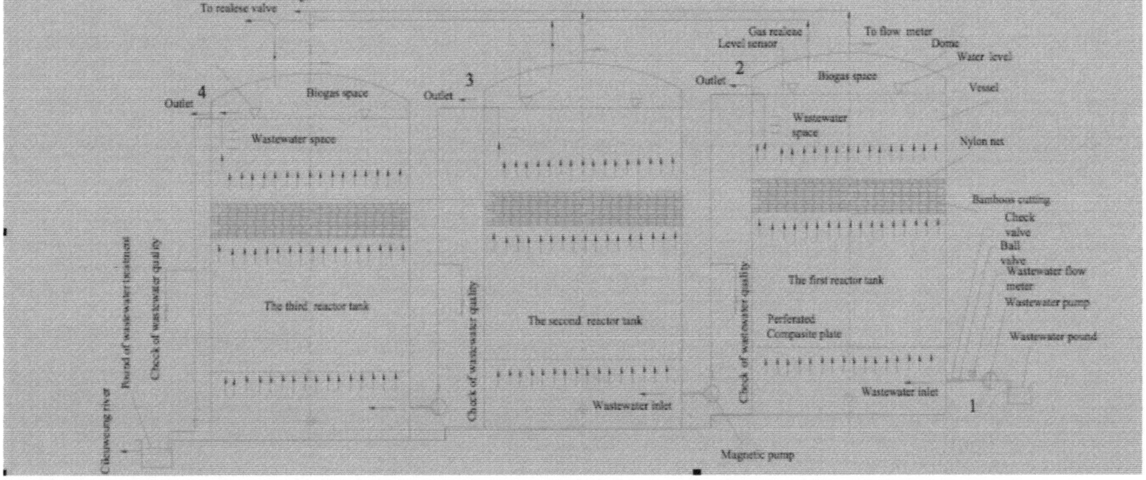

Figure 3. New arrangement of reactor tanks.

Notation: 1. Wastewater inlet to the first reactor tank, 2. Wastewater outlet of the first reactor tank and inlet for the second reactor tank,3. Wastewater outlet from second reactor tank and inlet for the third reactor tank, 4. Wastewater outlet of third reactor tank.

4. Conclusion

When the number of bamboo cutting was 1400 pieces and placed tightly at horizontal position, head losses occurred at each reactor tank was $8.42x10^{-1}$ m and it was higher than head reactor tank of $1.5x10^{-1}$ m. The first reactor completed by a pump at capacity of 1.39 $L.sec^{-1}$ and head of 30 m, therefore wastewater at the first reactor could flow well. On the contrary, the second and the third reactors without pump installation caused the wastewater could not flow optimally, therefore as a consequence the wastewater filled the biogas space. With pumps installed in every reactor, there is no resistance of wastewater flow inside reactors. The quantity of bamboo cutting, however, could be added on every reactor to increase the bacteria growth media, so the biogas produced optimally.

Acknowledgment

Authors thank to Head of Research Unit for Clean Technology - Indonesian Institutes of Sciences, Nanyang Technological University (NTU-Singapore), LIEN Foundation, and Association of Tofu Entrepreneur in Kebonjati Vilage-Sumedang District for their financial supports and facilities during the research. Appreciation is also addressed to Dr. S. Sentana for reading the manuscript.

References

[1]. Anonym, "Water Pump", Groundfos, 1-2.
[2]. Anonym 1990 *Design and Construction of Vessels and Tanks in Reinforced Plastic British Standard BS4994,* 12 – 13.
[3]. Naratnam S 1984 *Fluid Mechanics,* Khana Publishers Delhi, 383-393.
[4]. Robert W, Alan T. and McDonald F 1994 *Introduction to Fluid Mechanics*, John Wiley & Sons Inc New York, 32-34.

IOP Conf. Series: Earth and Environmental Science **60** (2017) 012024 doi:10.1088/1755-1315/60/1/012024

Determination of carbofuran in a river water sample using LC-MS/MS

R Yusiasih[1*], W C Nugraha[1] and A Hudayya[2]
[1]Research Unit for Clean Technology, Indonesian Institute of Sciences (LIPI)
Jalan Cisitu 21/154 D, Bandung 40135, Indonesia
[2]Indonesian Vegetables Research Institute
[*]Email : retno.yusiasih@lipi.go.id

Abstract. A study on the presence of carbofuran was conducted in Cikapundung river from an agricultural zone of the Lembang, West-Java, Indonesia. The present study aimed to determine the trace carbofuran in river water using LC-MS/MS. Extraction and purification of carbofuran was carried out simultaneously by Solid Phase Extraction (SPE) system with C18 sorbent. The procedure was evaluated by carbofuran recovery determination. Recovery was studied by spike technique with the addition of 2 and 5 ng/g carbofuran in river water sample. The recovery were of 83.87 and 115.88 % with relative standard deviation (RSD) of 6.28 and 0.47 % respectively. Carbofuran contained in Cikapundung river water samples was of 0.2898 ng/g.

1. Introduction

Indonesia is one of the agrarians country. It is well known that the use of large quantities of pesticides in agricultural activities is one of the main causes of surface and ground waters pollution. Consequently, very strict program is required to control and monitor the levels of these contaminants.

One source of drinking water for the rural areas around Lembang generates from Cikapundung river. However, it may contains carbofuran from Furadan insecticide commonly used by farmers. Therefore analyzing carbofuran contained in Cikapundung river water is indispensable. Based on Ministry of Health Regulation No.492/Menkes/Per/IV/2010 of drinking water quality requirement, carbofuran contained in drinking water should be less than of 7 ng/g, which was the legal maximum concentration permitted [1].

Since the concentrations of the compounds was normally low in ng/g or less as well as due to the low detection levels required by regulatory bodies and the nature of the matrices in which the target compounds was present, efficient sample preparation and trace-level detection and identification are important aspects of analytical methods. The analytical methods with less quantity of solvents, easy and rapid to execute are required to monitor the widespread distribution of trace levels of pesticides [2].

Most carbamates are thermally unstable chemical. Hence, it is necessary to analyze by liquid chromatography (LC) rather than by gas chromatography (GC). LC with variety of detection system can be done such as LC-post column fluorescence detector system and LC-diode array UV detector system. LC-UV detection low sensitivity compare with fluorescent derivates. Derivatization with o-phtalaldehyde (OPA) is the basis of the LC-fluorescence method which has been accepted as a standard protocol by several official organizations, including Environmental Protection Agency (EPA) and Association of Official Analytical Chemist (AOAC) [3,4]. However, post column flourescence requires o-phtalaldehyde (OPA) for derivatization [5]. Another LC with post column chemiluminescence (CL) detection also applies luminol for reaction. This method is based on the enhancing effect of these analytes on the CL emission generated by the oxidation of luminol with potassium permanganate in alkaline medium. Nowadays, the on-line combination of liquid chromatography and mass spectrometry (LC-MS) occupies an important place in the analysis of pollutants since this method provides for identification of thermally labile and polar pesticides at trace levels [6,7].

Extraction and purification were crucial step for sample preparation. Solid-phase extraction (SPE) used off-line or on-line is useful for trace determination of pesticides in water samples. SPE is already a well-established and routine technique. Presenting itself as a good alternative for the traditional techniques based on liquid–liquid extraction (LLE). LLE was needed a chemicals such as dichloromethane (DCM), methylene chloride, which was high consumption of solvent, which are often hazardous, which render them expensive or time consuming and organic solvent waste problem [5-9].

The concept of green chemistry emerged in the 1990s with the aim of minimizing the environmental impact of chemical activities. The basic principles to which a chemical procedure should correspond in order to be recognized as environmentally friendly (green) are reviewed. An important part of the green chemistry philosophy is the need to develop and adopt green analytical techniques and procedures. The key goals to be achieved in greening analytical methods are elimination or reduction of the use of chemical substances (solvents, reagents, preservatives, additives for pH adjustment and others); minimization of energy consumption; proper management of analytical waste; and increased safety for the operator. Most of these issues require reductions e.g., sample number, reagents, energy, waste, risk and hazard [10-15].

The objective of this work was to study of sample preparation and analysis of carbofuran in water sample based on analytical green chemistry (GAC). The analytical method then was applied to sample water from Cikapundung river as a source for drinking water.

2. Experimental

2.1. Standard and reagents

Carbofuran was supplied from Hunan Research Institute of Chemical Industry, HRICI – Changsha, PR. China, its purity of 99.7%. Stock standard solution was prepared by dissolving 10 mg of carbofuran in 10 g methanol. The working standard solution was prepared by dilution of the stock solution in methanol at concentration from 2 ng/g to 40 ng/g. The LiChrosolv-grade methanol and acetonitrile were supplied by Merck. SPE C18 cartridge was used for this study.

2.2. Source, collection and sample treatment

Water samples were collected from Cikapudung river, Eest Java, Indonesia on March which was rainy season. Freshly water sample was spike with carbofuran standard solution so that the concentration analyte in sample was 2 and 5 ng/g, respectively. To remove the suspended particulate matter, the spike sample was filtered trough Whatman filter paper No.1 with a Buchner funnel and vacuum. Prior to analysis the filtered spike sample was stored in the dark glass bottle at 4°C.

2.3. Sample preparation

Extraction and purification of carbofuran was done simultaneously using Solid-phase extraction (SPE) with C18 as a sorbent. SPE cartridges was conditioned with 3 ml of methanol and 3 ml of deionized water. As many as 5 g of the spiked river water samples was applied to the cartridge, then was eluted by 5 g methanol. The eluent then was filtered trough 0.45 µm membrane and analysis using LC-MS/MS. The blank sample was prepared with the same procedure without spiking with 3 replications.

2.4. Instrument and chromatographic condition

The instrument was LC-MS/MS 3200 QTRAP Triple Quadropole. Analysis carbofuran was performed on a Zorbax XDB-C8, 4.6 x 50 mm, 3.5 µm particle. The total flow rate of eluent A (0.1% formic acid in water) and B (0.1% formic acid in acetonitrile) was 0.35 ml/min. The profile of gradient was 0-1.5 min/90% A/10% B, 1.5 – 5 min/10% A/90% B, 5-8 min/10% A/ 90% B, 8 – 8.5 min/90% A/10 B, 8.5 – 10 min/90% A/10% B. The analysis was performed at ambient temperature and injection volume was of 10 µl.

IOP Conf. Series: Earth and Environmental Science **60** (2017) 012024 doi:10.1088/1755-1315/60/1/012024

3. Result and discussion

3.1. Qualitative analysis

Molecular formula of carbofuran was $C_{12}H_{15}NO_3$ and its molecular weight is of 221. Fragmentation schemes for carbofuran is shown in Figure 1 whereas ion chromatogram peak which retention time (tR) of 6.57 min is shown in Figure 2. The Multiple Reaction Monitoring (MRM) mode for carbofuran was precursors ion (m/z, Q1 mass) of 222.1 and products ion (m/z, Q3 mass) were 165.2 and 123.0, respectively. The isolated precursor and product ion in MRM using energy profile were De-clustering Potential (DE, volt) of 30.0, Entrance Potential (EP, volt) of 10.0, Collision Energy (CE, volt) of 17 and 29, Collision Exit Potential (CXP, volt) of 3.0.

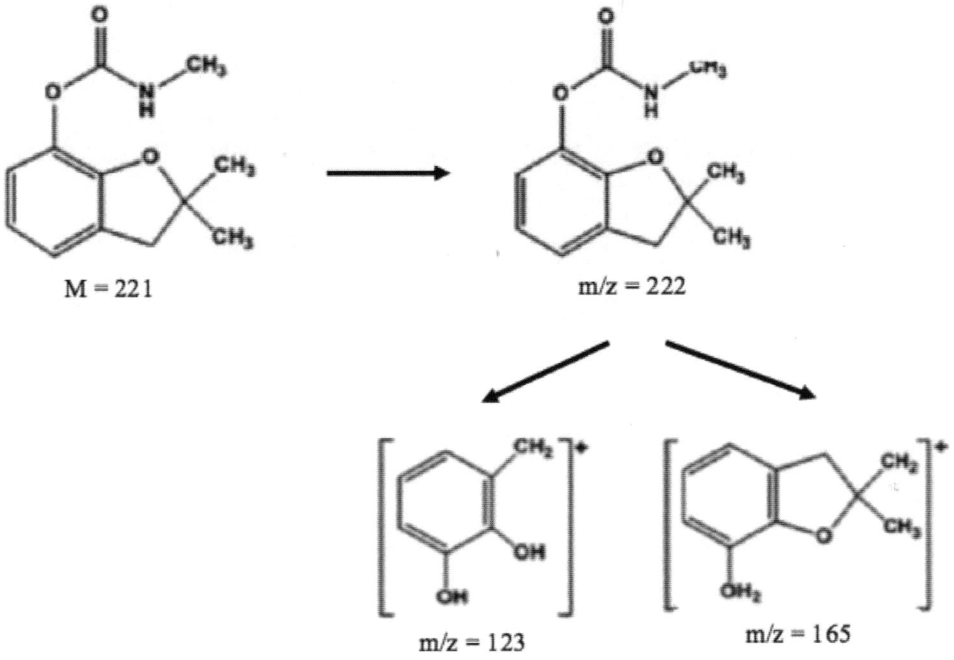

Figure 1. Fragmentation schemes of carbofuran [8].

3.2. Extraction and purification by SPE- C18

Spike river water sample with carbofuran was applied to the SPE. Filtrate from SPE was then analyzed using LC-MS/MS. The result of the analysis showed that no carbofuran was identified. The results indicated that carbofurans were absorbed to C18 as an absorbent.

3.3. Recovery and accuracy

Accuracy was calculated of the known concentration level by comparing the measured concentration with the spike concentration as the reported method [10]. Sample spike at 2 and 5 ng/g concentration with number replicate of 3. The average result of recovery obtained was of 83.87 and 115.88%. These values was considered acceptable and good result since a 70 – 130% recoveries were approved [4,5]. Precision was evaluated in term of repeatability and expressed as relative standard deviation (RSD, %). The result of precision were 6.28 and 0.47% respectively, indicating good repeatability.

IOP Conf. Series: Earth and Environmental Science **60** (2017) 012024 doi:10.1088/1755-1315/60/1/012024

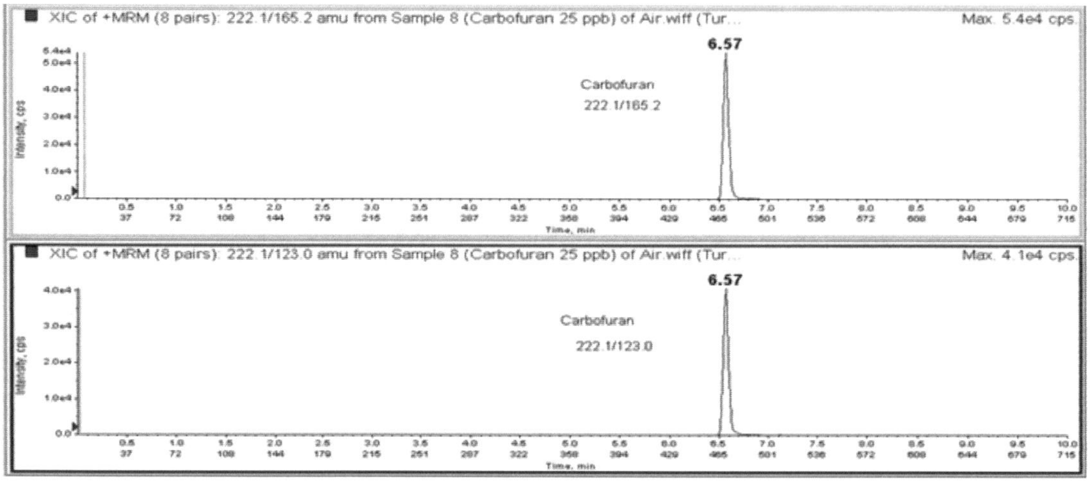

Figure 2. Extracted ion chromatogram (XIC) carbofuran standard solution of 25 ng/g.

Table 1. Experiment summary of recovery and precision.

Replication	Spike conc, ng/g	Recovery, %	Spike conc, ng/g	Recovery, %
1	1.9840	89.95	4.9581	115.2579
2	1.9824	80.88	4.9580	116.1623
3	1.9849	80.78	4.9555	116.2205
Average		83.8678		115.8803
Standard Deviation, SD		5.2643		0.5397
Rel. SD, %		6.28		0.47

3.4. Application of a qualitative analysis

The method was applied to water sample which was collected from Cikapundung river in rainy season on March of 2016. Concentration of carbofuran obtained was of 0.2098 ng/g was lower than requirement for water drinking of 7 ng/g. Analytical method was sensitive enough to detect the carbofuran. Water from Cikapundung river was safe for a source of drinking water. In this study, the concept of green analytical chemistry (GAC) has been done. Analysis of carbofuran using LC-MS/MS required mobile phase of 0.35 ml/min. Extraction and purification were required only as much as 5 g methanol. Flow rate and gradient system of mobile phase as described in section 2.4 were chosen as the optimum performance of LC-MS/MS which was usually used for carbofuran analysis.

4. Conclusion

Trace determination of carbofuran in water sample was achieved by using LC-MS/MS. Extraction and purification of carbofuran with SPE-C18 was very fast and simple. Recovery with spike technique of 2 and 5 ng/g was 83.87 and 115.88%, respectively and assuming as good result. Result for precision were of 6.28 and 0.47%, indicating good repeatability. The method was sensitive enough to detect carbofuran. Application of these analytical methods to water samples from Cikapundung river was found of carbofuran at 0.2098 ng/g, that was safe as a source for drinking water

Acknowledgement
This work was financial supported by Indonesian Institute of Sciences by "Mandiri" project on 2016. The authors wish to thank Mr. Efendi for his cooperation in collecting samples and Mrs. Evita Boes for her encouragement in this work.

References

[1]. Peraturan Menteri Kesehatan Republik Indonesia No.492/Menkes/Per/IV/**2010**.
[2]. Sergiane Souza Caldas, Fabiane Pinho Costa, Ednei Gilberto Primel 2010 *Analytica Chimica Acta.* **665** 55–62.
[3]. José Fernando Huertas-Pérez, Ana María García-Campana 2008 *Analytica Chimica Acta.* **630** 194–204.
[4]. Sergiane S. Caldas, at all. 2010 *J. Braz. Chem. Soc.* **21(4)**.
[5]. Garci'a Dellasera, M.P. and Bernal-gonza'lez, M. 2011 *Wat. Res.* **35(8)** 1933–1940.
[6]. Sonia C.N. Queiroz, Karel Lazao, Pat Sandra, Isabel C. S. F. Jardim 2004 *Curitiba.* **14** 53-60.
[7]. Nikolina Dujaković´, at all. 2010 *Analytica Chimica Acta.* **678** 63–72.
[8]. SOP#, Carb – Water, Revision #2 2006.
[9]. George Z. Tsogas, at all. 2006 *Analytica Chimica Acta* .**573–574** 354–359.
[10]. Sukesh Narayan Sinha 2011 *American J. Analytical Chem.***2** 511-521.
[11]. Armenta S., Garrigues S., De la Guardia M. 2008 *Trends in Analyt. Chem.* **27(6)** 497-511.
[12]. Marek Tobiszewski, at all. 2009 *Trends in Analyt. Chem.* **28(8)** 943-951.
[13]. Marinella Farre, at all. 2010 *Trends in Analyt. Chem.* **29(11)** 1347-1362.
[14]. Ivanova E. H., Detcheva A. K. 2012 *Bulgarian Chem. Comm.* **44(1)** 5 – 10.
[15]. Agnieszka Gałuszka, Zdzisław Migaszewski, Jacek Namieśnik 2013 *Trends in Analyt. Chem.* **50** 78–84.

Enhancing faecal sludge management in peri-urban areas of Lusaka through faecal sludge valorisation: challenges and opportunities

J M Tembo[1*], E Nyirenda[1] and I Nyambe[2]
[1]University of Zambia, School of Engineering, Lusaka, Zambia
[2]University of Zambia, School of Mines, Lusaka, Zambia
[*]Email : wazatemboj@gmail.com

Abstract. Lusaka, the capital city of Zambia, has two million inhabitants with 70% residing in peri-urban areas. Ninety (90) % of this population employ pit latrines for excretion generating approximately 22,680 tons of faecal sludge per annum. This sludge is inadequately managed hence of the generated amount, over 60% remains within the residential environment thereby compromising both the environment and public health. To foster a solution to this problem, a study was commissioned to assess faecal sludge valorisation potential and how it would impact on Faecal Sludge Management. The study evaluated policy, institutional and regulatory frameworks, sanitation practices including latrine construction and usage aspects and also characterised the faecal sludge for selected parameters relevant to valorisation. Four peri-urban areas were adopted as study sites. Policy issues together with existing institutional and regulatory frameworks were assessed through literature review. Sanitation practices were evaluated through physical observations, focus group discussions, interviews and questionnaire administration. Faecal sludge characterisation was through sampling and analysis. It was observed that there are policy gaps in fostering faecal sludge valorisation. Sanitation practices and latrines construction also do not favour valorisation. The quality of the raw sludge has potential for valorisation though again, some parameters like solid waste content require drastic changes in sanitation practices in order not to compromise the reuse potential of the sludge. It was concluded that if faecal sludge management is to be enhanced through valorisation, there is need to have policies promoting pit latrine faecal sludge reuse and strengthened regulatory and institutional frameworks in this respect.

1. Introduction

Faecal Sludge Management (FSM) has continued to be a world-wide challenge to sanitation provision especially in developing countries. Between 65 and 100 % of the urban population of most Sub-Sahara African countries access sanitation exclusively through onsite systems [1]. For these countries, pit latrines prevail especially in peri-urban areas [2]. However, pit latrines have one drawback in that their capacity is limited. They fill up after some time making them unusable [3][4]. When a latrine gets full, two options are available for its ultimate fate: It can be buried and a new one dug to replace it; or it can be emptied and put back into service [5][6]. In high density areas where space is a challenge, the latter option is more appealing.

The success of pit emptying in any locality requires the availability of a system which adequately addresses all aspects of FSM including safe faecal sludge emptying, transportation, treatment and disposal [7]. However, these systems are rarely available in low cost areas of most developing countries. It has been observed that lack of incentives is one of the factors contributing to the absence of adequate FSM systems in low cost areas of developing countries [8]. The perception that faecal sludge is a waste and not a resource is one major reason for the observed lack of incentives in FSM service provision. It is suggest that one way in which FSM can be improved is through valorisation [8]. Initial steps towards faecal sludge valorisation include creation of an enabling environment [9]and identifying potential areas for its reuse which needs to be coupled with an understanding of the quality aspects. The quality aspects need to be understood to know exactly what needs to be done to enhance the valorisation potential.

Lusaka, the capital city of Zambia, is home to over two million inhabitants [10]. About 70 % of this population resides in peri-urban areas [11][12]. It is reported that over 90% of people residing in peri-urban areas of Lusaka use pit latrines [13]. This translates to 1.26 million people employing pit latrines as a means of excreta disposal in the city. With an excreta generation rate of 1.5 litres per capita per day [14], this population generates about 1,890m^3 of fresh excreta every day which converts to over 22,000 tons of sludge per annum when biodegradation aspects are factored in. Most of this sludge remains within the residential environment where it continues to compromise the environment and public health. The major reason for this is the absence of effective and efficient FSM systems in these areas. This paper looks at faecal sludge management in peri-urban areas of Lusaka; the existing policy, legal and institutional frameworks, practices, quality and associated challenges and opportunities for faecal sludge valorisation with regards to improving FSM.

2. Experimental

2.1. Study areas

The project identified four peri-urban areas from within Lusaka. These included Chaisa, George, Kanyama and Madimba compounds (Figure 1).

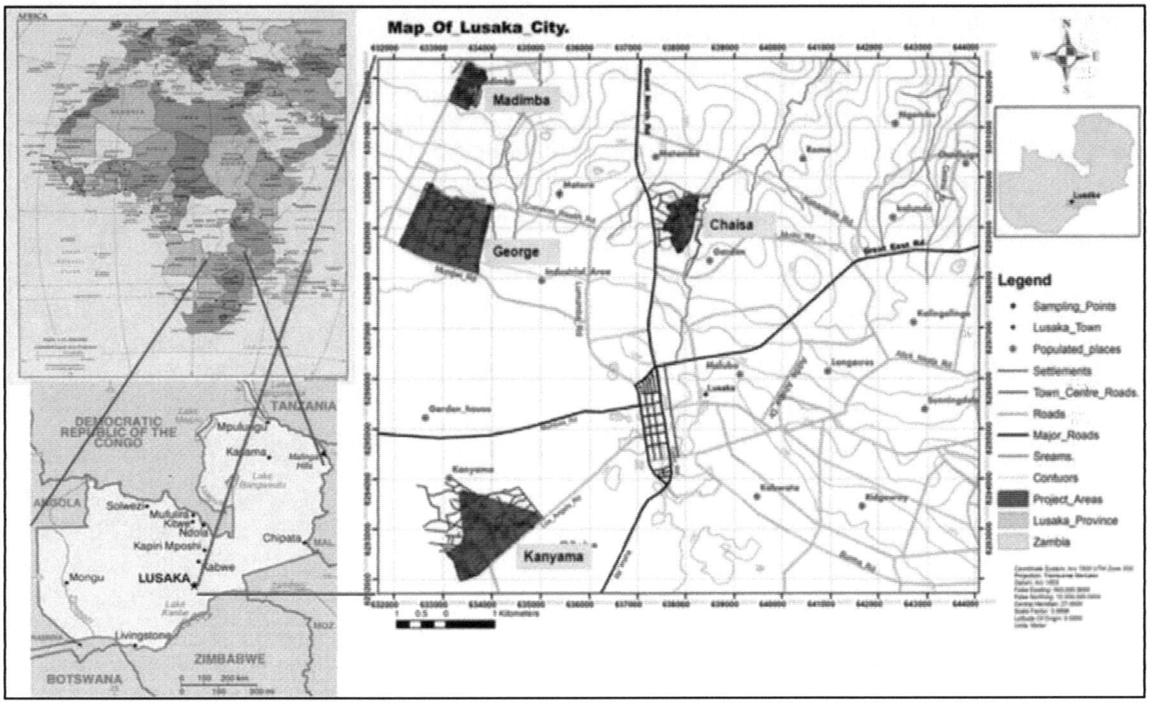

Figure 1. Map showing location of project sites.

The four selected study sites represented a full spectrum of the geophysical environments within the city. Kanyama and Madimba compounds are predominantly on dolomite. George compound sits on dolomite in the south but the central part including the north eastern parts are sited on schists. Chaisa is exclusively on schist formation. Within the compounds, areas are affected differently in terms of drainage. Madimba is mostly waterlogged throughout the year. Other compounds have areas that are water logged throughout the year as well as those that only get affected during the rainy season. The degree to which they are affected also differs depending on nature of the terrain. This selection was in order that effects of the hydrogeological aspects on faecal sludge quality could be captured.

2.2. Desk study
Desk study was cardinal in the assessment of the effectiveness and adequacy of policy, legal and institutional frameworks as they relate to faecal sludge management in Zambia. Desk study was also important in the garnering of information and knowledge available on the aspects of sanitation practices and faecal sludge management in relation to beneficiation prospects in low income areas.

2.3. Field work
Field work involved Focus Group Discussions (FGDs), questionnaire interviews and in-depth interviews. These were used to gather data on sanitation practices. Characteristics and nature of the latrines was assessed through physical surveys and measurements as well as interviews. Faecal sludge characterisation was through sampling and analysis of samples. Samples were collected from the top and bottom layers of each of the pits used as sampling points. From each layer, three samples were collected and then put in a container to formulate a composite sample. Three samples were then collected from the composites for analysis at the laboratory. Sample analyses were in accordance with the Water Research Commission analytical protocols which are based on the American Standard for Water and Wastewater Analysis [15].

3. **Result and discussion**
3.1. Policy, legal and institutional framework
Policy, together with legal institutional frameworks are critical ingredients to fostering reuse prospects of faecal sludge. Effective and responsive policies and regulations are critical in formulation of effective FSM systems [9]. However, the study revealed that there are policy, legal and institutional gaps relating to FSM in peri-urban areas. There are policies and legislative instruments that guide the provision of sanitation services in peri-urban areas of Zambia. However, the study showed that most of them apply to FSM in the general context of sanitation. Some of the applicable policies include: the 1994 National Water Policy; the 2004 National Decentralisation Policy; and the Sixth National Development Plan (SNDP). Legislative instruments in place which apply directly or indirectly include: the Environment Management Act of 2011; the Water Resources Management Act No, 21 of 2011; the Water Supply and Sanitation Act No. 28 of 1997; the Public Health Act (CAP 295); the National Health Services Act (CAP 315); and the Local Government Act (CAP 281). However, there is no a single policy or statute specific to management of faecal sludge from pit latrines especially with respect to reuse.

Several institutions are responsible for sanitation provision in peri-urban areas. However, until recently, these institutions have not been very visible in these areas. For example, through the Water Supply and Sanitation Act of 1997, commercial utilities were formed to take up the water supply and sanitation service provision to areas within the jurisdiction of the local authorities. However, most utilities only took up the reticulated sewerage systems leaving peri-urban areas to fend for themselves. Even where interventions are being designed in these areas, the approach has mainly been to provide reticulated systems. However, due to the nature of some of the peri-urban areas in terms of the layout of housing units and geological conditions which are rocky, it is clear that many areas will remain on pit latrines for a long time to come hence if nothing is done to enhance FSM like incentivising the faecal sludge chain, these challenges will continue in perpetuity.

Clear, effective and responsive policies are paramount to addressing the challenges. Statutes specific to FSM should also be enacted which clearly spell out and enforce aspects related to faecal sludge which are relevant to faecal sludge valorisation. Examples here include statutes that spell out latrine designs and usage. Formulation of Institutions with a clear mandate on enhancing FSM from the perspective of valorisation is important. In the current arrangements, aspects of FSM, especially in peri-urban areas, are just one of the main responsibilities of responsible institutions and this has seen them being relegated into oblivion hence exacerbating poor sanitation.

3.2. Sanitation practices

Over 90% of the people in peri-urban areas of Lusaka used pit latrines [13]. The study showed that the designs of facilities were not regulated. Most of the facilities were used as receptors of both excreta and grey water which impacted of the faecal sludge quality from these units.

Generally, there were no formal desludging services in the project areas. Hence, where it was impossible to dig a new latrine when the old ones filled up, the only solution was informal desludging. Due to limited capacity to appropriately dispose of the sludge removed from the latrines, or simply due to the desire to cut down on costs, informally desludged pit contents still ended up within the domestic environment. However, much as it was acknowledged by residents in these areas that latrine emptying was the solution to sustained access to the facilities, not a single latrine from the more than 100 pits that were assessed was constructed with provision for desludging. It was only after the pit got full that issues on how to appropriately desludge arose. In most cases, desludging was done by enlarging the squat hole or by breaking the side wall to access and scoop out the sludge. However, as the latrines were usually constructed from low quality materials, this process compromised the structural integrity of the latrines. Cases of latrine collapses due to desludging had been recorded. This was an aspect that could be addressed by imposing standards on the design and construction of facilities.

The study revealed that the high solid waste in the latrines was as a result of several reasons. Firstly, most latrine designs were not user-friendly to young children. Because of this, children usually defeacated on the ground within the premises of the household after which the excreta was picked and deposited into the latrine. The picking up of the excreta was through a scooping tool like a shovel or hoe. Therefore, when the excreta was picked, an appreciable amount of soil would also be collected which would also end up in the latrine. Secondly, the socio-cultural perceptions and stigma attached to menstruation in most of the Zambian culture, leads to disposal of menstruation waste in a manner that ensures maximum secrecy. Unfortunately, in peri-urban areas, pit latrines serve as one of the most appropriate means of disposal that ensures secrecy. Lastly, the lack of solid waste management systems leads to deliberate disposal of the waste in pit latrines. Solid waste negatively impacts on the beneficiation potential of the sludge. If not addressed, even when latrines are designed with a provision for desludging, the solid waste could still complicate the desludging and treatment processes.

3.3. Cost of desludging

Adequate FSM requires that desludging aspects are addressed. In Lusaka, to respond to the challenges of inappropriate desludging, formal desludging means were piloted in some of the peri-urban areas. However, the cost of desludging was high for most people resident in these areas. Desludging was through 60 litter barrels and the costs were US$40, US$ 60 and US$ 70 for 12, 24 and 32 barrels respectively. A survey on household income levels revealed income levels of less than 1US$ per capita per day with the majority living on a hand to mouth basis. This implied that faecal sludge emptying systems, although a solution to the FSM challenges faced in these areas, was out of reach for most residents and required subsidization.

The study revealed that the setup with respect to desludging was such that the whole cost of desludging, transportation and disposal was borne by the client. It is suggest that one way in which financing of the faecal sludge chain could be subsidized is to allow for income generation through resource recovery from waste derived products [16]. This is because the faecal sludge would generate income at the various stages of reuse like biogas generation. The generated income can then subsidise desludging thereby contributing to enhanced FSM in peri-urban areas.

3.4. Faecal sludge quality

Faecal sludge was characterised from the perspective of parameters that impact on valorisation. These included solids and moisture content, microbiological parameters, chemical parameters including metals and organic matter. No clear variations in quality were observed with respect to study sites.

IOP Conf. Series: Earth and Environmental Science **60** (2017) 012025 doi:10.1088/1755-1315/60/1/012025

The average solid waste content in the faecal sludge from the pit latrines was around 20% expressed as percent of unit weight of solid waste per unit weight of dry solids. This definitely has negative impacts on valorisation. In the production of biogas, high waste content would result in rapid filling up of the reactors hence increasing on the cost of operation. This had been a challenge at a biodigester that was being piloted in one of the project sites. At the reuse stage, the high waste content would make the sludge objectionable and would thus reduce its value. Therefore, regulation of latrine usage is imperative for effective faecal sludge valorisation as elaborated in section **Error! Reference source not found.**.

Moisture content ranged between 50.4±2.8 and 99.2±2.8% indicating high variability. This showed that whilst some faecal sludges were thick and could possibly not be desludged using suction action, others were watery and similar to wastewater.

Microbiologically, the faecal sludge, as expected, had high prevalence of microorganisms with Total Coliforms averaging $29.1\pm9\times10^9$ and $28.7\pm9\times10^9$ CFU/100ml of sample in the top and bottom sludge respectively. Prevalence of *Ascaris Lubricoides* was at 56% with viability at 27%. *Cryptosporidium Parvum* had a 90% prevalence rate in numbers averaging 22±5 and 14±6 in the top and bottom sludges respectively. The presence of all these microorganisms was indicative of the high risk the sludge posed with respect to potential for diseases transmission. This implied that in valorising faecal sludge, measures to guard against transmission of diseases need to be instituted.

The concentration of heavy metals was found to be very low ranging from zero to a few micrograms per gram of the dry solids (Table 1). Analysis of the faecal sludge for base metals also indicated low concentrations (Table 2).

Table 1. Average concentration of heavy metals in latrine sludge.

Element	Source of Sample	Range (μ/gTS)	Mean (μ/gTS)	Mean STD Dev
Copper	Top Sludge	0-300	17.2	±1.4
	Bottom Sludge	0-345	18.8	±2.0
Cobalt	Top Sludge	0-124.3	5.5	±4.0
	Bottom Sludge	0-27.7	3.4	±2.0
Cadmium	Top Sludge	0-23.4	1.9	±2.0
	Bottom Sludge	0-67.6	5	±3.7
Zinc	Top Sludge	0-134	8.7	±4.4
	Bottom Sludge	0-90	8.2	±5.7
Chromium	Top Sludge	0	0	0
	Bottom Sludge	0	0	0
Lead	Top Sludge	0-4	0.08	±0.1
	Bottom Sludge	0-13	0.25	±0.3

The low concentrations of heavy metals are indicative of the potential for reuse of the sludge in both agricultural activities as well as a biogas generation. Copper, lead, chromium and zinc inhibit anaerobic digestion by inactivating enzymes within microorganisms [17]. However, all these were low with concentrations ranging from not detectable to only a few micrograms per gram of the total solids (Table 1). Therefore, anaerobic digestion of this sludge cannot be impeded by the presence of heavy metals. Similarly, heavy metals were well below the threshold for reuse in agriculture. European Union Standards prescribe 20-40μg/gTS for Cadmium, 1,750μg/gTS for copper, 750-1,200μg/gTS for Lead and 2,500-4,000μg/gTS for Zinc. All these were above the concentrations recorded in Table 1 indicating that heavy metals would not be a problem if the sludge was used in fertigating crops.

Concentration of base metals is an important aspect that needs consideration where the sludge is to be used in agriculture especially from the perspective of the Sodium Adsorption Ratio (SAR). When sodium is present in water in higher proportions than calcium and magnesium, it may cause permeability problems. To this effect, SAR was computed using proportions of Na. Ca, and Mg in the sludge. The results showed that the metals in the sludge were not in proportions that would promote sodicity problems (Table 3).

Table 2. Average concentration of light metals in latrine sludge.

Element	Source of Sample	Range (μ/gTS)	Mean (μ/gTS)	Mean STD Dev
Sodium	Top Sludge	2.4-164	17.6	±5.1
	Bottom Sludge	1.1-182	17.8	±5.6
Potassium	Top Sludge	1.9-195	24	±7.7
	Bottom Sludge	1.4-115	20	±5.2
Calcium	Top Sludge	1.4-140	15.6	±4.9
	Bottom Sludge	1.5-136	16.5	±5.2
Manganese	Top Sludge	2.6-117	13.5	±4.9
	Bottom Sludge	0.3-70	11.2	±4.8

Table 3. Computation of sodicity based on average values.

Element	Average Weights		Milliequivalents	
	Top Sludge	Bottom Sludge	Top Sludge	Bottom Sludge
Na	17.6	18.0	0.8	0.8
Ca	15.6	16.5	0.8	0.8
Mg	13.5	11.2	1.1	0.9
SAR			1.1	1.2
STD			<3	<3

Organic matter measured as COD revealed high concentrations averaging 1.5 gCOD/g TS. Biodegradability measured through volatile solids averaged around 65% g VS/g TS indicating more than 50% biodegradability. Based on these two parameters, it is evident that the sludge is a potential raw material for biogas generation. The organic matter is also an important constituent in soil conditioning properties of sludge.

4. Conclusion and recommendation

From the study, it was concluded that FSM has continued to pose a challenge in peri-urban areas of Lusaka. This was because of the lack of incentives in sanitation service provision in low income areas coupled with absence of policy to address the problem. The study showed that it is imperative to incentivise FSM in the peri-urban areas which incorporates desludging and reuse due to i) The environment in these areas does not favour burying of pits when they get full making desludging the most suitable option for addressing the problem of full latrines. This is because of lack of space and also hostile geology which makes excavation impossible; ii) The quality of the sludge has reuse potential save for a few parameters like microorganisms and solid waste content that require interventions for effective and safe reuse; and iii) Promoting reuse would subsidies the desludging costs and hence contribute to enhanced sanitation in the service areas. The study also revealed that for faecal sludge to be effective, it is imperative that the following are addressed:

i) Formulation and implementation of specific, effective and responsive policy to deal with FSM in peri-urban areas from the perspective of reuse;

ii) Regulation of the design and construction of the latrines so that desludging when required can be facilitated; and

iii) Regulation of latrine usage to avoid usage that negatively impacts on the faecal sludge's potential for valorisation. This can be through statutes or/and bye laws.

References

[1] Strande L Ronteltap M and Brdjonovic D 2014 *Faecal Sludge Management: System approach for implementation and operation* (London: IWA Publishing) 1.

[2] Michael R T Acile S H Adrian P B Laura B Julie-Anne F Johanna G Moussa B Serigne F and Jean P J 2015 *J. Env. Sci.* **2(2)** 302-313.

[3] Still D A, Salisbury R H, Foxon K M, Buckley C A and Bhagwan J N 2005 *The challenges of dealing with full VIP latrines* (WRC: South Africa) 1.

[4] Murungi C and Van Dijk P 2014 *J. Habitat Int.* **42** 69-75.

[5] Jenkins M W, Cumming O and Cairncross S 2015 *Int. J. Environ Res Public Health Mar* **12(3)** 2588–2611.

[6] Pickford J and Shaw R 1997 *J. Waterlines* **16(2)** 15-18.

[7] Strauss M and Montangero A 2002 *FS Management – Review of Practices, Problems and Initiatives* (EAWAG/SANDEC, Switzerland) 10.

[8] Daniel B and Reed B 2010 *Conference Proceedings on 'From Small Steps to Giant Leaps...putting research into practice'* (Engineers Without Borders, UK).

[9] Hasan M, Udin N and Parison J 2004 *30th WEDC Int. Conf. Proc.* (Vientiane, Lao PDR) 86-90.

[10] Central Statistics Office 2011 *A Report on Population and Housing* (Lusaka, Zambia).

[11] Devolution Trust fund and National Water and Sanitation Council 2005 *Reaching the Millennium Development Goals for Water Supply and Sanitation in Zambia - The Urban Perspective* (Lusaka, Zambia).

[12] Ministry of Local Government and Housing 2014 *National Urban Sanitation Strategy* (Lusaka, Zambia).

[13] World Bank AFTU 1 & 2 2002 *Upgrading low-income urban settlements: Country Assessment Report, Zambia.*

[14] Koné D and Peter S 2008 *Sandec Training Tool 1.0* (EAWAG/SANDEC, Switzerland).

[15] APHA 1998 *Standard Methods for the Examination of Water and Wastewater* (Washington DC, USA).

[16] Murray A and Ray I 2010 *J. Plann Educ Res.* **30 (1)** 94–102.

[17] Hussein I A Mona S M 2014 *Egyptian J. of Petroleum.* **23**, 409–17.

Influence of voltage input to heavy metal removal from electroplating wastewater using electrocoagulation process

D R Wulan*, S Cahyaningsih and Djaenudin
Research Unit for Clean Technology, Indonesian Institute of Sciences (LIPI)
Jalan Cisitu 21/154 D, Bandung 40135, Indonesia
*E-mail : diana.rahayuning.wulan@lipi.go.id

Abstract. In medium capacity, electroplating industry usually treats wastewater until 5 m^3 per day. Heavy metal content becomes concern that should be reduced. Previous studies performed electrocoagulation method on laboratory scale, either batch or continuous. This study was aimed to compare the influence of voltage input variation into heavy metal removal in electroplating wastewater treatment using electrocoagulation process on laboratory-scale in order to determine the optimum condition for scaling up the reactor into pilot-scale. The laboratory study was performed in 1.5 L glass reactor in batch system using wastewater from electroplating industry, the voltage input varied at 20, 30 and 40 volt. The electrode consisted of aluminium 32 cm^2 as sacrifice anode and copper 32 cm^2 as cathode. During 120 min electrocoagulation process, the pH value was measured using pH meter, whereas the heavy metal of chromium, copper, iron, and zinc concentration were analysed using Atomic Absorption Spectrophotometer (AAS). Result showed that removal of heavy metals from wastewater increased due to the increasing of voltage input. Different initial concentration of heavy metals on wastewater, resulted the different detention time. At pilot-scale reactor with 30 V voltage input, chromium, iron, and zinc reached removal efficiency until 89-98%, when copper reached 79% efficiency. At 40V, removal efficiencies increased on same detention time, i.e. chromium, iron, and zinc reached 89-99%, whereas copper reached 85%. These removal efficiencies have complied the government standard except for copper that had higher initial concentration in wastewater. Kinetic rate also calculated in this study as the basic factor for scaling up the process.

1. Introduction

Electroplating is an electro-deposition process for producing a dense, uniform, and adherent coating, usually of metal or alloys, upon a surface by the act of electric current deposits metal coating onto metal or plastic by electrolytic process [1]. On their process, electroplating industries need raw water that usually comes from groundwater which later result wastewater with high heavy metals concentration, i.e. copper, nickel, chrome, and zinc diluted in cyanide, acid, alkaline and phosphate. They must install wastewater treatment plant to comply government standard before they discharge the wastewater into water body. In medium capacity, each industry could discharges wastewater until 5 m^3 per day. From environmental prospective, this condition results not only waste water with high heavy metals concentration, i.e. copper, iron, chrome, and zinc, but later also facing the lack of ground water.

Heavy metal in wastewater could be reduced by electrochemical, hydroxide precipitation and extraction [2]. Previous researches applied hydroxide precipitation process for metal wastewater treatment that usually requires different optimum pH. This precipitation might cause some problems of bulky low-density-sludge and different optimum pH that probably dilute the other metals present in wastewater [2]. Electrochemical is applicable for heavy metal removal with high concentration, as well as wastewater from electroplating industry, and more selective compare to hydroxide precipitation.

Electrocoagulation, as one of the electrochemical processes, could minimize sludge that usually exists as the end product from any conventional hydroxide precipitation process [3]. Electrocoagulation produces less sludge than conventional precipitation process for removing chromium [4].

Electrocoagulation combines electrochemical and coagulation-flocculation process that apply cell with metal anode (usually steel and aluminium), and requires electric supply to produce chemical reaction. Anode in this system would function as "sacrificial electrode ion". Ion Al^{3+} and OH^- resulted on electrode surface would react with metals in wastewater and later forms aluminium hydroxide. Main chemical reactions on electrocoagulation process are:

Anodic reaction [3]: $Al \rightarrow Al^{3+} + 3e^-$

Cathodic reactions that may occur might be expressed as [3,5] :

$$2H_2O + 2e^- \rightarrow H_2 + 2OH^-$$

$$O_2 + 4\,H^+ + 4e^- \rightarrow 2H_2O$$

$$2H^+ + 2e^- \rightarrow H_2$$

Further, Al^{3+} and OH^- ions generated at electrode surfaces react in the bulk wastewater to form aluminium hydroxide:

$$Al^{3+} + 3OH^- \leftrightarrow Al(OH)_3$$

Base on Faraday law, the amount of sacrifice anode dissolved depend on electricity pass through electrolytic solution, so the heavy metal removal by electrocoagulation is influenced by direct voltage input. The optimum voltage input could produce less sludge production on maximum removal efficiency [6]. Previous studies performed electrocoagulation process on heavy metal or organic removal from synthetic or industrial wastewater on laboratory-scale, either batch [3,6,7,8,9] or continuous [7]. This study present work would determine studied the optimum voltage input base on removal efficiency on batch laboratory-scale to determine kinetical rate for scaling-up into pilot-scale reactor.

2. Experimental

2.1. Materials

This study was conducted in 1,5 L batch-reactor with aluminium-plate anode and copper-plate cathode with 2 cm width x 8 cm length x 1 mm thickness respectively. Wastewater sample was taken from rinsing wastewater of degreasing and metal plating process in electroplating industry PT.DSB near Bandung, West Java.

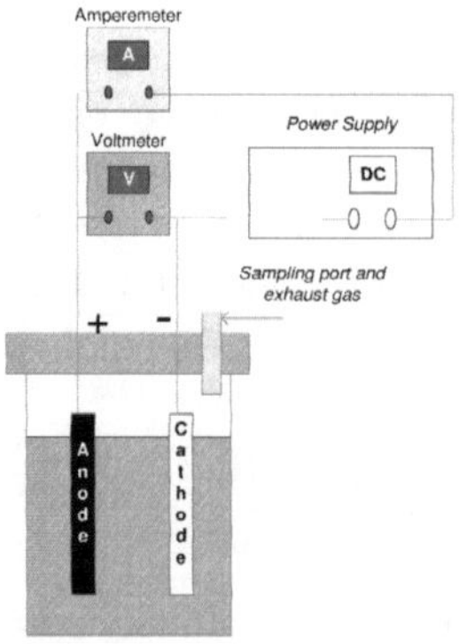

Figure 1. Scheme of electrocoagulation process.

2.2. Experimental setup

The laboratory-scale experiment was performed, as shown in Figure 1., on room temperature ($\pm 25^{o}C$), and pressure 1 atm. The electroplating wastewater was placed in the chamber. Previous studies showed optimum reactor performance on 12V [3] and 30V voltage input [6], or current density of 3A dm^{-2}[8] and 4.8 A dm^{-2}[9], depend on initial concentration. Considering the initial concentration of wastewater, in this study the direct voltage from power supply was varied at 15, 20, 30 and 40 volt during 120min. Optical density and pH were measured overtime and every 10 min sample was taken for metal concentration measurement. Using parameters from laboratory-scale experiment, pilot-scale reactor would be constructed and tested using further detention time calculation.

2.3. Analysis

The soluble metal concentration of Cr(VI), Cu(II), Fe(II), Zn(II) was analysed using Flame Atomic Absorption Spectrophotometer Varian GBC 903, the pH was measured using pHmeter Lutron PH-208, and Optical Density/OD was measured using Spectrophotometer LW Scientific V325XS on 464nm. OD measurement was performed to describe the turbidity fluctuation on reactor which was assumed to be correlated with metal concentration.

2.4. Calculation

Electrocoagulation process performance is described by the different of initial anode and cathode mass (m_0, g) with the mass of each electrode at time t (m_t, g):

$$Electrode\ mass = m_0\text{-}m_t \tag{1}$$

To calculate kinetic rate from batch experiment, the equation below was used [10]

$$y = kx + a \tag{2}$$

where k is the slope of metal ion concentration, y, during reaction time, x.

The reaction rate would help the calculation of the laboratory-batch scale into scaling-up continuous reactor. Using continuous equation on continuous stirred tank reactor (CSTR)

$$V.\frac{dC_i}{dt} = F.(C_i - C_0) + R_A.V \tag{3}$$

where V = reactor volume (liter), C_i = wastewater concentration (mg.L-1); C_o = outlet concentration (mg.L-1); F = debit (L.second^{-1}), R_A = $-k.\ C_o$, and k is the value from batch reaction. In steady state condition,

$$V.\frac{dC_t}{dt} = \text{-}k.C_t = 0 \tag{4}$$

So that:

$$k.C_0.V = F.(C_i - C_0) \text{ or } V = \frac{F(C_i - C_0)}{k.C_0} \tag{5}$$

Heavy metal removal (%) was calculated based on initial metal concentration (C_0, mg.L^{-1}) and metal concentration at time t (C_t, mg.L^{-1}):

$$Removal = \frac{C_0\text{-}C_t}{C_0} x100\% \tag{6}$$

3. Result and discussion

3.1 Electrode mass and removal

Electrocoagulation process occurred including reaction on anode and cathode. Al^0 from sacrifice anode is oxidized into Al^{3+} ion, reacts with ion hydroxide in cathode, tend to flock in $Al(OH)_3$ form and sweep the other metal in electroplating wastewater. Base on equation (1), during 120 min for each voltage input, anode and cathode mass change occurred, as shown in Figure 2.

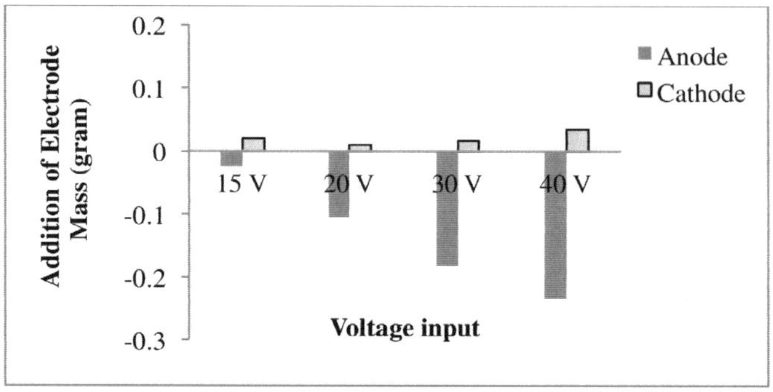

Figure 2. Mass change on anode and cathode.

Positive mass change on cathode means an addition on surface area, whereas negative change on anode means a mass reduction on anode or anode has sacrificed its metal to dilute in electrolyte solution. Base on Figure 2, electrocoagulation process occurred as expected. Higher voltage input gives higher anode mass reduction rate compare to cathode mass addition. On electrolysis cell (as electrocoagulation process), cathode has negative charge and anode has positive charge. It means oxidation rate of Al^0 anode into Al^{3+} was dominant that allows other ion metals in wastewater (positive charge) swept by $Al(OH)_3$ presipitat before attract to cathode surface (negative charge).

The metals in wastewater could present as colloid. Although initial pH of wastewater is around 5.2 until 7.5, at the end of the process it tends to increase for every voltage input variation, as shown in Figure 3. This condition allows the metals that present in colloid are swept by $Al(OH)_3$ precipitate, since aluminium hydroxide species from sacrifice anode are transformed into insoluble as the pH increase [11] and provide advantage for metal removal by hydroxide precipitation. This condition could be described by optical density as shown in Figure 4.

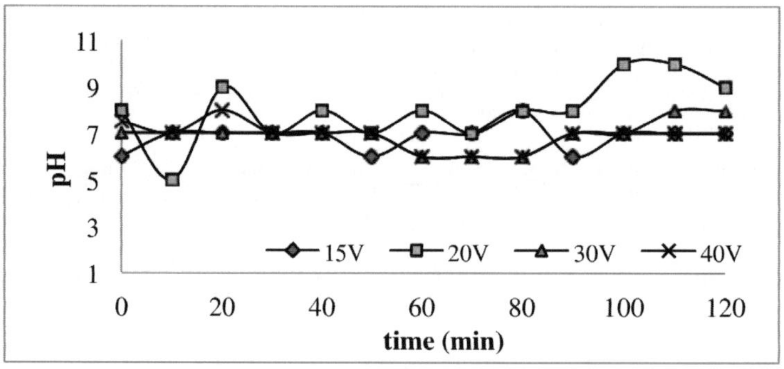

Figure 3. pH during reaction time.

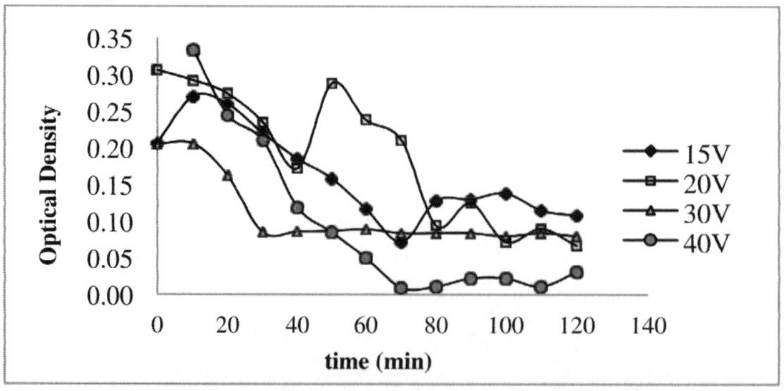

Figure 4. Optical density during reaction time.

Base on Figure 4, OD decreased during electrocoagulation and higher voltage input give faster OD decreasing. On 30V, at t = 30min the process has reach its lowest OD value, whereas on 40V it needs 70 min to reach its lowest OD value and afterward the value relatively constant. The OD decreasing on 30V and 40V give better result than 15 and 20V voltage input. However, even OD decreasing also occurred on 40V, but there was no significant difference between 30V input voltage. Further detail for every heavy metal removal would be analysed on 30V and 40V.

3.2 Kinetic rate

To determine the reaction kinetic rate, regression with polynomial approach was made on first order 30V and 40V input [11, 13]. Initial copper concentration was quite high compare to other metals concentration, so that copper concentration plotted in second axis. Using 30V input, iron, zinc and copper concentration drastically decreased at first 10 min. Chromium required 40 min until reach constant concentration. It means no significant metal removal after t = 40 min. As shown in Figure 5., based on the equation (2), the slope during electrocoagulation process is 0,035/min until 0,067/min, with $k_{average}$ is 0,051/min.

Using 40V input, all metals decrease significantly at t = 10min. After that, the removal showed insignificant decreasing. Based on equation (2), the decreasing results slope between 0.003/min until 0.032/min, with $k_{average}$ is 0,018/min.

Due to equation (5) k value affects to volume dimension of the reactor where higher k gives less volume of reactor when debit and concentration are constant.

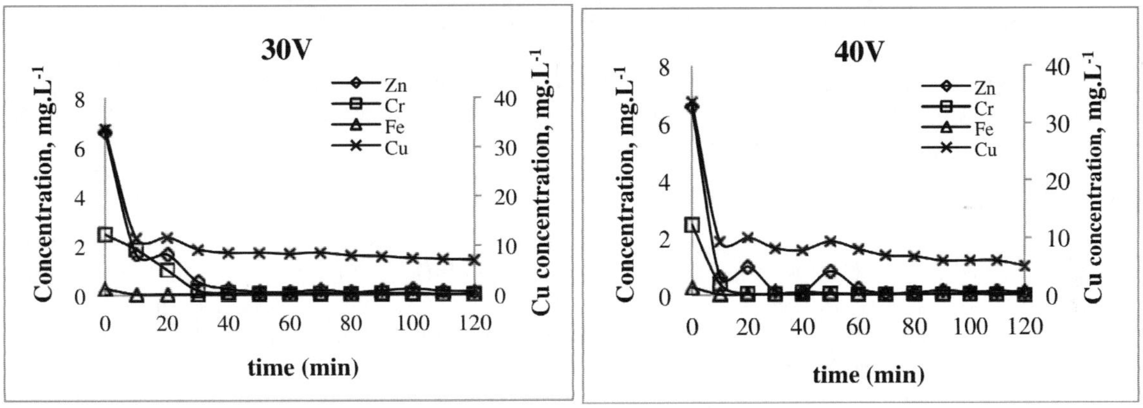

Figure 5. Heavy metal concentration at voltage input variation.

3.3 Pilot-scale

Electrocoagulation process on laboratory-scale of 1.5 L on batch system results kinetic rate as the basic factor for pilot scale calculation. Base on equation (5), it assumed that wastewater debit (F) 3 m^3 on 6 hours working period or equal with 500 $L.hour^{-1}$, and the metal concentration in the effluent fulfilled to standard regulation, so the pilot-scale reactor volume is 915 L. Detention time on the reactor, t, is calculated by dividing reactor volume by debit [13]. It results 100 min detention time.

Pilot-scale was constructed and tested using 30 and 40V input, due to the optimum OD removal efficiency on laboratory-scale, then the removal efficiency was calculated using Equation (6). The removal of metal concentration during 100 min detention time that shown in Table 1, compared with regulation standard, met the requirement except for copper. It might be caused by copper initial concentration that higher than other metals, which impact into longer reaction time to remove all concentration.Value of pH may influence copper removal since copper removal using hydroxide precipitation occurs on high basic condition, around pH 10 [14].

In electrocoagulation process, heavy metals may react in some different ways, as [3]:

- a reduction at the cathode, as in the case of chromium (VI):

$$Cr_2O_7^{2-} + 6e^- + 7H_2O \rightarrow 2Cr^{3+} + 14OH^-$$

- or oxidation, as in the case of iron:

$$Fe^{2+} \rightarrow Fe^{3+}$$

- or hydroxide precipitation of heavy metals as the reaction below:

$$M^{n+} + n(OH)^- \leftrightarrow M(OH)_n\downarrow$$

The reduction of chromium (VI) occur in cathode into Cr (III) then precipitate as hydroxide.

Table 1. Removal efficiency.

Heavy Metal	Influent Conc. ($mg.L^{-1}$)	Standard* ($mg.L^{-1}$)	30 V		40 V	
			Effluent Conc. ($mg.L^{-1}$)	η (%)	Effluent Conc. ($mg.L^{-1}$)	η (%)
Cr	2.450	0.5	0.054	98	0.063	97
Cu	33.670	0.5	7.052	79	5.000	85
Fe	0.265	5.0**	0.030	89	0.030	89
Zn	6.600	1.0	0.108	98	0.077	99

*Regulation of Ministry of Environmental no. 5 year 2014
**Decree of Governor of West Java no. 6 year 1999

Based on Table 1, voltage input between 30 and 40V does not give significant differences. Since operation cost also being a consideration factor when deciding the optimum condition [15], so 30V should be the optimum voltage input that suggested to fulfil the regulation standard.

4. Conclusion

Electrocoagulation using aluminium anode and copper cathode on electroplating industry wastewater on laboratory and pilot-scale has been studied. Voltage input influence the removal efficiencies. Due to efficiency, the optimum voltage input to remove chrome, copper, iron, and zinc from electroplating wastewater in this study was 30V. On this optimum voltage, OD removal efficiency on laboratory-scale is 60% in 30min while on pilot-scale removal efficiency ranges between 79-98%, depend on initial heavy metal concentration. Series installation of electrocoagulation might be an alternative to overcome high concentration of copper in the effluent.

Acknowledgment

We would appreciate Ministry for Research and Technology Republic of Indonesia for funding support and PT.DSB electroplating industry for the collaboration.

References

[1] H. H. Lou and Y. Huang 2006 *Electroplating.*
[2] F. Fu and Q. Wang 2011 *J. Environ. Manage.* **92(3)** 407–418.
[3] B. Lekhlif, L. Oudrhiri, F. Zidane, and J. F. Blais 2014 *J.Mater.Environ.Sci.* **5(1)** 111–120.
[4] A. K. Golder, A. N. Samanta, and S. Ray 2007 *Sep. Purif. Technol.* **53(1)** 33–41.
[5] Z. Gu, Z. Liao, M. Schulz, J. R. Davis, J. C. Baygents, and J. Farrell 2009 *Ind. Eng. Chem. Res.* **48(6)** 3112–3117.
[6] N. Beyazit *Int. J. Electrochem. Sci.* 2014 **9** 4315–4330.
[7] M. Sarkhosh, Z. Atafar, E. Ahmadi, S. Nazari, and Y. Fakhri 2016 *Int. J. Curr. Microbiol. App. Sci.* **5(4)** 615–625.
[8] K. Dermentzis, E. Valsamidou, A. Lazaridou, and N. C. Kokkinos 2011 *J. Eng. Sci. Technol.*

IOP Conf. Series: Earth and Environmental Science **60** (2017) 012026 doi:10.1088/1755-1315/60/1/012026

Rev. **4 (2)** 188-192.
[9] N. Adhoum, L. Monser, N. Bellakhal, and J. E. Belgaied 2004 *J. Hazard. Mater.* **112(3)** 207–213.
[10] H. D. Tchobanoglous,G; Burton, F.L.; Stensel 2004 4th ed. McGraw Hill
[11] N. Marriaga-cabrales and F. Machuca-martínez 2014 *Res. Sig.* **661(2)** 1–16.
[12] M. Al-shannag, Z. Al-qodah, K. Bani-melhem, and M. Rasool 2015 *Chem. Eng. J.* **260** 749–756.
[13] V. Kuokkanen, T. Kuokkanen, J. Rämö, and U. Lassi 2013 *Green Sustain. Chem.* **5** 89–121.
[14] A. Negrea, M. Ciopec, and L. Lupa 2008 *Bull. Politeh. Univ.* **53(67)** 1–2.
[15] C. F. Rajemahadik, S. V Kulkarni, and G. S. Kulkarni 2013 *Int. J. Sci. Res. Publ.* **3(10)** 1–5.

Heat transfer characteristics of coconut oil as phase change material to room cooling application

M Irsyad* and Harmen
Mechanical Engineering Department, Engineering Faculty, University of Lampung
*Email : irsyadmuh@eng.unila.ac.id

Abstract. Thermal comfort in a room is one of human needs in the workplace and dwellings, so that the use of air conditioning system in tropical countries is inevitable. This equipment has an impact on the increase of energy consumption. One method of minimizing the energy use is by using the phase change material (PCM) as thermal energy storage. This material utilizes the temperature difference between day and night for the storage and release of thermal energy. PCM development on application as a material for air cooling inlet, partitioning and interior needs to be supported by the study of heat transfer characteristics when PCM absorbs heat from ambient temperature. This study was conducted to determine the heat transfer characteristics on coconut oil as a phase change material. There are three models of experiments performed in this research. Firstly, an experiment was conducted to analyze the time that was needed by material to phase change by varying the temperature. The second experiment analyzed the heat transfer characteristics of air to PCM naturally convection. The third experiment analyzed the forced convection heat transfer on the surface of the PCM container by varying the air velocity. The data of experimental showed that, increasing ambient air temperature resulted in shorter time for phase change. At temperatures of 30°C, the time for phase change of PCM with the thickness of 8 cm was 1700 min, and it was stable at temperatures of 27°C. Increasing air temperature accelerated the phase change in the material. While for the forced convection heat transfer, PCM could reduce the air temperature in the range of 30 to 35°C at about 1 to 2°C, with a velocity of 1-3 m/s.

1. Introduction

The need for electricity in Indonesia increase every year, in which on average the increase each year is 6.6%. The using of electricity is dominated by household sector (41%), industrial sector (34%), and commercial sector (24%) [1]. In commercial sector, the use of AC dominates the use of electrical power at approximately 65% [2]. The same case applies in household sector, that is shown by increase in use of split AC.

What encourages people in using AC is to experience room thermal comfort, either for doing activities, working and for taking a rest. Based on Indonesian National Standard (SNI) 03-6390-2011, room thermal comfort for occupants is at the temperature of 25°C ± 1.5°C and relative humidity of 60% ± 5% [3]. The air temperature condition in Indonesia on average at daytime is higher than the thermal comfort temperature, such as 28-28.5°C in Jakarta, 29°C in Makassar, Surabaya and Yogyakarta [4]. This has caused the use of AC to be the primary need. On the other hand, the difference in daytime and night time temperatures is quite significant. The average minimum temperature at night time is quite low at 22°C, or it could even get lower than 20°C. This potential may be used to cool room air at daytime by using phase change material (PCM) as storage of thermal energy. Cool air at night time can take the energy from PCM thus phase change takes place. While at daytime, PCM can take the energy from the air which is passed until the phase change takes place again. Compared to one phase material, the energy that could be absorbed by PCM at phase change range is much higher because of using energy in the form of latent heat. The superiority of this PCM high stored energy as well as able to work in low temperature range at its phase change temperature.

The study related to the use of PCM as thermal energy storage for room cooler has been conducted by researchers. There are three forms of PCM application for room cooling: air cooling by channeling into PCM container, as thermal energy storage (TES) to increase efficiency in heating ventilating air conditioning (HVAC) system, and using PCM on wall, floor as well as ceiling, by mixing into the material, or installed in the form of container. Kange et.al. [5,6] conducted a study on the utilization of PCM in which cool air was channeled at night time and room air was circulated at day time known as night ventilation with a PCM packed bed storage (NVP) system. Yamaha et.al. [7] used PCM HVAC system as thermal energy storage, that is used when load condition is at the highest. Indartono et.al. [8] used micro encapsulation phase change material (MPCM), one of the materials in the making of bricks which functions as daytime thermal energy storage. PCM material that has been studied for the application of room cooler is summarized in the Table 1. This study aimed to evaluate characteristics of coconut oil as phase change material to room cooling application.

Table 1. Phase change material to room cooling application.

PCM	Melting temperature (°C)	Heat fussion (kJ/kg)	c_p (kJ/kg.K)	k (W/m.K)	ρ (kg/m^3)	Ref
s19	19	160	1.9	0.43	1520	[9,10]
S21	22	170	2.2	0.54	1530	[9,10]
ClimSel C21	21	144	3.6	0.5	1380	[9,11]
S23	23	175	2.2	0.5	1530	[9,10]
CimSel C24	24	126	3.6	0.5 – 0.7	1380	[9,11]
S25	25	180	2.2	0.54	1530	[9,10]
A22	22	145	2.22	0.18	785	[9,10]
A23	23	145	2.22	0.18	785	[9,10]
A24	24	124	2.22	0.18	790	[9,10]
A25	25	150	2.26	0.18	785	[9,10]
SP 22A17	22 -24	150	-	0.6	1490	[9,12]
Paraffin C13-C24	22 -24	189	2.1	0.21	900	[9,13]
Mystiric acid–Capric acid (34%–66%)	24	147	-	0.164	888	[9,14]

2. Experimental

In this PCM research, Barco brand coconut oil was used. From the initial test, the freezing process took place from the temperature of 24°C. While the melting process took place at the temperatures of 22-24°C. The characteristics of this material from several references are shown in Table 2.

Table 2. Properties of coconut oil.

Density (kg/m^3)	Thermal conductivity (W/m.K)	Melting temperature (°C)	Heat of fusion (kJ/kg)	References
916	=0.62+0.1006*T	22 – 24	103.25	[15]
-	0.321	26.78	110.4	[16]

To determine characteristics of heat transfer at this PCM, three tests were conducted: time of phase change, characteristics of heat transfer when coming in contact with ambient air, and characteristics of heat transfer with forced convection. The schemes of the three test equipment are shown in Figure 1.

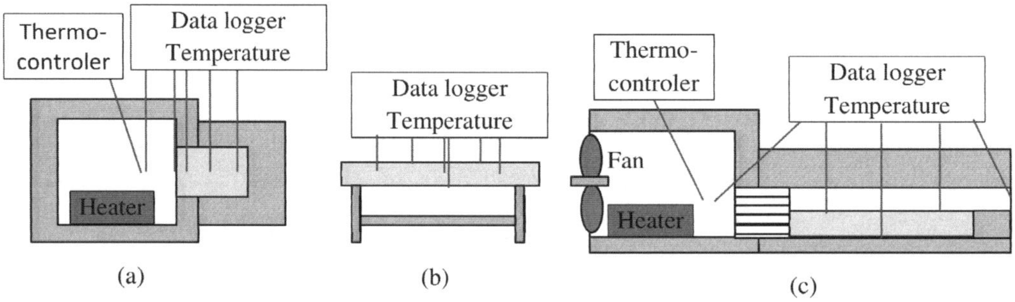

Figure 1. Scheme of test equipment: a) phase change time, b) characteristics of heat transfer with natural convection by the surrounding air, c) characteristics of heat transfer with forced convection.

In the test of melting time, PCM was placed in container of 4 x 6 x 10 cm (Figure 1a). The surface of heat transfer was only found on one side of the surface with the size of 4 x 6 cm, made of aluminum of 0,2 mm in thickness. PCM thickness from the surface of heat transfer was 10 cm. Thermocouple was applied in conditioned air, surface of heat transfer, and at PCM thickness of PCM 1 cm, 5 cm and 9 cm. Hot air was controlled by using a thermo-controller. Data were taken at air temperatures of 30, 40 and 50°C. Data was recorded until the temperature of the PCM was not increased again. At the test of heat transfer with the surrounding air, PCM was placed inside an aluminum profile, with the size of 0.5 x 1 inch and length of 10 cm as shown in Figure 1, section b. Data collecting was started at a PCM temperature of 20°C. Thermocouples were installed on the top and bottom surfaces of PCM location and inside the PCM. Data was recorded until the PCM temperature exceeding the melting temperature. On the other hand, test with forced convection was conducted on one surface of PCM container made of three aluminum profiles with the size of 0.5 x 1 x 10 cm as shown in Figure 1, section c. Air velocity was varied at 1 and 1.5 m/s. Air temperature was varied at 30 and 33°C. Thermocouple was placed on the top and bottom surfaces of the container, also in air before and after test section. Temperature test data was recorded by using temperature data logger Lutron 452SD. The velocity of air flow was measured using anemometer Lutron 22SD.

The result was analyzed for each test. In the first test, time needed for phase change in several air temperature condition was analyzed. In the second test model what was analyzed was the process of phase change by obtaining thermal energy from ambient air. While in the third, test absorption of thermal energy by PCM from accelerated air through convection heat transfer on the surface of PCM container was analyzed.

3. Result and Discussion

Initial study was conducted testing of environmental temperatures at during the day. Measurement did in Bandung city on 18-19 August 2016, in which in the daytime from 9.00 AM-5.00 PM, where the surrounding air temperature exceeded the thermal comfort temperature, as shown in Figure 2.

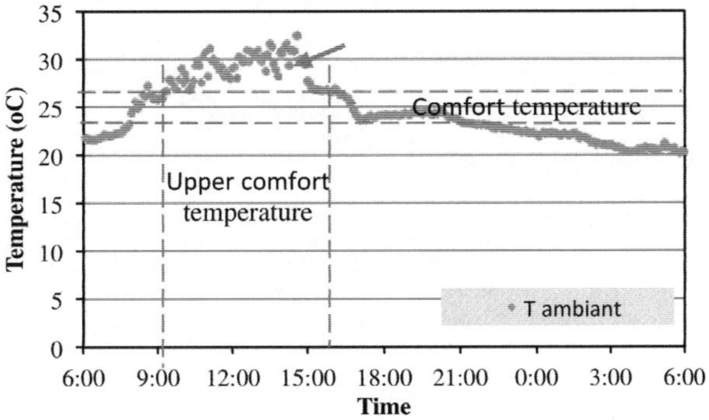

Figure 2. Temperature of ambient air in Bandung on 18-19 August 2016.

The process of PCM melting was started from the part nearest to heat transfer surface. Time of phase change to take place for air temperature of 30°C was quite long starting from 700 min for a distance of 1 cm to 1700 min for a distance of 9 cm from the surface of heat transfer, as shown in Figure 3. One of the causes was that thermal conductivity PCM was not as high compared to water. Besides that also the temperature differences of air and PCM were not too great. A high energy for phase change process also contributed to the time taken for phase change. From Figure 3, sloping curve showed phase change process of PCM at temperatures of 22 to 25°C. Differences in air temperature affected of phase change period as shown in Figure 4. At temperature of 50°C, period of phase change was shorter at 228 min for a distance of 5 cm from heat transfer surface, compared to temperature of 30°C at 1260 min.

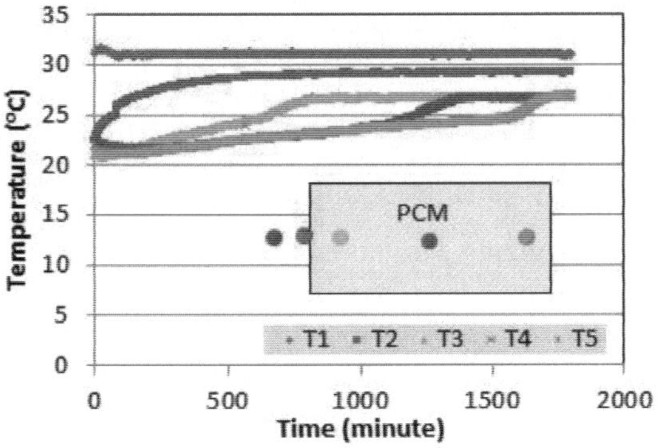

Figure 3. Period of phase change at air temperature of 30°C.

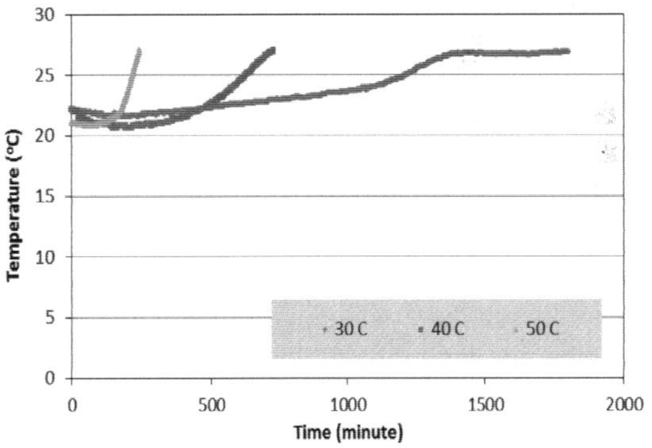

Figure 4. Period of phase change at PCM thickness of 5 cm.

The process of thermal energy absorption from ambient air is shown in Figure 5. At air temperature higher than PCM temperature, heat transfer to PCM took place. The process of phase change took place at the same time as the increase in ambient air temperature. Heat transfer was also caused by radiation heat transfer to surface of PCM container and then continued to the PCM. This was shown starting from the 220 to the 270th min.

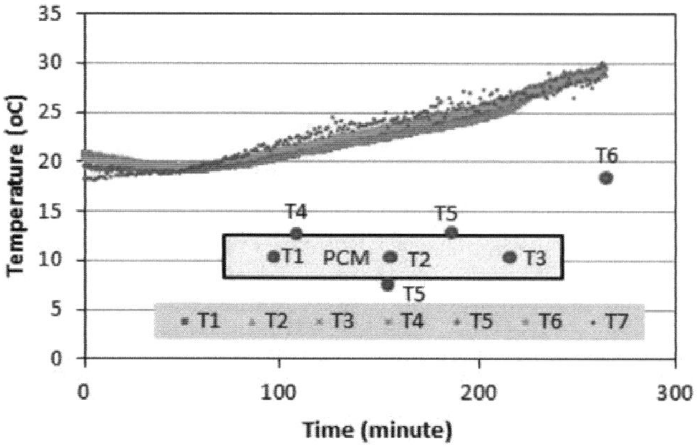

Figure 5. Process of PCM phase change in ambient air.

For forced convection heat transfer, there are two variables tested: air temperature and flow velocity. The process of phase change started from the closest area to incoming air and on the upper side. The process of change moved to the area of air outside and lower part of PCM, as shown in Figure 6. The time for the process of phase change for the entire PCM in the container at the speed of 1m/s and temperature of 30°C was 65 min. Average decrease of temperature was 0.83°C. Thermal energy absorbed from the air was 9.6 kJ. The impact of air velocity and temperature towards reduction of air temperature and time of phase change was shown in Table 3. Increasing the air velocity and temperature difference of air and PCM could increase of reduction of air temperature. Reduction of air temperature was range between 0.83 and 1.29°C. For the application of surrounding air cooling still possible for improve in temperature decrease by increasing flow speed and expanding heat transfer surface.

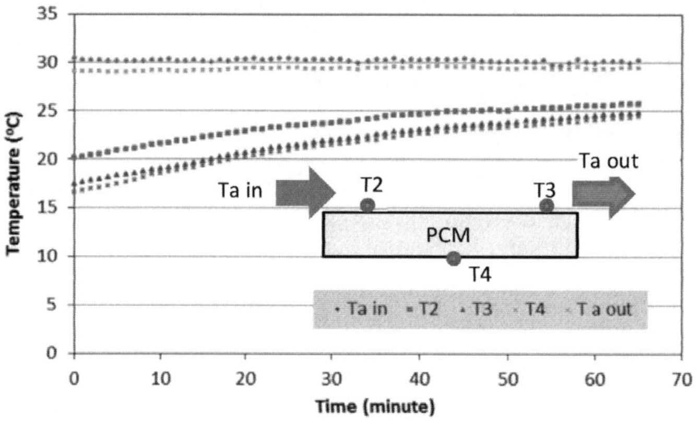

Figure 6. Surrounding air temperature decrease by using PCM.

Table 3. Heat transfer in ambient air using PCM.

Velocity (m/s)		ΔT (°C)	Time (min)	Heat transfer (kJ)
1.0	30°C	0.83	65	9.58
	33°C	1.19	32	6.60
1.5	30°C	0.82	48	10.24
	33°C	1.29	24	8.05

4. Conclusion
The utilization of PCM with high thermal conductivity needs to consider the thickness of container as it gives an impact on time of phase change. By natural convection heat transfer the time of phase change takes a long time. PCM in container that comes in contact directly to the surrounding air experiences phase change process with the increase in surrounding air temperature. Heat transfer through forced convection on PCM container surface can decrease the ambient temperature of 0.82-1.29°C, with air velocity of 1-1.5 m/s and air temperature of 30-33°C

Acknowledgment
Authors would like to acknowledge the Indonesian Ministry of Research, Technology and Higher Education for Fundamental Research Project 2016.

References
[1] Indonesia Energy Outlook 2015, Agency for the Assessment and Application of Technology, Indonesia
[2] JICA Study on Energy Efficiency and Concervation Improvement in Indonesia 2007 -2008 http://eneken.ieej.or.jp/en/data/pdf/491.pdf, access 8 – 10 – 2012
[3] Badan Standardisasi Nasional (BSN). Standar Nasional Indonesia (Indonesian National Standardization)-SNI 6390:2011, 2011.
[4] Karyono TH, 2015, *Buildings*, **5**, 802-813
[5] Kang YB, Jiang Y, Zhang YP. 2003, *Energy and Build.* **35** (4):417–25.
[6] Zhang Y, Zhou G, Lin K, Zhang Q, Di H, 2007, *Build. and Environ.***42** 2197–2209
[7] Yamaha M, Misaki S. 2006, *HVAC & R Res* **12** (3c): 861–9.
[8] Indartono YS, Suwono A, Pasek AD, Christantho A, 2012, *ASEAN Eng. J. Part A*, **3** (2), 46 – 53
[9] Tatsidjodoung P, Pierres NL, Luo L, 2013, *Renew. and Sustainable Energy Rev.* **18** 327–349
[10] Environmental Process Limited. Available from: /http://www.pcmproducts.net/S
[11] Climator. Available from: /http://www.climator.com/en/climsel/S
[12] Rubitherm, 2010, Available from: /http://www.rubitherm.de/ english/index.htmS.
[13] Zhou D, Zhao CY, Tian Y, 2012, *Appl. Energy.* **92** 593–605
[14] Lane GA, 1980, *The Int. J. of Ambient Energy.* **1** 155–68
[15] Mettaweea ES, Ead AI., 2013, *Int. J. of Thermal & Environ. Eng.* **5** (1) 21-30.
[16] Wi S., Seo J, Jeong SG., Chang SJ., Kang Y., Kim S., 2015, *Solar Energy Mat. & Cells.***143** 168–173

Heat transfer characteristics of building walls using phase change material

M Irsyad[1,2*], A D Pasek[2], Y S Indartono[2] and A W Pratomo[2]
[1]Mechanical Engineering Department, Engineering Faculty, University of Lampung
[2] Faculty of Mechanical and Aerospace Engineering – Institut Teknologi Bandung
[*]Email : irsyadmuh@eng.unila.ac.id

Abstract. Minimizing energy consumption in air conditioning system can be done with reducing the cooling load in a room. Heat from solar radiation which passes through the wall increases the cooling load. Utilization of phase change material on walls is expected to decrease the heat rate by storing energy when the phase change process takes place. The stored energy is released when the ambient temperature is low. Temperature differences at noon and evening can be utilized as discharging and charging cycles. This study examines the characteristics of heat transfer in walls using phase change material (PCM) in the form of encapsulation and using the sleeve as well. Heat transfer of bricks containing encapsulated PCM, tested the storage and released the heat on the walls of the building models were evaluated in this study. Experiments of heat transfer on brick consist of time that is needed for heat transfer and thermal conductivity test as well. Experiments were conducted on a wall coated by PCM which was exposed on a day and night cycle to analyze the heat storage and heat release. PCM used in these experiments was coconut oil. The measured parameter is the temperature at some points in the brick, walls and ambient temperature as well. The results showed that the use of encapsulation on an empty brick can increase the time for thermal heat transfer. Thermal conductivity values of a brick containing encapsulated PCM was lower than hollow bricks, where each value was 1.3 W/m.K and 1.6 W/m.K. While the process of heat absorption takes place from 7:00 am to 06:00 pm, and the release of heat runs from 10:00 pm to 7:00 am. The use of this PCM layer can reduce the surface temperature of the walls of an average of 2°C and slows the heat into the room.

1. Introduction

Room thermal comfort is a human need in which to rest and do activities. In Indonesia room thermal comfort standard is based on Indonesian National Standard (SNI) 03-6390-2011 which is at the temperature of 25°C±1.5°C and relative humidity of 60%±5% [1]. There are several factors which influence room thermal comfort, namely thermal loads from occupants, equipment, lights/lamps, and thermal load coming from sun rays. In tropical area, the main thermal load is due to the average temperature of ambient air exceeding the thermal comfort temperature, as well as sun shine which hits buildings continues into the rooms. Solar radiation which reaches the wall surface partly continues through the wall and a part is reflected as shown in Figure1. In Indonesia, cooling load of which generates from thermal load from outside the building, reaches 70% [2]. With this condition, the use of air conditioning system is highly necessary to acquire room thermal comfort.

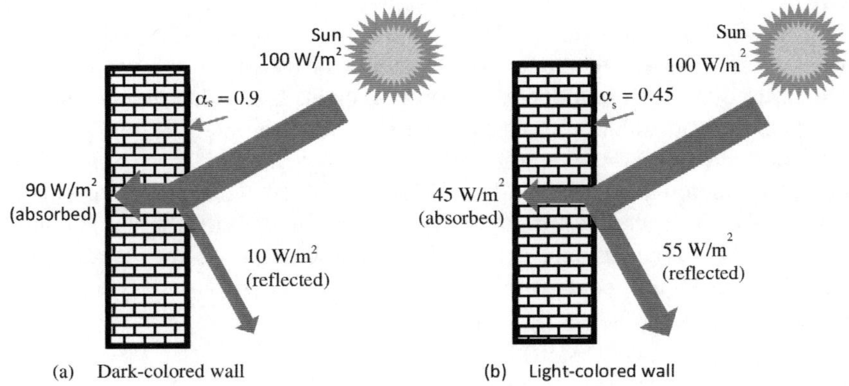

Figure 1. Distribution of thermal energy on walls that receive solar radiation [3].

Thermal load which enters the room is solar radiation energy received by wall surface and then through conduction is continued to the surface of the inside part of the wall. Solar radiation energy which can be continued to the inside surface of the wall can reach 90%, depending on the darkness level of the wall [3]. The heat transfer on the wall is also influenced by wall thermal conductivity. The varied thermal conductivity of wall material is as shown in Table 1.

Table 1. Properties of wall materials [4].

Material	Thermal conductivity [W/m.K]	Specific heat capacity [kJ/kg K]	Mass density [kg/m^3]
Tuff	0.630	1.300	1500
Concrete	1.263	1.000	2000
Brick	0.500	0.840	840
Plaster	0.900	0.910	1800
Insulating material	0.170	0.920	1200

Besides the choice of material, to slow down heat transfer in walls, nowadays there has also been a development on the use of phase change material (PCM) on walls. This PCM is functioned as thermal energy storage, thus the transfer energy in the wall is collected in the PCM. The utilization of PCM on walls according to Baetensa et al. [5] is grouped into three types namely the use in wall board, the use in concrete, and the use as insulation material. A number of research results of PCM use in wall showed positive results on the decrease of thermal energy which enters a room. Diaconu et al. [6] used PCM made of composite on building walls and can decrease cooling load up to 35.4%. Indartono et al. [7] used micro-encapsulation phase change material (MPCM) from coconut oil as material in brick mixture and can decrease room temperature and can delay the incident of peak load. Barrenech et al. [8] used PCM DS5001 Micronal as material of Portland cement mixture and gypsum with mass fractions of 5 and 15%. The use of this material can reduce wall thermal conductivity up to 50%. Catsell et al. [9] used MPCM which has been commercialized as type RT27 and SP25. The use of this material can decrease building energy consumption of 15%. Arce et al. [10] used Micronal, BASF product PCM in portable building walls and can decrease peak temperature of 6%. Besides that it can also slow down the occurrence of peak temperature of 36%. For the application of PCM in building material in Indonesia, study on heat transfer characteristics in materials is needed, either laboratory test or application test on building type.

2. Experimental

In this study, heat transfer characteristics were tested for bricks added with encapsulated PCM, and building wall made of bricks with the inner side coated with PCM were conducted. In brick test, temperature was measured at several points by using thermocouple and data logger lutron type 224SD either for standard brick or brick containing encapsulated PCM. Scheme of test tool (Figure 2a). Heat source was from heater and surface temperature of 80°C was maintained by using thermo controller. Besides that thermal conductivity test in standard brick and brick containing encapsulated PCM was conducted (Figure 2b). While in the building model wall test, points with thermocouple are outer and inner wall surfaces, PCM container surface, ambient temperature. At the same time, a standard building model has been tested as a comparison. Test equipment is shown in Figure 3. PCM used was Barco brand coconut oil.

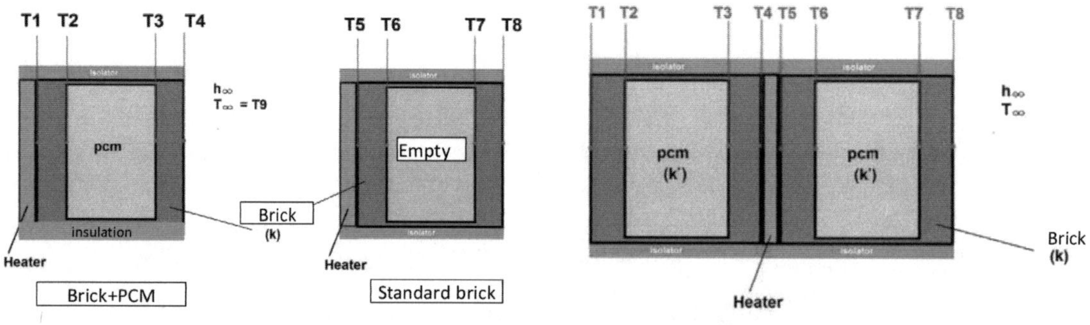

(a) Heat transfer time (b) thermal conductivity

Figure 2. Test equipment scheme of heat transfer characteristics in brick.

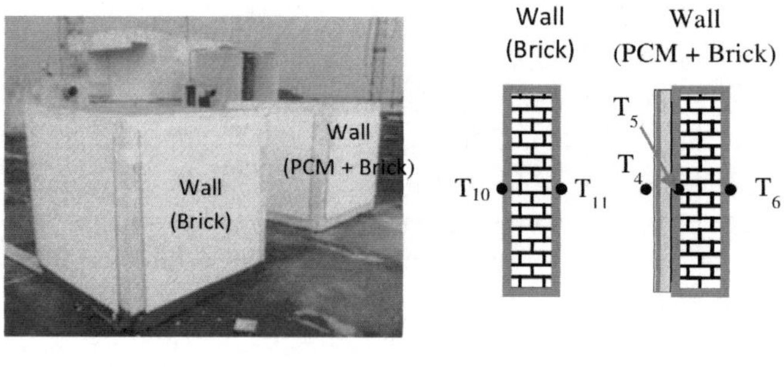

Figure 3. Heat transfer test on walls in building models.

3. Result and discussion

In this part heat transfer in bricks, temperature change in each part, thermal conductivity, as well as PCM coated wall heat transfer characteristics to the surrounding air are discussed. Data of test result were compared to bricks and standard wall.

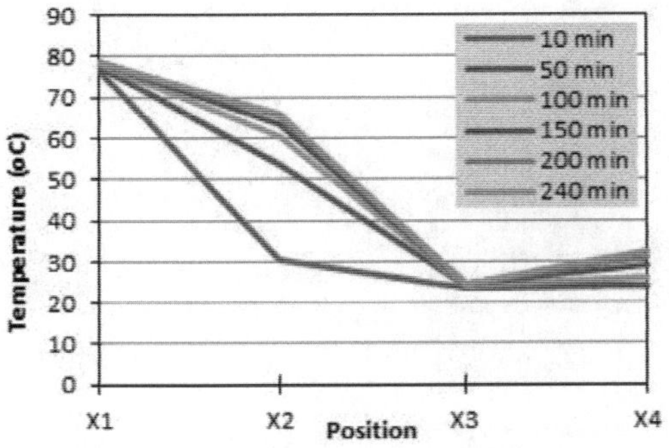

(a) Brick+PCM.

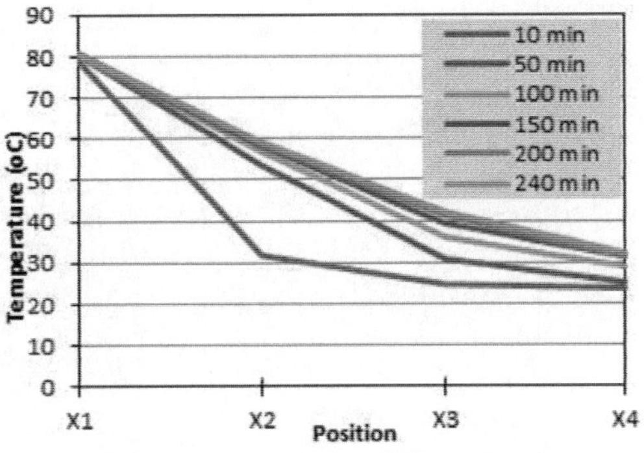

(b) Hollow Brick (standard brick).

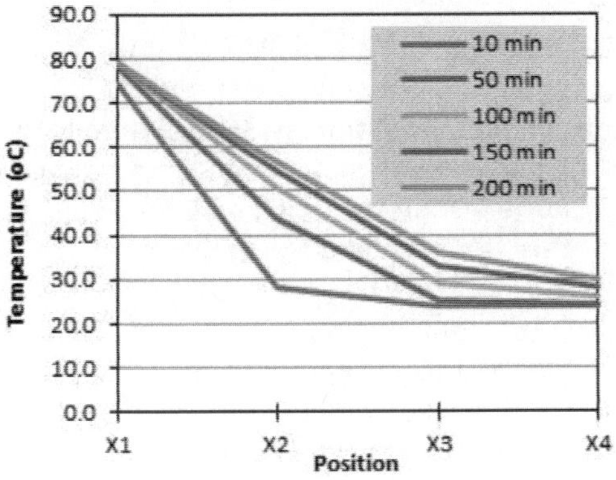

(c) Cement + brick.

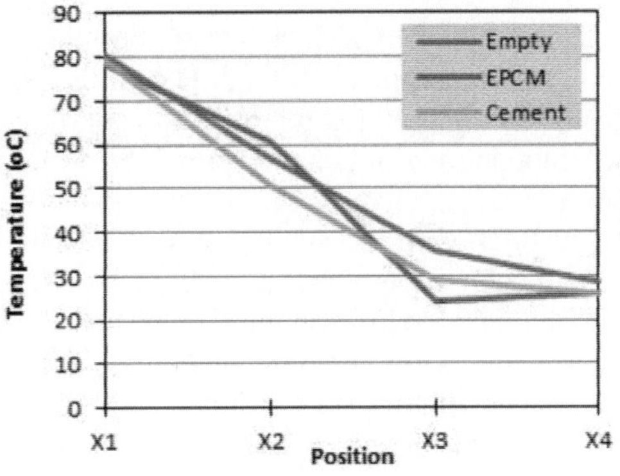

(d) Temperature of 200th min.

Figure 4. Temperature distribution in brick and brick + PCM.

In the test of thermal transfer in bricks, the use of encapsulated PCM in bricks could slow down heat transfer into inner surface of brick, thus the time to increase surface temperature was longer (Figure 4). The thermal energy entering through PCM was used for the process of phase change. The use of thermal energy did not result in significant change in temperature. Energy required was great, compared to using sensible heat. After 240th min, brick surface received heat, position 3 did not experience significant temperature change, but in position 4 which is the back surface of the brick, there was a significant increase in temperature. This increase is predicted to come from heat transfer on the sides of the brick.

The result of thermal conductivity test is shown in Figure 5. Brick with hollow inside has higher thermal conductivity compared to brick with encapsulated PCM in the hollow part itself. The thermal conductivity is for one piece of brick. Observed from each brick material, such as sand and cement, thus coconut oil has lower thermal conductivity. The thermal conductivity decrease is also made possible as a result of the encapsulated layer.

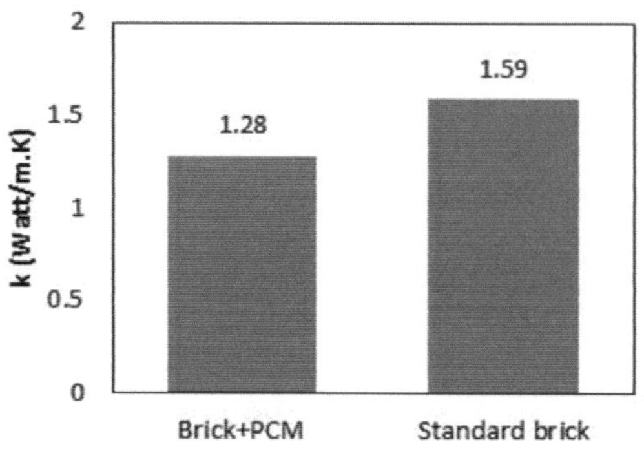

Figure 5. Thermal conductivity in brick and brick + PCM.

The result of PCM use test on building model wall from 15 to 19 August 2016 is shown in Figure 6. The temperature of standard building wall outer surface as well as that with PCM container was affected by ambient temperature and solar radiation. In peak condition, its temperature higher than ambient temperature. On the other hand, the inner surface temperature there was a delay when ambient temperature change took place. This is due to the occurrence of thermal energy storage in brick material. In brick wall installed with PCM container, energy stored increased thus the delay became longer. For peak temperature, it took place at 12:00. Peak temperature for inner surface of standard wall was at 17:00. While on the inner surface of the wall containing PCM, it was at 21:00. The decrease in thermal conductivity in PCM caused heat flow part to collect in brick wall. Thus, this is what causes temperature on the inner side of the brick to increase compared to standard brick. The placement of PCM container on the inner side of the wall can slow down heat transfer into the room (Figure 6d). When ambient temperature is lower than wall temperature, energy release occurs. In standard wall, outer and inner surface temperature can reach the ambient temperature. While in wall + PCM container, energy release to the surrounding still results in higher wall inner surface temperature than the ambient temperature. This is due to energy release stored in PCM in phase change process did not provide great temperature change effect. The use of coconut oil as PCM for the application on walls in the city of Bandung is still suitable because PCM temperature is still below its phase change temperature.

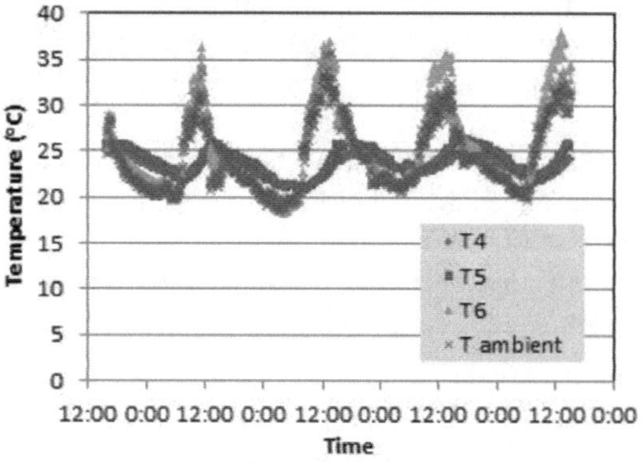

(a) Wall with PCM layer on inner surface.

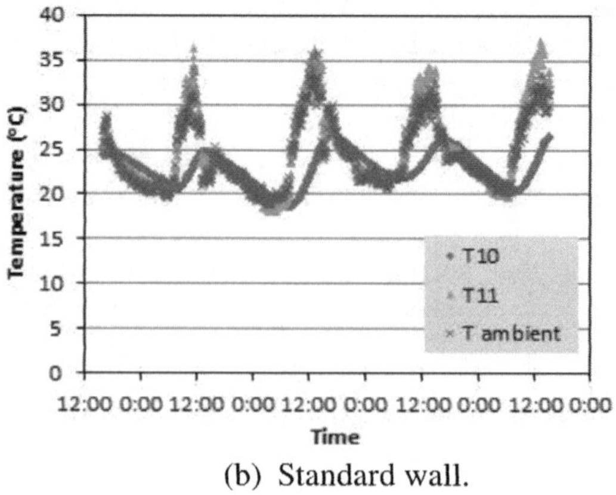

(b) Standard wall.

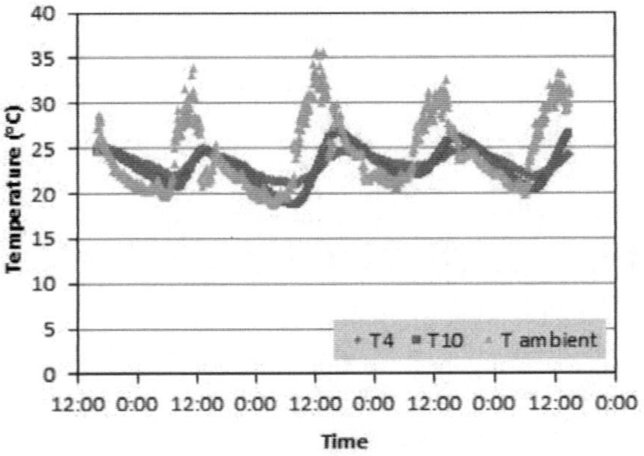

(c) Comparison of temperature on inner wall inner surface of wall.

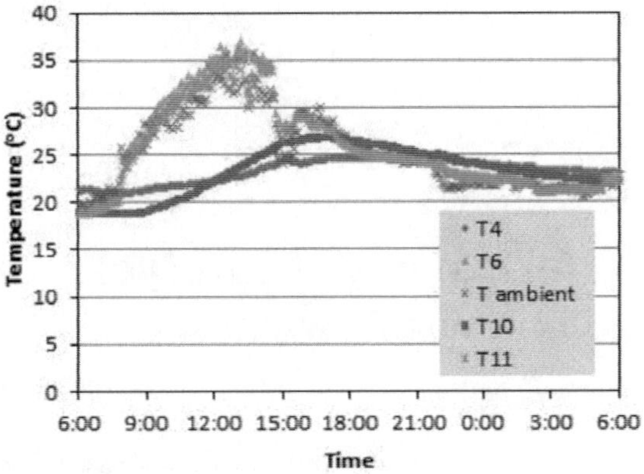

(d) Comparison of temperature on surface for 24 hr.

Figure 6. Temperature distribution on building model wall (5 – 19 August 2016 in Bandung).

4. Conclusion

The use of PCM could delay heat transfer from outer wall surface to inner surface due to increase in material thermal capacity. The addition of PCM in the hollow inside the brick slightly reduced thermal conductivity compared to hollow brick. The utilization of PCM from coconut oil applied on wall inner surface in the container is suitable for minimum ambient temperature ranging between 18 – 20°C, because PCM can still undergo complete phase change process at 22°C, thus the use of the PCM is still suitable with the weather in the city of Bandung. PCM wall inner surface temperature can be lower on average 2°C compared to standard wall. The process of heat absorption in PCM wall takes place from 7:00 to 18:00, while the process of heat release then takes place at 22:00 to 6:00.

Acknowledgment

This work was financially supported by International Research Collaboration and Scientific Publication 2016 from Indonesia's Ministry of Research, Technology and Higher Education.

References

[1] Standar Nasional Indonesia (Indonesian National Standardization)-SNI 6390:2011 BSN: Jakarta, Indonesia.

[2] N.A. Utama N.A., Ishihara K.N., Tezuka T., Gheewala S.H., Zhang Q., 2011, *J. Sustainable Energy & Environ.* **2** 21-25

[3] Kreider, J.F., and Rabl. A., 1994, Heating and Cooling of Buildings. New York: McGraw-Hill.

[4] Vollaroa R.D., Evangelistia L., Carnieloa E., Battistaa G., Goria P., Guattaria C., Fanchiottia A., 2014, *Energy Procedia.* **45** 372 – 378

[5] Diaconu BM, and Cruceru M, 2010, *Energy and Build.* **42** (10) 1759–1772.

[6] Baetensa R., Jelle B.P., Gustavsen A., 2010, *Energy and Build.* **42** 1361–1368.

[7] Indartono YS, Suwono A, Pasek AD, Budiman AC, 2011, *Proceeding of 7th Fluid and Thermal Energy Conversion (FTEC)*, 24-27 September 2011, Zhengzhou, China.

[8] Barreneche, C., Navarro, M.E., Fernandez, A.I., Cabeza, L.F., 2013, *Appl. Energy.* **109** 428–432

[9] Castell, A., Martorell, I., Medrano, M., Perez, G., Cabeza, L.F., 2010, *Energy and Build.* **42** 534–540

[10] Arce, P., Castellon, C., Castell, A., Cabeza, L.F., 2012, *Energy and Build.* **44** 88–93

Performance of PEM fuel cells stack as affected by number of cell and gas flow-rate

A Syampurwadi*, H Onggo, Indriyati and R Yudianti
Research Unit for Clean Technology, Indonesian Institute of Sciences (LIPI)
Jalan Cisitu 21/154 D, Bandung 40135, Indonesia
*Email : anung.syampurwadi@lipi.go.id

Abstract. The proton exchange membrane fuel cell (PEMFC) is a promising technology as an alternative green energy due to its high power density, low operating temperatures, low local emissions, quiet operation and fast start up-shutdown. In order to apply fuel cell as portable power supply, the performance investigation of small number of cells is needed. In this study, PEMFC stacks consisting of 1, 3, 5 and 7-cells with an active area of 25 cm^2 per cell have been designed and developed. Their was evaluated in variation of gas flow rate. The membrane electrode assembly (MEA) was prepared by hot-pressing commercial gas diffusion electrodes (Pt loading 0.5 mg/cm^2) on pre-treated Nafion 117 membrane. The stacks were constructed using bipolar plates in serpentine pattern and Z-type gas flow configuration. The experimental results were presented as polarization and power output curves which show the effects of varying number of cells and H_2/O_2 flow-rates on the PEMFC performance. The experimental results showed that not only number of cells and gas flow-rates affected the fuel cells performance, but also the operating temperature as a result of electrochemistry reaction inside the cell.

1. Introduction

Global demand for new and renewable energy has lead to several alternative energy resources & technology. One of them is fuel cell (FC) which convert chemical energy of hydrogen and oxygen (or air) directly to electrical energy. There are many kind of fuel cell technology, such as alkaline FC (AFC), proton exchange membrane FC (PEMFC), phosphoric acid FC (PAFC), molten carbonate FC (MCFC), and solid oxide FC (SOFC) [1].

Proton exchange membrane FC (PEMFC) is a promising technology due to its high power density, low operating temperatures, low local emissions, quiet operation and fast start up-shutdown. One of PEMFC application is as mini/portable power supply. For this purpose, the PEMFC is usually assembled as a stack of many cells. Investigation of PEMFC performance was conducted by assembling 1, 3, 5 and 7 cells, then evaluated the polarization curves yielded by fuel cell test device [2].

2. Experimental

2.1. Materials

Components of PEMFC stacks consisted of membrane electrode assembly (MEA), aluminum end-plates, gold coated copper current collectors, silicon rubber gaskets (sheet thickness 0.5mm), FU4369 graphite bipolar plates and other accesories (Swagelok tubing fitting, rods, stainless steel nuts and bolts, ring, heat-resistants shells, silicone tubing). MEA consisted of Nafion 117 membrane, gas diffusion electrodes (GDE) with 0.5 mg/cm^2 Pt loading of 60% Pt/C and 5 wt.% Nation 117 solution. The reactant gases were hydrogen UHP (99.99% purity) and oxygen, with nitrogon as purging gas. In this assembly, the reactant gases flow configuration was Z type, as showed at Figure 1.

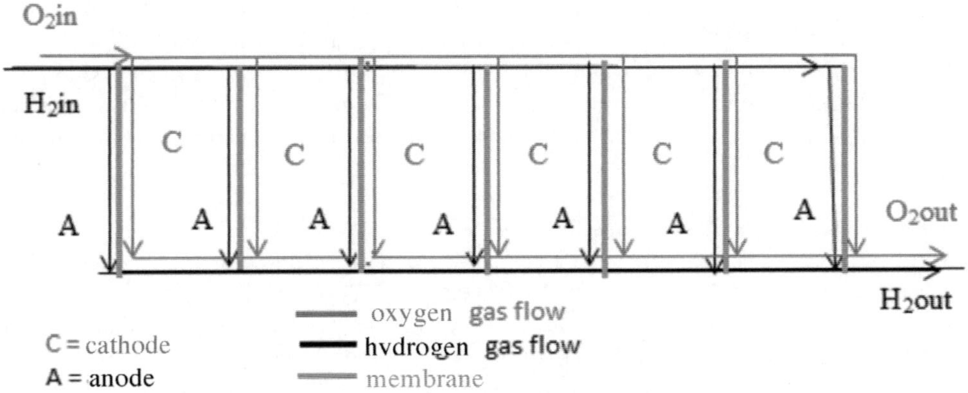

Figure 1. The Z type of reactant gases flow configuration.

2.2. Fabrication of some stack components

The bipolar plates were fuel cell grade FU 4369 graphite bipolar plates, dimension of 62 mm x 95 mm x 5 mm, with a 6-channel serpentine pattern in the hydrogen flow field and in oxygen flow field. Depth and width of the channels were 1mm, and width of the rib were 0.8 mm (Figure 2). The curent collector plates were gold coated (0.2 μm)-copper plates, dimension of 62mm x 120 mm x 1.0 mm). The aluminum end-plates had dimension of 80 mm x 115 mm x 16 mm.

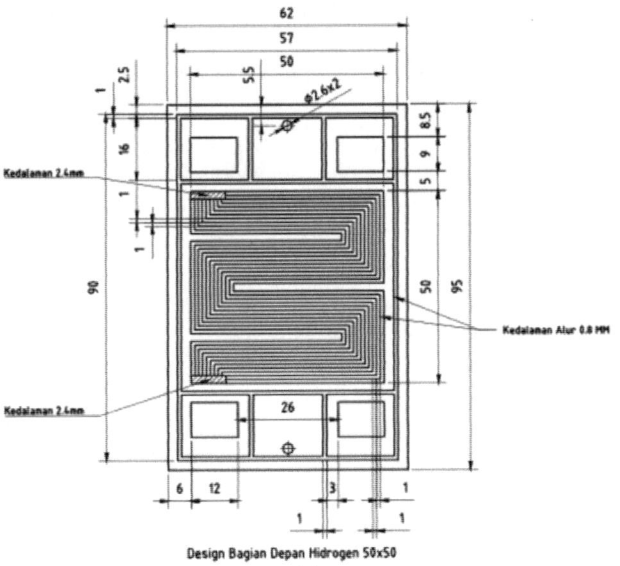

(a)

(b)

Figure 2. Bipolar plates design, (a) schematic, (b) photo.

2.3. Preparation of membrane electrode assembly
The membrane electrode assemblies (MEAs) used for the experiment were prepared by hot-pressing of two commercial gas diffusion electrode (GDE) (60 wt% Pt/vulcan XC72 on carbon cloth, Pt loading 0.5 mg/cm^2) on either side of a DuPont™ Nafion 117 membrane at 120°C, 20 kgf for 1 min. For effective contact between the GDE and the membrane, the Nafion 117 solution (5% solution of 1.0 mg cm^{-2} (Neuroflash PTE Ltd)) was applied on catalyst surfaces of two GDE by screen printed methode[3]. Nafion 117 membrane was previously treated with 3% hydrogen peroxide (H_2O_2) at 70-80°C for 1 h to remove organic impurities. To get fully H-form, the membrane was boiled at 70-80 °C for 1 min 0.5 M H_2SO_4 , after that the membrane was rinsed twice repeatedly with distilled water at 70-80 °C for 1 h to remove H_2SO_4 that remain on the membrane surface.

2.4. Assembling single cell, 3-cells, 5-cells and 7-cells
The PEMFC stack consists of single cell, 3-cells, 5-cells and 7-cells with active area of 25 cm2. The MEAs, gasket, two monopolar plates (for hydrogen and oxygen flow), bipolar plates and current collectors were sandwiched between two aluminum end plates with eight threaded stainless steel rods, bolts and nuts.The gaskets are placed between electrolyte membrane and bipolar-plate which should have good elasticity, durability and corrosion resistant. During the stack assembling, two positioning rods were used for alignment. All components were aligned and stacked one by one. The sizes of SS 316 bolts are M6 x 0.8 mm with 58mm, 70mm, 90mm and 100mm length for the single cell , 3-cells, 5-cell and 7-cell stacks respectively. Equal clamping torque (20 kgf) was applied on each bolt to assemble the PEMFC single cell, 3-cells, 5-cells and 7-cells.

2.5. Measuringf PEMFC performances
In the cell testing, high purity hydrogen (99,99%) and oxygen gases were used as fuel and oxidant, respectively. Smart2 PEM/DM hybrid fuel cell test system (WonATech Co., Ltd.Korea) was used to test the performance of the stacks. The test station consisted of two main system: a gas feeding and a controlling system. A gas feeding system supplied and treated the fuel and oxidant gases fed to the cell, while the controlling system measured and controlled the operation conditions of the FC stack. The back pressure of the fuel cell could be varied with the help of mechanical valves in the test station, for each reactant.

2.5.1. Varying gas flow rate for single cell
At first step, number of cell was kept constant at single cell. Gas flow rate variation was applied, then each of its performances were evaluated. As addition, back pressure was also applied to show its effect. Table 1 shows each condition.

Table 1. Variation of gas flow rate & back pressure for single cell.

	Condition 1	Condition 2	Condition 3	Condition 4
H_2/O_2 volumetric flow rate (mL/min)	100 / 140	60 / 140	100 / 100	100 / 100, with ¾ psi back pressure

2.5.2. *Varying number of cell.* At second step, number of cell was varied. In the other hand, gas flow rate were kept proportional to number of cells. Because of the fuel cell test device limitation, back pressure was not applied in this step. Each of its performance were then evaluated. Table 2 show each condition.

IOP Conf. Series: Earth and Environmental Science **60** (2017) 012029 doi:10.1088/1755-1315/60/1/012029

Table 2. Variation of number of cells.

	Single cell	3-cells	5-cells	7-cells
H_2/O_2 volumetric flow rate (mL/min)	100 / 100	300 / 300	500 / 500	700 / 700

3. Result and discussion

Figure 3 showed that for single cell, the best performance was achieved by condition of 100/100 with back pressure, followed by 100/100, 100/140, and 60/140, respectively.

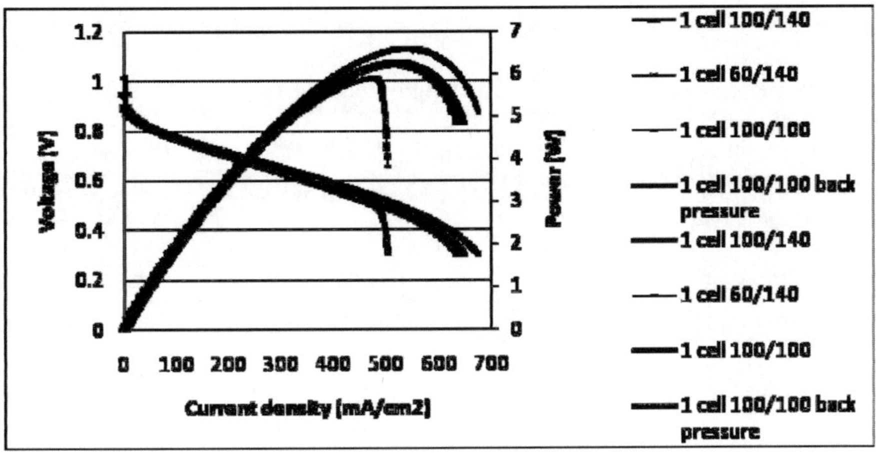

Figure 3. Polarization and power output curves for single cell.

First, it showed that back pressure have positive effect on fuel cell performance, as already confirmed by many scientific books and journal papers [4][5][6]. Second, it indicated that 100/100 mL/min not only showed the saturated flow rate for the single cell, but also indicated that 1 : 1 ratio is the best stoichiometry ratio for these reactants. Theoretically, the mole ratio of hydrogen and oxygen gases for water formation is 2 : 1. The discrepancy between the experimental and theoretical ratios need to be investigated at futher research.

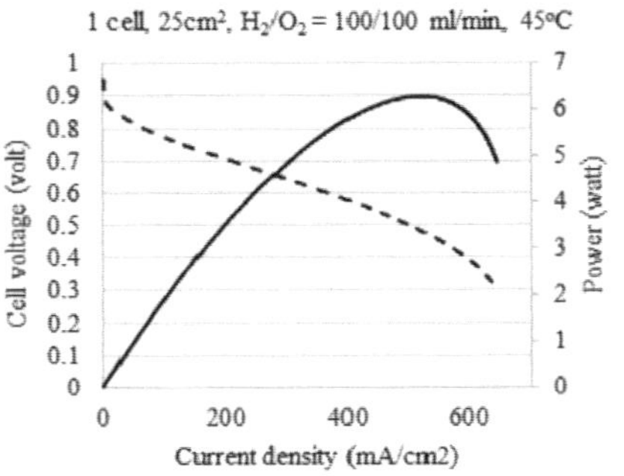

(a)

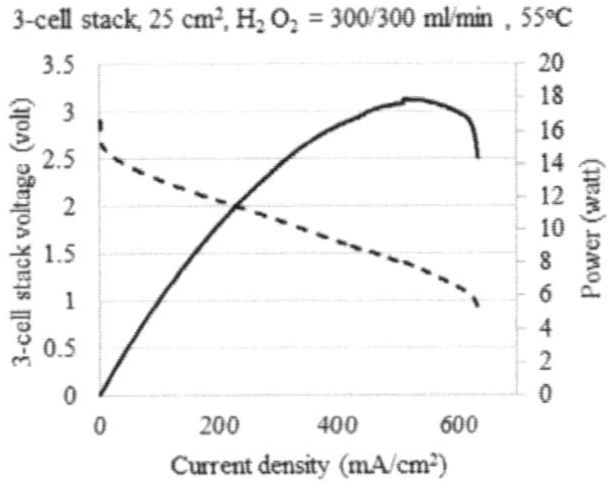

(b)

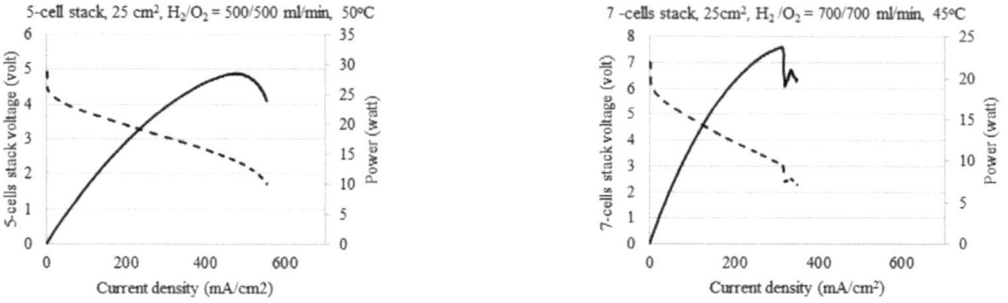

(c) (d)

Figure 4. Polarization and power output for (a) single cell, (b) 3-cells, (c) 5-cells, (d) 7-cells.

Figure 4 shows the performance test results for single cell, 3-cells, 5-cells and 7-cells. From these curves, it could be obtained data of Table 3 as follow.

Table 3. Extracted data from polarization and power output of 1-cell, 3-cells, 5-cells, 7-cells.

Parameter	1-cell	3-cells	5-cells	7-cells
H_2/O_2 volumetric flow rate (mL/min)	100/100	300/300	500/500	700/700
Open circuit voltage, OCV (volt)	1.0	2.9	4.9	7.0
Operational voltage, V_{op} (volt)	0.6	1.8	3.0	4.2
Operational current density, J_{op} (mA/cm^2)	379	297	316	165
Operational power density, P_{op} (mW/cm^2)	227	535	948	693

Table 3 shows that open circuit voltage values were normal for each cells, as it was predicted from theoretical study and confirmed by many research papers [7]. Operational voltage was determined by definition[8]. The results of operational current density were rather unexpected. The 7-cells indicated a fail or broken condition on operation/testing. It is normal if the expected values for J_{op} were same or similar for all of the 4 combination. But the results was different. The best J_{op} were 1-cell, followed by 5-cell, 3-cell, and 7-cell, respectively. This lead to irregular patern for operational power density P_{op}. There are many factors that could contribute to this unexpected results, from the quality of MEA assembling (hot-pressing), leakage of reactant gases, etc. The risk of defect is increased as number of cells increasing.

Figure 5 shows the correlation between cell temperature and current density for 1-cell, 3-cells, 5-cells and 5-cells.

(a) (b)

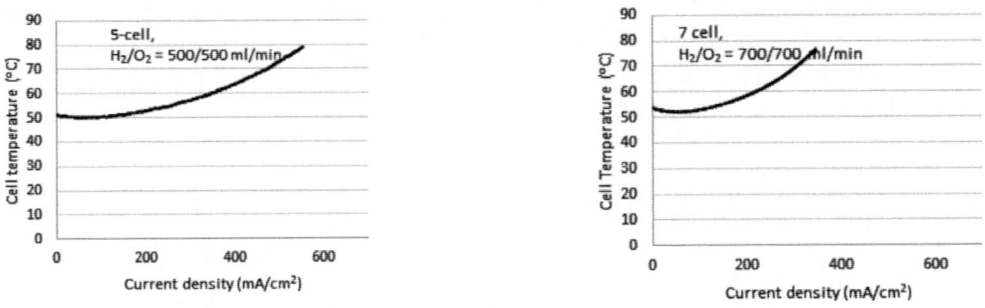

(c) (d)

Figure 5. Correlation of cell temperature and J_{op}, (a)1-cell, (b)3-cells, (c)5-cells, (d)7-cells.

From Figure 5, it is clear that there exist some trends between cell temperature and current density. For 1-cell, the temperature was relatively steady at 45°C. For 3-cells and 5-cells, temperature increased as current density was growing up, but not exceed 90°C (usually known as temperature limit for Nafion membrane)[9]. Instead, by reaching 70-80°C, the stack could improve its performance, as this temperature range is the best values for Nafion performance[10]. For 7-cells, the increasing temperature could exceed 80°C, if current density was above 400mA/cm^2. This means that the stack need a cooling sytem to prevent from possibility of broken. On the other side, this results suggested that there was a possibility that the 7-cells temperature had exceeded 80°C, so triggering of Nafion defective during testing.

3. Conclusion

The reactant flow rate of ratio 1:1 yielded best stack performance. The polarization curves showed that before electronic loading, the OCV of each cells have normal values. This means that all stack worked normally at start point. As current density was increasing, some of the cells was going underperformance. Some considerations could be taken such as hot-pressing that based on determined spacing (thickness) rather than pressure gauge, cooling system for increased number of cells, changing value of clamping torque, etc. The risk of fail is increased as number of cell is increasing.

More number of the cells and higher current density had influence on the increasing of cell temperatures. On the other hand, this increasing temperature could contribute to higher performance of the stack if could be maintained between 70-80°C.

Acknowledgement

Authors would like to acknowledge Nanik Indayaningsih, Yuyun Irmawati, Dita Rama Insiyanda, and Achmad Subhan from Research Center for Physics, Indonesian Institute of Sciences for supporting PEMFC performance test.

References

[1] Hoogers G 2003 *Fuel Cell Technology Handbook*.
[2] Hirschenhofer J H, Stauffer D B, Engleman R R and Klett M G 1998 *Fuel Cell Handbook*, 4th edition.
[3] Litster S and McLean G 2004 *J. Power Sources*. **130(1–2)** 61–76.
[4] Reshetenko T V, Bender G, Bethune K and Rocheleau R 2011 *Electrochimica Acta*. **56(24)** 8700–10.
[5] Amirinejad M, Rowshanzamir S and Eikani M H 2006 *J. Power Sources*. **161(2)** 872–5.
[6] Sun H, Liu H, Guo L 2005 *J. Power Sources*. **143(1–2)** 125–35.
[7] Zhang J, Tang Y, Song C, Zhang J and Wang H 2006 *J. Power Sources*. **163(1)** 532–7.
[8] Yu P T, Gu W, Makharia R, Wagner F T and Gasteiger H A 2006 *ECS Transactions* **3(1)**797-809

[9] Jung G, Weng F, Su A, Wang J, Yu T L, Lin H, Yang T and Chan S 2008 *International Journal of Hydrogen Energy* **33(9)** 2413–7.
[10] Williams M V, Kunz H R and Fenton J M 2004 *Journal of Power Sources* **135(1–2)** 122–34.

IOP Conf. Series: Earth and Environmental Science **60** (2017) 012030 doi:10.1088/1755-1315/60/1/012030

Gas production in anaerobic dark-fermentation processes from agriculture solid waste

L Sriwuryandari*, E A Priantoro and N Sintawardani
Research Unit for Clean Technology, Indonesian Institute of Sciences (LIPI)
Jalan Cisitu 21/154 D, Bandung 40135, Indonesia
*Email : lies.sriwuryandari56@gmail.com

Abstract. Approximately, Bandung produces agricultural solid waste of 1549 ton/day. This wastes consist of wet-organic matter and can be used for bio-gas production. The research aimed to apply the available agricultural solid waste for bio-hydrogen. Biogas production was done by a serial of batches anaerobic fermentation using mix-culture bacteria as the active microorganism. Fermentation was carried out inside a 30 L bioreactor at room temperature. The analyzed parameters were of pH, total gas, temperature, and COD. Result showed that from 3 kg/day of organic wastes, various total gases of O_2, CH_4, H_2, CO_2, and $C_nH_{n,}O_2$ was produced.

1. Introduction

Domestic solid waste is produced by people activities and serves as problem if there is no available space and technology for further handling and treatments. Many cities in the world have to face the problem to handle and treat their domestic waste in line with the increase of population. A proper technology is needed to manage and to treat the domestic waste, but unluckily the lack of funding for a proper infrastructure and management make this all more difficult.

Problems of waste can be generated as a result of population increases every year, the infrastructure is reduced, the development of the region urban, human resources are insufficient, the management system of waste management has not been good, limited land for waste disposal, lack of environmental education in the community, particularly the issue of waste and lack of understanding of the importance of protecting the environment and others [1].

Data in the Ministry of Environment (MOE) in 2010 states, the average volume of garbage in Indonesia reached 200,000 tons per day. Urban areas accounted for most trash. This is due to many factors, including population growth and urbanization. If the waste problem is not immediately addressed then in 2020 the volume of waste in Indonesia will increased to be five-fold or 1 million ton pile of garbage a day. In developed countries of whom people more aware for environment and supported by modern technology, solid waste is managed properly.

General garbage can be divided into organic and inorganic waste. The organic waste can rapidly decompose, such as food scraps, vegetable and fruits, and some part can be degraded slower like dried leaves, twigs and so forth. Inorganic wastes such as rubber, paper, cans, plastic, food wrappers, and glass are difficult to unravel or not easily decompose naturally. The dispose of waste will increase continuously. In order to reduce pollution, waste management is indispensable. Alteration of organic waste for energy resources sounds very interesting and beneficial.

Using anaerobic process, domestic solid waste produces biogas which is useful as energy source for cooking, heating or electricity. More hydrogen gas as alternative energy sources can be produced by anaerobic modification. Biogas derived from wet vegetable and fruits waste has been studied [2,3]. Hydrogen production by dark hydrogen fermentation is increasing [4]. Bio-hydrogen production has been done using biomass as sources for fermentation from sweet potatoes starch residues [5], molases [6], tofu solid waste [7], sweet buckwheat [8], sweet melon [9], wheat straw [10], food wastes [11,12,13,14], starch [15], and waste and wastewater [16,17,18]. It was reported that mixed culture microorganisms is more valuable rather than single or pure culture for bio-hydrogen production. A controlled environment condition such as pH and temperature is strictly required [19] as well as hydraulic retention time (HRT) to reduce H2 – consuming bacteria [20,21,22]. The current study aimed to use agricultural organic waste and mixed culture bacteria of which naturally available in waste as well as invulnerable condition for high rate bio-hydrogen production.

2. Experimental

2.1. Material

The material used for media solution was purchased by Merck, such as yeast extract, buffer solution, Chemolithotropic Growth H-3 (DSM Catalogue), and trace element SL-6. The mix culture microorganism was available in the laboratory and re-activated.

2.2. Method

2.2.1. Biomass waste

Biomass waste was collected from traditional fresh market garbage in Caringin Bandung. Organic waste was crushed and blended by mechanical shredder to be smaller size as the feeding sludge. Biomass sludge was used for substrate in anerobic digestion bio-reactor. Feeding of organic agricultural sludge to bioreactor was of 3 kg per 10 days intermittently during 70 observation days.

2.2.2. Bioreactor

Bioreactor capacity was 30 L consisted of feeder and digester of 0.5 HP and electric Pump (40 L/min capacity). This pump serves to feed substrate into bio-reactor properly as well as to circulate substrate and to optimize the process. Bioreactors operated intermittently.

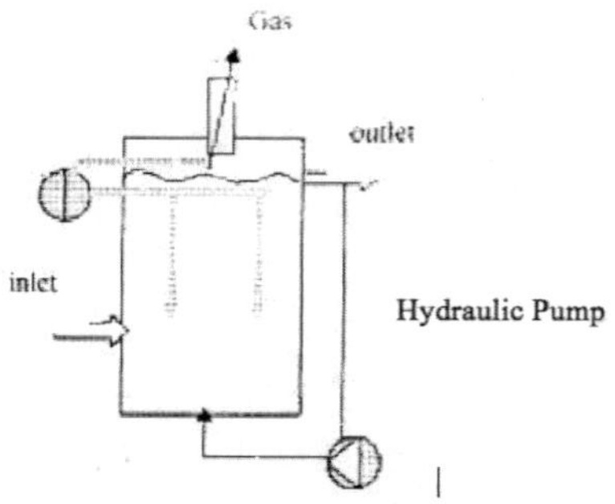

Figure 1. Schematic diagram of bio-reactor.

2.2.3. Mixed culture bacteria and growth media

As much as 100 ml of bacterial suspension from stock culture of mixed bacteria was inoculated in Erlemeyer containing of 1 L sterilized growth media. Growth media consisted of K_2HPO_4 750; KH_2PO_4 = 850; $(NH_4)_2SO_4$ = 40; $MgSO_4.7H_2O$ = 200; $CaCl_2.2H_2O$ = 0.75; Co $(NO_3)_2.6H_2O$ = 290; Fe $(NH_4)_2SO_4.6H_2O$ = 10; $MnCl_2.4H_2O$ = 2.1; $Na_2MoO_4.2H_2O$ = 0.75; H_3BO_3 = 2.8; Cu $(NO_3)_2.3H_2O$ = 0.04; $Na_2EDTA.2H_2O$ = 2.0; Nicotinic acid = 0.025; and Yeast Extract = 0,003 $mg.L^{-1}$. Trace element SL-6 = 5 ml. Total usage of inoculum in bioreactor was 10% (v/v). In order to optimize the production of biohydrogen, temperature was maintained (25-30°C, pH 5.6-6.5). Addition of NaOH 3N was carried out to increase pH whereas the addition of HCL 3N was required to decrease pH.

2.2.4. Sample analysis

As much as 100ml per day of sample recovered from bioreactor outlet was required for analysis Parameter analysis in this study was as of :

- *pH and Temperature :* pH of sample was measured by using pH-meter (LT Lutron pH-207) whereas temperature was measured by using Hg thermometer. Bioreactor was operated at room or ambient temperature about 25-28°C without setting. pH valued was adjusted to 5.6-6.5.

- COD was measured based on the standard methods for water and wastewater (APHA, 2005) with aUV-VISSpectrophotometer (Shimadzu GC14-A, Japan). The measured sample was on the filtrate COD.

2.2.5. Gas analysis
Gas composition and its analysis were carried out using gas analyzer of Cubic gasboard - 3100 P infra red syngas RRC.

3. Result and discussion
During the experiment, the feeding solid concentration was constant at 2.4%. Figure 2 shows the fluctuation of dry matter in the reactor. Average total solid concentration in reactor increased slowly from 2 to 3% in the I-IV batch (10 days for each batch), but it increased to 4.5% in batch VI and then decreased to 3.5%. It showed accumulations of organic solid material in reactor during experiment.

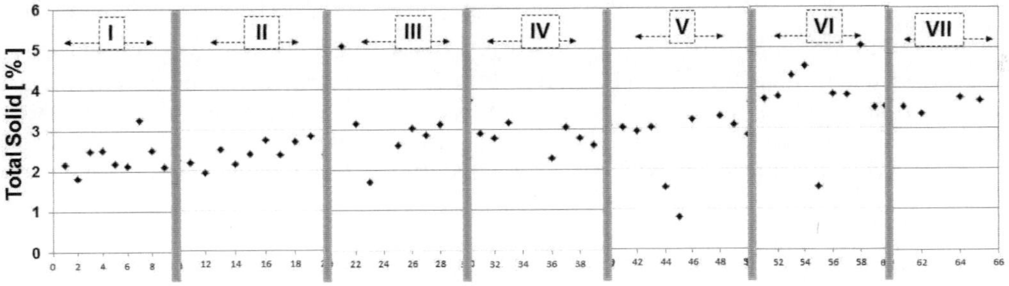

Figure 2. Total solid concentration in the reactor during observation days.

The expectation in anaerobic hydrolysis process is the liquefaction of organic solid material and production of hydrogen gas. Liquefaction of organic material to liquid could be observed in the concentration of filtrated or dissolved COD (Figure 3). Dissolved COD was increased during observation days from initial concentration of 62 to 114 g/L in the first batch. The concentration was fluctuated in the second batch, but then it tended to increase up to 200 g/L. After second batch the trend of COD fluctuation was like a bacterial growth curve, especially in the fourth, fifth to sixth batches.

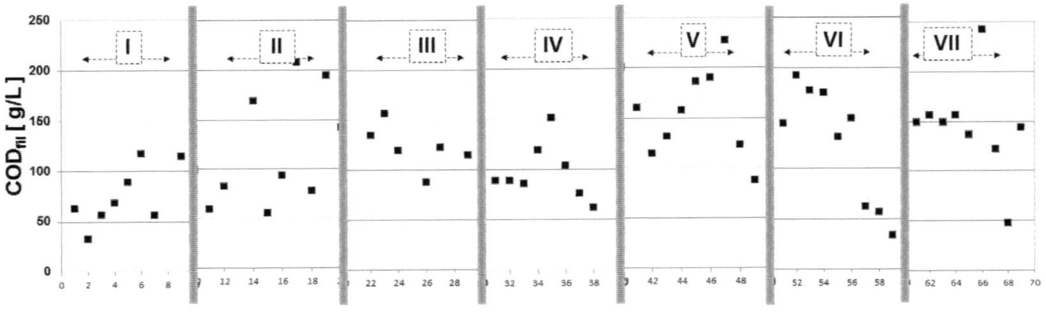

Figure 3. Concentration of filtrated COD during observation days.

Figure 4 shows the total solid and dissolved COD concentration. It is estimated that the first 30 days (3 batches) was the start-up process. In this period the anaerobic bacteria adapted to the feeding. After the third batch hydrolysis process became stable. The curve in each batch showed microbial activities from the log phase, a short stationary phase, and death phase. These estimations were supported by the result of hydrogen production per batch.

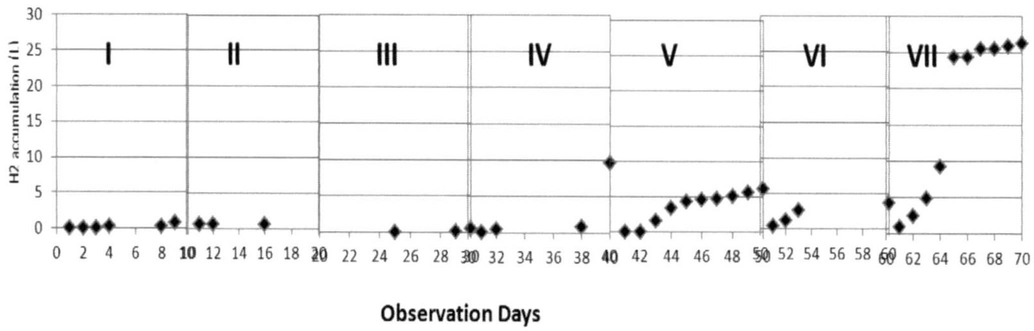

Figure 4. H_2 accumulation produced per batch during observation days.

Figure 5 shows concentration of hydrogen gas in each batch. In the first three batches, H_2 concentration was lower than 2% then increased in the next batches with the curve form looked like bacterial growth. H_2 concentration increased up to 14%. Gas production could be used as indicator for anaerobic microbial activities in the reactor. As the H_2 concentration was tended to increase during experiment days, it was measured in the batch V, VI and VII the methane production was increased to 4-6 L at the end of each batch. The methane content in the gas could indicate the activities of methanogenic bacteria.

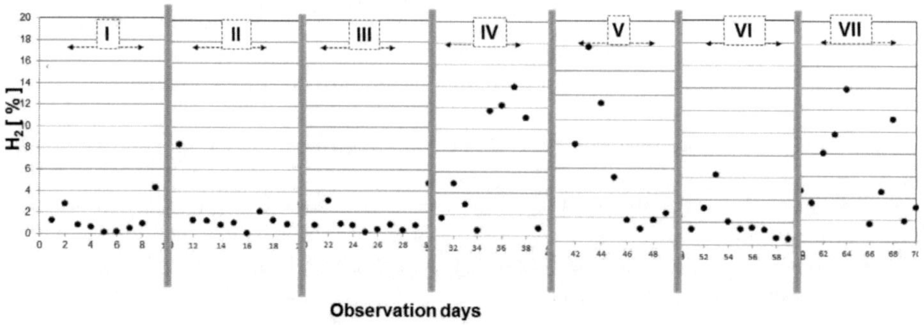

Figure 5. H_2 concentration during observation days.

It is important to adjust pH of 5.6 – 6.5 to produce effective conditions for microorganisms to produce hydrogen [5,17,19].

4. Conclusion

In current study, hydrogen production is not expected since the research was carried out in a few months and was of as early level. This onditions were in the process of acclimatization of which bacteria was adapting inside bioreactor. After lag phase period, the serial of batches hydrogen production increased after 30 days. After fourth batch, hydrogen was increased significantly.

Acknowledgement

We gratefully acknowledge the financial support from Riset Unggulan 2016, Sub Research *Pengembangan Energi Berbasis Biogas dan Bio-Oil* Project Research Center for physics, Indonesian Institute of Sciences.

IOP Conf. Series: Earth and Environmental Science **60** (2017) 012030 doi:10.1088/1755-1315/60/1/012030

6. References

[1] Widiatmoko, Aji 2011 *Scientific papers of Partnership*, BAPPENAS.

[2] Sirodjuddin, Ardan. 2011 *Pemanfaatan Sampah.* Suistaining Partnership, (IRSDP) BAPPENAS.

[3] PD. Kebersihan Kota Bandung 2015. *Condition of trash Bandung,* Final report JICA 2010

[4] Z. Rasdi., Tabassum Mumtaz, Nor'Aini A., M. Ali Hasan 2012 *Int. J. Hydr. Energy.* **37** p 17724-17730

[5] Fajar A. Pamungkas and Rudiana Agustini 2014 UNESA *J. Chemistry* **3**, No 1.

[6] Nanqi Rena, Jianzheng Lia, Baikun Lib,YongWanga, Shirui Liua. 2006 *Int. J. Hydr. Energy.* **31**. 2147 – 2157

[7] Amir Husin, Sarto, Siti Syamsiah, dan Imam Prasetyo 2014. *J. Rekayasa Proses*. **8**(2).

[8] Etty Tri Widyastuti 2011 *Thesis*. Sekolah Pasca Sarjana. Institut Pertanian Bogor.

[9] Nurkholis, Sarto, Muslikhin Hidayat 2016 *J. Inovasi Teknik Kimia*, **1**(2), 78-83

[10] Prasad Kaparaju , María Serrano, Anne Belinda Thomsen, Prawit Kongjan, Irini Angelidaki 2009 *J. Bioresource Technol.* **100** 2562–2568.

[11] H.M.Y Nazlina, A.R. Nor Aini, F.Ismail, M.Z.M Yusof and M.A. Hasan 2009 *Asian. J. Biotechnol.* **1(2)**; 42-50.

[12] Sun-Kee Han, Hang-Sik Shin 2004 *J. Hydrogen Energy* **29**. 569 – 577

[13] Hang-Sik Shina;Jong-Ho Younb, Sang-Hyoun Kim 2004 *Int. J. Hydr. Energy.* **29**. 1355-1363.

[14] StevenW.Van Ginkela, Sang-Eun Oha, Bruce E. Logan 2005 *Int. J. Hydr. Energy.* **30** 1535-1542

[15] Tong Zhang, Hong Liu, Herbert H.P. Fang. 2003. *J. Environ. Management*. **69**. 149-156

[16] Ilgi Karapinar Kapdan, Fikret Kargi 2006 *Enz. and Microbial. Technol.* **38** 569–582.

[17] S. Venkata Mohan, Y. Vijaya Bhaskar, P. Murali Krishna,N. Chandrasekhara Rao, V. Lalit Babu, P.N. Sarma 2007 *Int. J. Hydr. Energy* **32** 2286-2295

[18] Shihwu Sung, Dennis A. Bazylinski, Lutgarde Raskin 2003 *Hydrogen, Fuel Cells, and Infrastructure Technologies.* FY Progress Report

[19] Arief Wijaya, 2009 *Proceeding of National Conference of Indonesia Chemical Engineering*, Bandung. Pp. 19-20.

[20] D. Levin, R. Sparling, N. Cicek, R. Islam, 2005 *Proceedings International Hydrogen Energy Congress and Exhibition IHEC Istanbul, Turkey*, 13-15 July 2005

[21] Yao-Ting Fan, Ya-Hui Zhang, Shu-Fang Zhang, Hong-Wei Hou, Bao-Zeng Ren. 2006 *J. Bioresource Technol.* **97** 500–505.

[22] Mao-Lin Zhanga, Yao-Ting Fana, Yan Xinga, Chun-Mei Pana,Gao-Sheng Zhanga, Jiunn-Jyi Lay 2007 *J. Biomass and Bioenergy*. **31** 250–254.

IOP Conf. Series: Earth and Environmental Science **60** (2017) 012031 doi:10.1088/1755-1315/60/1/012031

Application of a micro-credit scheme to some ecological activities

F Hakoyama
International Peace Research Institute, Meiji-Gakuin University, Japan
Email : fhakoyama@hotmail.com

Abstract. Micro-credit schemes are expanding very rapidly worldwide in ecological activities. Providing gas-cooking equipments in Burkina Faso is a successful example in which the micro-credit system contributes to improve not only poor women's life but also ecological environment. In Bangladesh, a solar PV system program through micro-credit has been implemented widely and successfully: big NGOs act as equipment dealers and provide micro-credit loans to individual poor households. In contrast, there are very few cases which showed positive results in sanitation projects. Micro-credit schemes are, in principle, based on the income generated through the fund. But in usual cases, sanitation activities do not yield any income. High cost of latrine construction is another barrier. In this paper, we reviewed why we could not apply a micro-credit scheme to our "Améli-eaur project" in Burkina Faso. Common features for the success in ecological activities are 1) enough income yielded from the activity itself, 2) the strong needs from population side, and 3) established system support, both technically and administratively. If we find a way to fulfill these elements in a sanitation project, it can be a long, sustainable project.

1. Introduction

The "Ameli-Eaur Project on sustainable water and sanitation system" was a joint action/research project of Hokkaido University in Japan and 2iE (International Institute for Water and Environmental Engineering) in Burkina Faso. It started in 2009 and terminated in 2014, aiming at creation of a new recycling system of domestic gray water and human excreta as natural resources for agriculture. More precisely speaking, this project tried to invent a new technology on gray water treatment and compost toilet which can be utilized at a pilot site in a rural area in Burkina Faso. This study aimed to look for the conditions which facilitate to apply a micro-credit scheme to ecological projects as a means of ensuring sustainability.

2. Methodology

In this context, the author conducted a research on a possible micro-credit scheme which can be applicable for this new technology, so that a large number of farmers will be able to benefit from it after the termination of the project.

The author began the research in interviewing various financing groups dealing with micro-credit schemes in Burkina Faso, such as banks, ministries and NGOs/private organizations. Facing some difficulties in application of micro-credit scheme to sanitation technologies, the author looked for other examples worldwide by reviewing documents of sanitation projects, and then moved to some cases of various ecological aspects.

3. Result and discussion

3.1. Application of a micro-credit scheme to a sanitation project in Burkina Faso

3.1.1. Existing micro-credit schemes in Burkina Faso

In Burkina Faso, three types of financial organizations deal with micro-credit systems: banks, ministries and NGOs/private organizations.

Among banks, "Caisse populaire" is the biggest and most experienced in the area of micro-credit. It has a special scheme called "caisse villageoise" targeting village women, providing thus a potential for a future cooperation with our Améli-eaur project. The borrowers opened their account with the bank, and then they can get a micro-credit for their income-generating activities. The loan periods differ from 6 to 8 months (exceptional maximum is one year.) according to the nature of the activity. The interest rate is around 9.75%, and the reimbursement rate is as high as 98%.

One of the characteristics in Africa is that national ministries try to help population to undertake a small business. In Burkina Faso, the Ministry of Finance has the FAARF(Fonds d'appui aux activités rémunératrices des femmes) through which female agents address to rural women. The credit system is almost the same as the Caisse populaire: the interest rate is 10%, with monthly reimbursement, and the period is 8 to one year according to the activity. The reimbursement rate is 97%. The Ministry of Education has its special organization AME (Association de Mères Educatrices) which provides small funds to pupil's mothers through school directors.

The third type of financial organization dealing with micro-credit is NGOs, international organizations and private organizations. NGOs are functional if they are supported by some international organizations, but it seems difficult if they are only locals. Private organizations are small but active and they take action especially at market places in towns, targeting working urban women. We visited three private organizations which deal with small micro-credit schemes in the Ouagadougou market places. AGD (Action Grâce Divine) is a religion-based organization. The amount of each activity is 5,000 to 2,000,000 CFA (of which up to half can be credited), with loan period of around 4 months. Interest rate is 8%, plus 1% of insurance. The reimbursement rate is around 75%. In the case of LAAFI SIRA KWIEOGO (it means, in the local languages of mollé and djoula, "the intimacy is the way to health"), micro-credit is given to informal sector workers who are doing savings in their account of the organization. The interest rate of the credit is 13% per year for social credit, and 15% per year for commercial credit. FINACOM (Finance communautaire) is also working actively for small female entrepreneurs.

3.1.2. Application of micro-credit to sanitation projects in Burkina Faso

Micro-credit system seems to work well in many activities in Burkina Faso. However in sanitation projects there were only very few examples, moreover none of them showed positive results. The reasons for this failure is the same as the cases described in 2.2.1. In almost all the sanitation projects of international organizations, toilet construction is financed by donors, and no scheme was foreseen after its termination. Only one micro-credit example was found with Caisse Populaire. It was realized in collaboration with World Bank and an international NGO: CREPA (Centre Régional pour l'Eau Potable et l'Assainissement à faible coût), but when the World Bank project terminated (that means, financial input finished), reimbursements from the farmers were not expected high enough to continue the micro-credit scheme. So it was in failure. Another sanitation project which we visited in Koupela was supported by CREPA. It did not adopt a micro-credit scheme, and therefore not expandable after the termination of the project.

The case of our "Améli-eaur" project is not the exception, and we could not even start a micro-credit scheme: the reasons are many: 1) the construction cost (although it is still under improvement) was far beyond the limit of small farmers economical capacity. (It was calculated for 100,000 -500,000 CFA/toilet) 2) The agricultural benefit expected from the use of compost and purified gray water was too low to cover the investment. 3) A strong support system did not exist technically nor administratively (a support system was planned but not realized before the end of the Améli-eaur project). And 4) the people did not feel so much impact of using sanitation devices comparing to big financial expenses (as shown in the results of interviews of pilot families: "Report of the Sociological Survey" conducted from March 11 to 14, 2014, at Améli-Eaur project pilot sites).

IOP Conf. Series: Earth and Environmental Science **60** (2017) 012031 doi:10.1088/1755-1315/60/1/012031

3.2. Review worlwide

3.2.1. Application of micro-credit to sanitation systems worldwide

After the interviews and research in Burkina Faso, we reviewed documents worldwide, but the results were the same. "La question du recouvrement des fonds est plus problématique. Nombreux sont des projets où le recouvrement des sommes prêtées a éte nul ou faible." (Arnaud Assié: "Le micro-crédit pour l'accès à l'assainissement autonome", Revue documentaire d'expériences dans les pays en développement) Micro-credit schemes are, in principle, based on the income generated through the fund. But the borrowers of sanitation micro-credit have to reimburse the loan through their own savings. High cost of latrine construction is another barrier: poor families have often difficulties in reimbursement, especially when there is some natural calamity such as drought or flood. In usual micro-credit projects the initial capital is used as revolving funds. In the case of sanitation projects, the reimbursement rates are too low for such a revolving system to work properly. Moreover running costs such as excreta disposal have not necessarily been considered. This also affects the sustainability. As a result, micro-credit projects in sanitation are in many cases not sustainable, and expansion of latrine construction after termination of donor inputs cannot happen.

3.2.2. *Micro-credit in ecological activities*

In contrast, in the area of ecology, micro-credit schemes are expanding very rapidly worldwide. Providing gas-cooking equipments in Burkina Faso is a successful example in which the micro-credit system contributes to improve not only poor women's life but also ecological environment. The borrowers reimburse their debt from the difference of expense between gas cost and eventual purchase of woods or carbon. For example according to AGD scheme, for 28,000 CFA of a gas bottle cost, the borrower pays 10,000 CFA at first, and 6,000 CFA per month for three months. The eventual purchase cost of woods is estimated 9,000 CFA/m, while the expense for gas is 3,000 CFA/m. So the borrower can pay the difference of 6,000 CFA/m. Another example is LAAFI SIRA KUIEOGO: since 2008, this organization lends 25,000 CFA for a bottle of gas, with daily reimbursement, small by small, and the loan period is flexible according to borrowers' situation. They can thus get a high reimbursement rate at the end.

In Bangladesh, a solar PV system program through micro-credit has been implemented widely and successfully: big NGOs act as equipment dealers and provide micro-credit loans to individual poor households. Up to January 2013 over 1 million systems have been installed with an average loan collection rate of 96%. (Green Growth Best Practice (GGBP) report: "Microfinance for Solar Home Systems") For example, starting with PV systems, Grameen Shakti (GS) is offering 24-month leasing financing to consumers to spread out the initial equipment cost. In parallel, GS is training a network of local retailer-technicians to provide on-going maintenance and customer support. (FAO report: "Micro-finance for Rural Energy Using PV Systems")

In Morocco, the NGO: Migrations et Developpement has helped electrify nearly 100 villages in a remote mountainous area through diesel-based mini-grids. Local operators provide electricity to a village cooperative on a fee-for-service basis. The NGO helps the villagers form an association to own, manage and finance the electrical systems. (FAO report: "Micro-finance for Rural Energy Using PV Systems")

In Sri Lanka, the Sarvodaya Economic Enterprises Development Services (SEEDS) is the most financially self sufficient microfinance institution within the NGO sector. SEEDS has embarked on programme to support rural electrification with the emphasis being directed to home solar and micro-hydro units. The programme of support operates under the World Bank RERED Programme. A minimum down payment of 15% is required. The balance is granted as a loan and repayments are made over 1-4 years. A field officer calls each month to collect the agreed repayment. This comprehensive checking on each loan together with frequent direct contact has led to a very low level of default (no more than 2%). (University of Greenwich report: "Micro-finance and Credit in the Development of Renewable Energy Sector")

In India, the Ministry of Non-Conventional Energy Sources (MNES) has supported establishment of first two demonstration biomass projects of size 6 MW each in the year 1998 & 2000 by extending capital grant to the extent of 15% of the capital cost. Private entrepreneurs were encouraged to set up further biomass power projects through suitable interest subsidies. (University of Greenwich report: "Micro-finance and Credit in the Development of Renewable Energy Sector")

3.3. List of Interviews on micro-credit system in Burkina Faso

- AME (Association de Mères Educatrices), Ministere de l'Education, Direction nationale de la promotion d'éducation des filles, at Ouagadougou, on 20 March 2012
- Caisse Populaire, at Ziniaré, on 22 March 2012
- Association pour la promotion et l'émancipation des femmes, at Ziniaré, on 22 March 2012
- FAARF (Fonds d'appui aux activités rémunératrices des femmes), Ministère de Finance, at Ziniaré, on 22 March 2012
- AGD (Action Grâce Divine)/VIME at Ouagadougou, on 27 March 2012
- LAAFI SIRA KWIEOGO ("the intimacy is the way to health" in the local languages of mollé and djoula), at Ouagadougou, on 27 March 2012
- FINACOM (Finance communautaire), at Ouagadougou, on 27 March 2012
- Caisse Populaire, at Bobodjulasso, on 28 March 2013
- AME (Association de Mères Educatrices), Direction provinciale de l'éducation de base et de l'alphabétisation, at Bobodjulasso, on 29 mars 2013
- FAARF (Fonds d'appui aux activités rémunératrices des femmes), Ministère de Finance, at Bobodjulasso, on 29 March 2013
- Association des femmes, Cooperative Zoor Tanga, at Manga, 3rd April 2013
- Caisse populatire, at Ouagadougou, on 5 April 2013
- Fédération Caisse Populaire, at Ouagadougou, on 13 August 2013

4. Conclusion

In Burkina Faso, our team has renounced a plan to apply a micro-credit scheme in the Améli-eaur sanitation project, due to very high cost of latrine production, very low financial capacity of rural households, lack of business chain of compost/agricultural products sales, and insufficient keenness from users to install sanitation devices in overcoming big financial barriers.

But more widely speaking, in ecological activities there are a plenty of success stories both in Burkina Faso and worldwide. Common features for the success are 1) enough income yielded from the activity itself, 2) the strong needs from population side, and 3) established system support, both technically and administratively. If we find a way to fulfill these elements in a sanitation project, it will have a big chance to be a long, sustainable project.

Acknowledgement

We would like to express our sincere thanks to Prof. Boureima Oueodraogo of University of Ouagadougou, and his graduate student, Ms. Coulibaly Konotoumou. Without their contribution, we could not undertake our interviews with private financial organizations at the Ouagadougou market places. We also express our gratitude to Mr. Arnaud Assié, who kindly sent to us his report, which helped us greatly in progressing our research.

References

[1] Arnaud Assié Le micro-crédit pour l'accès à l'assainissement autonome Revue documentaire d'expériences dans les pays en développement Août 2013 Document de travail n° 3 Edition 2013 Fondation pour l'agriculture et la ruralité dans le monde.

[2] University of Greenwich, Natural Resources Institute: "Micro-finance and Credit in the Development of Renewable Energy Sector", http://projects.nri.org/biomass/conference_papers/the_role_of_credit_in_the_development_of_renewable_energy.pdf
[3] Green Growth Best Practice (GGBP): "Microfinance for Solar Home Systems" . http://www.ggbp.org/case-studies/bangladesh/microfinance-solar-home-systems
[4] FAO corporate document repository
[5] Hokkaido University and 2iE: Report of the Sociological Survey at Améli-Eaur project pilot sites, conducted from March 11 to 14, 2014.

IOP Conf. Series: Earth and Environmental Science **60** (2017) 012032 doi:10.1088/1755-1315/60/1/012032

Development of sanitation technologies in African context : how could we make it more sustainable?

M S Dakouré*, M B Traoré, S K Sossou and A H Maïga

International Institute for Water and Environmental Engineering (2iE), 01 Rue de la science, P.O. Box 594, Ouagadougou 01, Burkina Faso

*Email : mariam.sou@2ie-edu.org

Abstract. Access to sanitation technologies remains one of the biggest challenges in sub-Saharan Africa. To overcome this gap, a sanitation project called "Ameli-EAUR" translated from French as improvement of water and sanitation in urban and rural areas, was implemented in Burkina Faso for 5 years (2010-2016). The technologies from the project were designed on the basis of agro-sanitation concept, leading to package containing a composting toilet, a grey water treatment facility and a set of urine collection and treatment. The study aimed to evaluate of Ameli-EAUR project, one year after the end, and identify some key factors of sustainability of technologies. As methodology, a survey and a technical diagnostic of implemented technologies were done. The results showed that, the pilot families stopped using all the technologies one year after the end of the project. However, two main lessons can be learnt: (1) in term of efficiency and effectiveness of the project the technology of composting toilet was not robust enough, leading to a rapid abandonment after the project (2) in term of impact and sustainability, the economic incentive of the resource oriented sanitation concept was very weak compared to the needed workload. The technologies development in this kind of project should be carried on and associated with a more inclusive system driven by economic incentive.

1. Introduction

The lack of adequate sanitary installations and proper sanitary waste disposal systems leads to serious implications on human and environment health [1]. Sahelian countries are still among those where sanitation access is very limited [2]. The target 7c of United Nations millennium development goals (MDG) was to reduce by halve people without access to potable water and basic sanitation by 2015 [3]. In 2016, none of the Sahelian countries reached this target, leading to a big challenge to fill the gap as we start the Sustainable Development Goals (SDG) which target a universal access to water and sanitation [4]. Burkina Faso is not left out from this situation, this typical Sahelian country located in western Africa still face open defecation by the majority of its rural communities [5].

The sanitation challenge makes sanitation projects very relevant, in order to increase the number of adequate sanitation facilities. Several projects were initiated during the two past decades in Burkina Faso. Ecological Sanitation (ECOSAN) project is one of the biggest one which was initiated in 2006 for 3 years and aimed to develop onsite sanitation technologies in Ouagadougou, the capital city of Burkina Faso including by-product reuse in agriculture. Among the 5 criteria commonly used to assess such kind of developmental project, that is relevance, effectiveness, efficiency, impact and sustainability, the mid-term evaluation highlighted a weak sustainability for two reasons (1) the strategy of the project to cover the major part of investment expenses and (2) the fact that incomes generated per month by the sanitation system represent only 9% of the total operating expenses [6].

The Ameli-EAUR project on the other hand as described in the present paper was launched in 2010 for 5 years in 2 villages of Burkina Faso. The project aimed to develop onsite low cost and low maintenance system including waste separation and treatment and reuse of sanitation by-products (urine, faeces based compost and grey water) in agriculture. The final evaluation concluded with a mixed results. Indeed, the project relevance was not doubted because water and sanitation was a priority both for Government of Burkina Faso and The Japan International Cooperation Agency which

IOP Conf. Series: Earth and Environmental Science **60** (2017) 012032 doi:10.1088/1755-1315/60/1/012032

financed the project. However, the efficiency and effectiveness of the project received a mixed opinion mainly due to a lack of maintenance system of facilities. Therefore, impact and sustainability were also judged as not reached.

2. Methodology

2.1 Presentation of Ameli-EAUR project

The project is named Ameli-EAUR project, the term “Ameli-EAUR” is a french acronym defined as "Improving Sustainable Water and Sanitation Systems in Sahel Region in Africa: Case of Burkina Faso". This project is an international development joint project on water, sanitation and hygiene between mainly the International Institute for Water and Environmental Engineering (2iE, Burkina Faso) and Hokkaido University (Japan) and funded by Japan International Cooperation Agency (JICA), Burkina Faso’s Government and Japan Science and Technology Agency (JST). The project, started in 2009 and ended in 2014 (5 years), aimed at proposing a sustainable agro-sanitation model in Burkina Faso based on recycling households waste (grey water, urine and human excreta) as natural resources for agriculture.

2.2. Description of the sanitation technologies

2.2.1. Composting toilet

The composting toilet is an urine diverting type toilet. A reactor (of composting) and a urine collection tank are placed in a hole and covered by a platform where urine bowl, faeces hole and washing bowl were designed (Figure1). The reactor must be rotated after defecation to provide oxygen and mix faeces with matrices (i.e. sawdust). So a mixing system connected to the reactor is available on the platform. One of the most important improvements of the toilet was this mixing system. It has been changed for each version to enhance its robustness (Figure 2). The other main modification was the reactor. Previously wood was used but has been replaced in version 3 by metal matter to ovoid its rapid degradation (Figure 3). In addition, pilot families were trained on the inactivation and using of the compost.

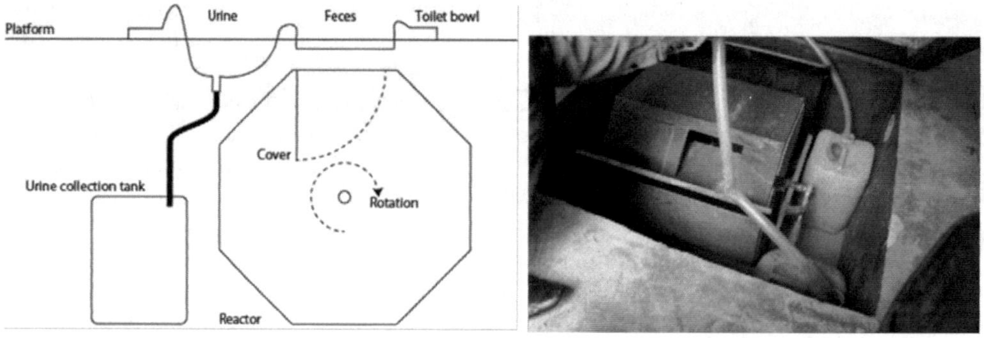

Figure 1. Composting toilet used in rural Burkina Faso (design and picture) [8].

First version Second version Third version

Figure 2. Composting toilet mixing system used in rural Burkina Faso) [8].

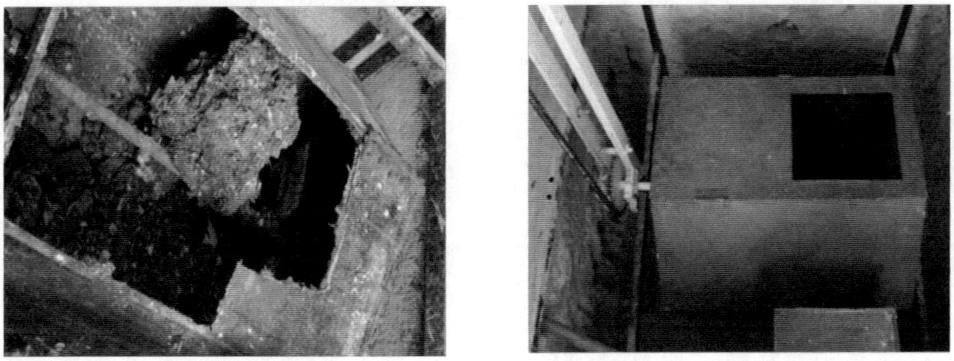

Figure 3. Composting toilet reactor made by wood and metal used in rural Burkina Faso [8].

2.2.2. Urine treatment set

The urine is collected in toilet as previously mentioned in the composting toilet description, but also in shower room where urinary bowl has been set to ovoid presence of urine in shower water. Figure 4 shows urine collection, packaging and treatment following the training pilot families received.

Figure 4. Collection of urine in PET bottles of 1.5 litres and exposition on roof during 7 -14 days in rural Burkina Faso.

2.2.3. *Grey water treatment facility*

Figure 5 presents the two versions of grey water treatment facility developed during the project. The system was made of an influent tank, an upper filter, a lower filter and a buried collection tank. The influent tank receives grey water from different activities like dishwashing, laundry etc. From version 1 to version 2, the influent tank was improved by adding a screen at the top to retain some solid waste (i.e. food waste). The upper filter is filled with gravel (1-9 mm). It receives grey water both from influent tank and shower room. The facility is placed just behind the shower room so that it can be directly connected by a pipe to collect the shower water. In version 2, the size of gravel in upper filter has been decreased (1-4 mm) to improve filtration while reducing the clogging time. The weight of the filter was also reduced to allow women to carry it during maintenance operations. The lower filter also filled with gravel (1-9 mm in version 1; 1-6 mm in version 2) receives grey water from the upper filter and discharge the treated grey water in the buried tank. Grey water is taken from the buried tank in a similar manner to a shallow well for irrigation purposes. The quality of treated grey water met WHO standard [8] for restricted irrigation but not allowed to be eaten as raw vegetable. This quality is due to grey water contamination by babies' bath water discharged in the influent tank.

Figure 5. Grey water treatment facility (first and second version) used in rural Burkina Faso.

2.2.4 *Household garden and by-product reuse*

The best urine to compost ratio was identified and published by Sangaré et al. [9]. The pilot families were trained based on this data. Each family had a small garden about 20 m² (Figure 6) near the grey water facility. Treated urine and water have been tested through cultivation of several crops (except for raw vegetables that are to be eaten) in accordance with the interest of families. Demonstration of compost application was limited to training with guidelines because the post-treatment developed to improve the microbial quality was too time consuming.

Figure 6. Household garden developed by a trained family in rural Burkina Faso.

2.3. Description of pilot families and sanitation technologies

Two pilot sites were identified at the beginning of the project. The first one was in rural area, within two villages near Ziniare, a small city situated at about 50 km from Ouagadougou. In each village, a preliminary survey was performed before the Chief of each village selected the pilot families. The survey helped to define some socio-cultural criteria to be taken into account in the final selection. Table 1 displays information about the pilot families. All pilot families received a package of sanitation technologies containing: a composting toilet, a grey water treatment facility with new or renovated shower room, a set of urine treatment. Family 3 (Table 1) received 3 packages due to the high number of residents, three brothers were living there with their wives and children.

Table 1. Description of pilot families in the two villages near Ziniare, Burkina Faso.

Location	Household	Number of person	Religion and occupation	Sanitation facilities before the project
Kologodjesse	Family 1	5	Christian Farmers	1 traditional toilet cement blocks + 1shower with local soil
	Family 2	18	Christian Farmers	3 shower rooms with local soil (1 was partly broken)
Barkoundba	Family 3	20	Muslim Breeders	1 toilet with collapsed wall + 1 collapse shower room
	Family 4	8	Muslim Breeders	1 toilet in local blocks (used as shower room because the shower room was collapsed

2.4. Survey method

A questionnaire was administrated to the pilot families of Kologodjesse. The survey was limited to this village because it is the only village where all facilities were working and employed at the end of the project. The questionnaire aimed to evaluate the level of satisfaction related to the use of the different facilities as well as the use of sanitation by-products i.e. treated urine, treated water and compost. The level of satisfaction is denoted in Table 2 with an "X" during the survey.

Table 2. Satisfaction questionnaire administered to pilot families of Kologodjesse, Burkina Faso.

Family Name	1	2	3	4	5	6
Composting toilet using & maintenance						
Shower room using						
Grey water facility maintenance						
Urine collection and treatment						
Reuse of treated urine in the garden						
Reuse of treated grey water in the garden						
Reuse of compost in the garden						
1. Unused, facility broken						
2. Unused, lack of interest						
3. Used, but not satisfied						
4. Used with mitigate satisfaction						
5. Used and satisfied						
6. Used and very satisfied						

2.5. Technical diagnostic of implemented sanitation facilities

The survey was followed by visiting the implemented sanitation facilities. The various components of each facility were checked for their well functionality as well as for any observations that will prevent their proper functionality.

3. Results

3.1. Result of the survey

Figure 7 summarize results of the satisfaction survey performed with pilot families of the Kologodjesse Village following the Ameli-EAUR project. As described in the legend of Table 2, the higher is the score is, the better is the satisfaction. For the shower room, the high score of 6 means that users appreciated the facility in term of easiness both for using and maintain. In contrast, urine treatment and reuse as well as composting toilets and compost reuse have very low scores (between 1 and 2). Such score reveal an abandonment of the facility using and/or the by-products reuse practice. The results show that the main reasons facilities remained unused was due to breakdown whereas none use of by-products was due to lack of interest.

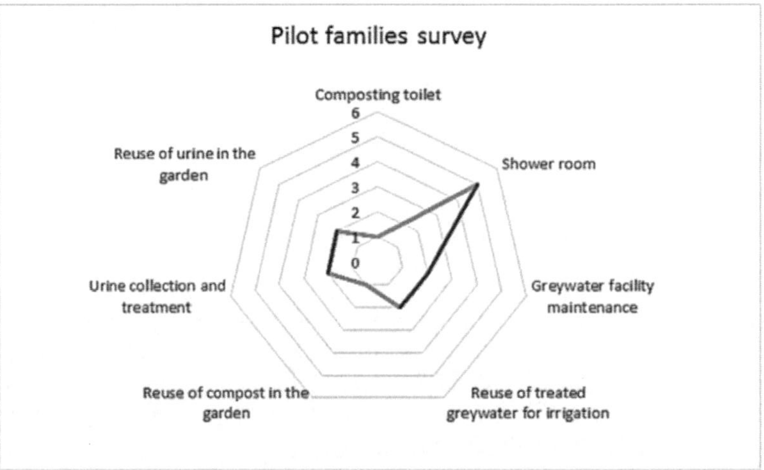

Figure 7. Results of pilot families' survey one year after the end of the project in Kologodjesse Village, Burkina Faso.

The mixing systems of composting toilets were all broken. During the project, this issue was frequent and several modifications were necessary to improve the crank. Lack of continued system maintenance after the project leads to the abandonment of broken facilities. It was learned through discussion with pilot families, that they are still waiting for project members to repair the toilet despite the agreement they signed to maintain all facilities after the end of the project. They justified the situation by lack of money to repair the crank but also by the fact that local trained technician to repair the toilet is too fair from the village. The filter media of grey water facilities were very clean in all pilot families. However, the buried tanks were full of grey water, implying that it was not used for gardening. They were no crops growing in the garden. By discussing with families, we realized that all filter media had been washed just before our visit. This highlights the progressive abandonment of the system. Some families attributed this to the fact that the garden is not well protected against livestock who systematically destroyed the cultivated vegetables. Urine is collected from the shower room urinary bowl but it is directly discharged into their traditional manure pit.

3.2. Technical diagnostic of implemented sanitation facilities
The technical diagnostic of implemented sanitation facilities showed that some components were not properly functional. At the shower, there is an accumulation of dirt on the cap of the urinal and piping; the presence of a lot of debris on the gravel; the tear of the filter and the drum cover is lost. The composting toilet is not functional and has been neglected in favour of the toilet which existed before the project because the reactor is no longer running when you move the handle. At the household garden, the family does not practice vegetable growing because of wandering of animals. Indeed, despite the protective grille, chickens, goats and pigs, raised in semi-open-air conditions, frequently penetrate inside the garden place to eat the vegetables. Pilot families did not find a way to avoid these intrusions so they finally abandoned the gardening.

3.3. Discussion
The survey highlighted a progressive abandonment of sanitation facilities one year after the end of the project. The main reasons of such abandonment are facilities breakdown and progressive lack of interest both to maintain/repair these facilities associated with a total disinterest for by-products reuse in agriculture. However, pilot families of Kologodjesse Village seem to have appreciated all the technologies and had expressed a very good commitment to carry on the by-product reuse. But as mentioned by Shove [10], technology appropriation in developing project is usually oversimplified by considering only the lack of correct technical device as the main cause of appropriation failure. The author mentioned the necessity to include several other constraints like markets, economic incentives and credits. Ushijima et al. [11] completed this analysis by proposing a post-modern sanitation concept, more inclusive than a simple development and implementation of technologies. This concept has particularly proposed to develop sanitation value chain driven by a business model and supported by innovative financial system.

4. Conclusion
To the question "what do you need for your own development?" people never ask for a toilet. Their priority are borehole for drinking water, school, health center etc. sanitation facilities are not among their priority. In the present project, the relevance to develop sanitation facilities for rural people in Sahelian country was expressed by the Government of Burkina Faso through its program for water and sanitation access. However, after 5 years of implementation followed by one year after the end of the project, two main lessons can be learnt: (1) in term of efficiency and effectiveness of the project the technology of composting toilet was not robust enough, leading to a rapid abandonment after the project (2) in term of impact and sustainability, the economic incentive of the resource oriented sanitation concept was very weak compared to the needed workload. So people progressively abandon the by-product reuse. In the perspective of further investigation, the technologies development should be carried on and associated with a more inclusive system driven by economic incentive

Acknowledgement
The authors would like to express their sincere thanks to Bachelor students of Water and Sanitation Department of the International Institute for Water and Environmental Engineering (2iE). Without their contribution, we could not use their interviews and technical diagnostic of implemented sanitation technologies.

References
[1] Owusu E and Adjibolosoo S K 2016. Int. *J. Curr. Res. Med. Sci.* **2(6)**, 8-18.
[2] Alagidede P and Alagidede A N 2016 *Public health.* **130**, 59-63.
[3] WHO and UNICEF 2012 Geneva: WHO Press. Available from: www.wssinfo.org
[4] WHO and UNICEF 2015 Geneva: WHO Press. Available from: www.wssinfo.org
[5] Galan D I, Kim S S and Graham J P 2013 *BMC public health*, **13(1)**, 1.
[6] CREPA 2010 Evaluation report www.washplus.org/sites/default/files/burkina_faso2010.pdf

[7] WHO 2006 *World Health Organization* (Ed.),Vol. 4, Geneva, Switzerland.
[8] Ito R, Ushijima K, Hijikata N and Funamizu N 2012 *Proceedings of 4th dry toilet conference, Tampere*, Finland.
[9] Sangare D, Sou/Dakoure M, Hijikata N, Lahmar R, Yacouba H, Coulibaly L and Funamizu N 2015. *Environ. Technol.* **36(10)**, 1291-1298.
[10] Shove E 2003 *The European J. of Soc. Scie. Res.* **16(2)**, 193-206.
[11] Ushijima K, Funamizu N, Nabeshima T, Hijikata N, Ito R, Sou M and Sintawardani N 2015 The *Water Policy*. **17(2)**, 283-298.

IOP Conf. Series: Earth and Environmental Science **60** (2017) 012033 doi:10.1088/1755-1315/60/1/012033

Ameli-EAUR project: which lessons for the promotion of agricultural value chains?

M B Traoré*, M S Dakouré and A H Maïga
International Institute for Water and Environmental Engineering (2iE), Burkina Faso
*Email : maimouna.bologo@2ie-edu.org

Abstract. Due to the lack of rain, dry spell and agricultural insufficient inputs, agricultural yields are very low and do not allow achieving food self-sufficiency in countries such as Burkina Faso. To fight against this situation in the context of climate change, several researches have emerged such as agronomic research with the promotion of crop varieties that require little water and short growth period. The Ameli-EAUR project is another example of research. This project aimed to promote hygiene and sanitation by the use of sanitation by-product in rural and sub-urban areas of Burkina Faso. Ameli-EAUR project was initiated and implemented within five years. Completed in 2015, it appears necessary to think over its impacts on production hence the question: which lessons for the promotion of agriculture, specifically for agricultural value chains"? Based on the result of a post-survey performed one year after the end of the project, the aim of this presentation is to show that project can support the promotion of agricultural Value Chains in spite of difficulties if we take into account some parameters. To do this, we will start by presenting briefly Ameli-EAUR; then we will take stock of the project; finally, we will examine the possibilities it offers for the promotion of agricultural value chains, taking as example the vegetable gardening. We will also try to show the limits of value chains which are presented nowadays as a solution to many problems of agriculture in Africa.

1. Introduction

The Ameli-EAUR project aims to recycle organic waste, and sanitation by-products such as feces, urine and grey water from showers, laundry and cooking. For that purpose, eight rural and suburban families were selected from 3 pilot villages of Barkoundba, Kolongdjessé and Kamboinsé around Ouagadougou, the capital city of Burkina Faso, to test a sanitation system that permits the use of recycled feces and urine in agriculture. However, the modes of provision and use of organic materials, which were applied in the project, occurred in a context where these wastes are generally considered repulsive or synonymous with taboo and contamination by people [1].

During five years, 8 families tested Ameli-EAUR infrastructures such as, bathrooms, toilets and little gardens. Our presentation will focus on the village of Kolongdjessé where post-survey has been realized one year after the completion of the project.

Before presenting the results of the surveys did by students, what is Value Chains?

“A value chain is a system of people, organizations and activities needed to create, process and deliver a product or service from input supplier to customer. Actors in the chains include primary input suppliers, producers, processors, packers, traders, distributors and service providers. They transform natural resources, raw materials and components into a finished product that is delivered to the end consumer. Value chains include the horizontal and vertical linkages between several actors that generate value in a product or service. They are interrelationship-driven and focus on governance and resilience while describing activities and actors related to bring a specific product to the market. Value Chain Governance is about the local mechanisms and power used to define or impose transaction parameters between the different actors. Value chains can present uneven power relations, barriers to entry or exclusion. “Good” value chain governance will strike a balance between a fair competitive advantage and an unfair dominance of lead firms.

Globalization and economic liberalization has exposed local farmers to new markets. Value chains show how smallholders and rural enterprises are linked to and are economically dependent on this wider environment. While markets could be a path out of poverty for many farmers, their potential is often left untapped [2]. The study aimed to show that project can support the promotion of agricultural Value Chains in spite of difficulties if we take into account some parameters. In our purpose, we don't focus on an example of Value chains. We insist that ecological sanitation can be a link in the agricultural value chain.

2. Methodology

In February 2016, we made a field trip with our third years' students in order to show them effectively what ecological Sanitation is. This field trip was focused on Kolongdjessé which is one pilot site of the Ameli-EAUR Project. Before the field trip, we explained to the students the aims of Ameli-EAUR projects; what was expected from the sociological section and how we collected data (surveys with questionnaire forms; Group focus, direct observation); and the relevant conclusions we attained. We also helped students to make questionnaire forms with engineering (type of technology) and sociological (behaviors of families) aspects. For a better consideration of the four pilot families, students were divided into four groups, with the goal for each one to administer a questionnaire to the head of the household and fill out an observation grid on the state of maintenance of the project's.

3. Result and discussion

One year after the Ameli-EAUR project, the assessment that we can make is as follows: The system of Composting toilet is broken; the mechanism to turn feces is not efficient. Families do not put urine systematically in bottle for sanitation process due to lack of bottles. The collected urine was thrown in a manure pit a traditional compost wall. Grey water collected during dry season was used to make bricks. This is another example of appropriating the AmeliEAUR project. The protection gloves and mask used by families to manipulate urine and feces were obsolete. The sustainability of sanitation practices and protection of people handling sanitation by-products threaten the sustainability of the infrastructure of Ameli-EAUR.

Figure 1. Area developed for Culture with crop during rainy season (Tomohiro GONDAIRA July 8, 2014).

Figure 2. Area developed for Culture without any crop, (Group 1, February 11, 2016).

Moreover, around houses, development of sanitation by-products by households proves difficult during the dry season because of wandering animals, the harsh climate and low water availability. Indeed, fencing of areas developed for the garden are not robust and during the dry season farm animals live outdoors out of the enclosure. The consequence of this situation is that no garden can survive once these animals in search for pasture pass by. That is why nobody maintained garden during dry season. Because there was no good protection for the garden, it is not possible to follow a good process for plants. On the other hand, there are possibilities of selling the yellow waters composed of urine contained in vegetable garden. Actually besides food crops that are cultivated in rainy seasons, most pilot families practice vegetable garden during dry seasons. Instead of chemical fertilizers used in the framework of their gardeners, these families could enhance the urine. That is, if the urine valuation is problematic around houses during the dry season, this difficulty would tend to fade at the market garden perimeters. This potential for reuse could lead to a systematic collection of urine, their sanitation, their direct valorization by households who practice gardening and / or marketing to gardeners. The marketing of urine will provide money for household and give it opportunity to maintain and/or renewal Ameli-EAUR infrastructure.

The assessment that can be drawn from the AmeliEAUR project is that, despite the difficulties regarding the operation of certain infrastructures set up by the project, pilot families tried to appropriate the technology through the use and maintenance of still functional equipment (shower), a collection of gray water and urine collection, part of which is put in manure pits waiting to be used as fertilizer during the winter campaign manure. This diversion function is a sign of technology ownership. However, to take into account the initially goal set to AmeliEAUR project , namely the use of by-products and in order to make a link with the green growth through the provision of agricultural inputs, it would be appropriate to develop grants to fund the construction and maintenance of sanitation facilities.

As part of the Value chain approach, the valuation of sanitation by-products as the urine reduced the costs of agricultural inputs while promoting productivity growth [3]. These solutions contribute to cleaner living environment for households and promote green growth. However, to meet these goals, it is important that certain limits of the value chains approach be taken into account

4. Conclusion

In conclusion, making of ecological sanitation a link in the agricultural value chain can be possible if incentives mechanisms such as subsidies to the adoption of ecological sanitation are planned. In a context where the primary concern is the satisfaction of basic needs such as food, ecological sanitation appears as an epiphenomenon. The opportunities it offers cannot be taken into account if a strong incentive based on:

- Highlighting the economic benefits related to technology;
- Mitigating the constraints linked to the nature of the materials to support (the repugnant nature of sanitation by-products)
- And alleviating or eliminating health risks related to the handling of sanitation by-products;
- Revising sanitation strategies and priorities and implement innovative funding mechanisms.

Acknowledgement

The Authors are grateful to JICA for AmeliEAUR Project funding. Our acknowledgements are addressed to Kolongdjessé families for their cooperation.

References

[1] Bologo/Traoré, M., Sou/Dakouré M., Maïga A.H., Nabeshima T., Hakoyama F., Dicko S. 2014 "Mobilizing Communities for use of by-products of Sanitation: Case study of the Ameli EAUR project", *9th* IWA International Symposium on Waste Management Problems in Agro- Industries 24-26.

[2] P Visser and al, 2011-2012 Value Chain for SNV Advisors
[3] D. Sangare, M. Sou Dakoure, N. Hijikata, R. Lahmar, H. Yacouba, L. Coulibaly & N. Funamizu, 2015 Toilet compost and human urine used in agriculture: fertilizer value assessment and effect on cultivated soil properties, Environmental Technology, DOI: 10.1080/09593330.2014.984774

Sanitation and income improvement by local community as sustainable participatory development

M Ikemi
Graduate School of Economics and Business Administration, Hokkaido University, Kita7 Nishi9, Kita-ku, Sapporo, Hokkaido 0600809, Japan
E-mail: mayu.ikemi@econ.hokudai.ac.jp

Abstract. Most people in low-income areas such as rural village in Africa and urban slum in Southeast Asia live with limited economic resources and poor sanitation conditions. In order to deal with the situation, many of the people have formed community-based organizations and joined the organization activities for the common purpose of improving these conditions as participatory development. This study attempts to examine and evaluate the impact of the activities in their respective local communities. From the case study in two villages of rural Senegal, self-help effort is considered essential to solve the people's income and hygiene problems through their community participation for sustainable development. For the implementation, the organizational solidarity, adequate water supply and water management are crucially needed. It is suggested to encourage community-based organizations effectively working for income improvement to also consider practicing for sanitation improvement.

1. Introduction

In general, many of low-income households in developing countries and emerging countries are settled in unstable economic conditions. Many of them are also not able to break away from the vicious circle of poverty: that is, low incomes force to keep low savings, low savings tend toward low investments, low investments result into low labor productivity, and low labor productivity results back to low incomes [1]. In addition, they are also typically living in poor sanitary environment. As the main reason, public hygiene control and management such as consistent water supply, sewage treatment and waste disposal are not adequately provided in their living areas.

However, many of the people in the spiral are making their self-help efforts for a better living environment in their local community on a daily basis. For example, most people in low-income areas such as rural village in Africa and urban slum in Southeast Asia form community-based organizations and participate in organization activities for their common purposes. The various purposes include income improvement, promoting economic activities, health maintenance, enhancement of child care and education, and even solving their personal problems [2,3,4]. These efforts can also be regarded as so-called participatory development or community initiative, which means a development implemented through participation of local residents. The participants incorporate the characteristics of their life environment, culture, customs, knowledge and wisdom into their development practice [5,6,7,8]. Learning

IOP Conf. Series: Earth and Environmental Science **60** (2017) 012034 doi:10.1088/1755-1315/60/1/012034

from the experiences and problems they face will provide us with a thought-provoking suggestion about the possibility to enable the acquisition of better life for people living in low-income areas in developing and emerging countries.

Based on the above background, the author has been conducting field research in rural area of Senegal and Burkina Faso, West Africa, and in urban slum of Indonesia, Southeast Asia. This paper focuses on the results of the field survey conducted in Senegal, and provides a case study of community-based organizations in the rural area. This case study seeks to provide a better understanding with respect to the effects and challenges encountered through the organizations' activities by the local residents. It is also considered to be a suitable way of participatory development for the low-income people in achieving their autonomous income improvement and sanitation management practice in their respective local communities.

2. Methodology

This study is based on field research in rural area of Senegal. The field surveys were conducted in two villages, named Fissel Deux and Tewrou Mbéyéne, located in Ouadiour county, in Fatick region. The surveys were carried out intermittently: July 2003-July 2005, February-March 2007, November 2009, and March 2016 respectively. The methods are interview and participant observation targeting adults aged 20 years or older who are the residents and members of community-based organizations in their villages. The contents of the surveys are about their standard of living, economic activities, infrastructure, sanitary condition, and group activities of the community-based organizations, including the awareness and evaluation of the residents about these matters. This paper deals with sanitation with particular focus on water supply and management.

3. Result and Discussion

3.1 Livestock breeding with microcredit in the village of Fissel Deux

The village of Fissel Deux is located about 4 km away from the national route, in rural area of Ouadiour county, in Fatick region. It belongs to the rural community of Patar Lia and this community was considered as the lowest standard of living and economic level among the total four communities in the county from the result of field research [9]. The population was 475 (232 men and 243 women) as of 2006. There is a community-based organization called Groupement d'Intérêt Économique (GIE) that works as an economic activity group. The GIE is composed of 36 members, 4 men and 32 women. In this village, livestock breeding project to improve income generation was carried out by the GIE with the support of Japan International Cooperation Agency (JICA) in July 2004 – July 2005. The Livestock breeding as a means of generating income was chosen by the GIE members and other villagers themselves after their group discussions and need investigation with a JICA volunteer. In the process of their practice for the project, firstly the residents decided the rules about the purchase, breeding and sale of livestock, specifically five cows on an autonomous basis. The 80% of the activity funds were contributed by a donation from Japanese general public, and the rest of 20% was funded by the GIE residents themselves which was loaned from JICA.

In this project, first, the GIE purchased five cows for a sum total of 500,000 CFA (about 1,000 USD), and in addition, spent 280,000 CFA (about 560 USD) for the tools and breeding. Thus a total of 780,000 CFA (about 1,560 USD) was spent at the start. After that, they bred the five cows for three months and sold them at a total price of 925,000 CFA (about 1,850 USD). They got a net profit of 120,000 CFA (240 USD) after deducting such extra expenses as transportation cost. The 20% part of the budget contributed by the GIE was made available through a microcredit scheme. In detail, the GIE received a loan of 160,000 CFA (about 320 USD) from JICA at the beginning, then every month they paid 500 CFA (about 1 USD) per each of total 36 members for 9 months. The repayment amount eventually was given as: 500

CFA × 36 members × 9 months = 162,000 CFA (about 324 USD), which amount was paid back to JICA. In addition, an interest of 10%, equivalent to 16,000 CFA (about 32 USD) on the amount was paid to JICA from the benefit of 120,000 CFA after selling the cows. This money that JICA received was used for the volunteer's work in support of the GIE activity in the village.

One and a half years later after the project completion, the first follow-up field work was conducted in February 2007. For the GIE in the village, the activity of livestock breeding project was not continued. However, it was confirmed that non-project-members in the village had imitated the project, and begun livestock breeding by themselves. Furthermore, about four years later after the project completion, the second follow-up field work was conducted in November 2009. The GIE had resumed livestock breeding project, and they had continued the project on a voluntary basis without original funders and any support by outsiders such as JICA.

From the above case, certain efforts of the project can be discussed. The first is the creation of cash income and learning the new economic means for the participating residents. The second is encouraging activities of the community-based organization and the pro-activeness. The third is strengthening the continuity of practical participation by local people themselves. And the fourth is that the impact of a development project contribution to the participants can be spread to other residents of the village. It is also suggested that the project impact can be sustained despite changing the content of the activity among the residents in their community. This case study examined some positive factors of the project from the point of view of income, community motivation, and sanitation of water supply. One of the factors is that livestock breeding is an economic activity not much influenced by weather conditions, and local residents of the community can earn income from the activity in a relatively short period of time. Another factor is that livestock breeding does not face serious problems of water supply as it is used only for watering animals. Therefore, it was not difficult for the residents to understand and participate in the project, start working, continue their efforts, and maintain their high motivation to the implementation.

However, as previously described, it must not be forgotten that this village belongs to the rural community that has been regarded as the lowest standard of living in the county. Therefore, it can be proposed that the villagers need not only to make efforts for improvement of cash income but also to consider sanitation such as hygiene and agricultural water management. In terms of circumstances of toilet and bathroom, for example, most households have one small place just surrounded by crop fences for common use for both urinating and bathing. There is no pit for excretion in the place, and in general the villagers go in the fields for excretion. Agricultural water use almost relies on only rainwater, and little is done other than rainy season cultivation. If the residents have a motivation to make the situation better or change, this challenge may also be addressed by the residents by taking advantage of the organizational solidarity and practical skills enhanced through the experience of the livestock breeding project.

3.2 Dry-season gardening in the village of Tewrou Mbéyéne

The village of Tewrou Mbéyéne is located 15 km away from the village of Fissel Deux in the county. It belongs to the rural community of Ouadiour which is a county seat and considered as the second standard of living and economic level in the county from the same research. The population was 207 (125 men and 82 women) as of 2007. Same as the village of Fissel Deux, there is a GIE organization and it is composed of 40 members, 17 men and 23 women.

In this village, dry-season gardening project to improve the villagers' nutrition and cash income was carried out for the community mainly targeting the GIE with the support of a Non-Governmental Organization (NGO) called Agence National de Conseiller Agricole et Rurale (ANCAR) in December 2003 – May 2005. The NGO local staff gave a seminar to the resident participants on the basic knowledge and technical guidance of vegetable cultivation at the beginning of the project. After the seminar, the staff demonstrated how to plant a garden practically making a pilot field with the participants. The practice was

IOP Conf. Series: Earth and Environmental Science **60** (2017) 012034 doi:10.1088/1755-1315/60/1/012034

proceeded by expanding the area of the field gradually, and increasing the type of crops grown such as tomato, eggplant, okra, and so on. The management and operation of the field were left up to the participants. The local staff visited the village regularly after the seminar to continue the observation and guidance of the field progress made by the participants themselves. Tools, seeds and fertilizer were provided by the NGO, while the other expenses and labor force were provided by residents of the village.

To summarize the results of the project implementation, some effects and challenges are discussed. First, many of the crops withered even before reaching maturity stage. It was caused by insufficient water supply to the vegetable garden due to undeveloped deep well in the village. Second, the residents who positively participated in the activity and worked hard for the project were limited to few people such as the family of the village headman who is the president of the GIE. During the project, only a few participants' efforts were noticeable, and group activities as a community-based organization could not be seen as very functional. The necessity of organizational solidarity, and of the role of human resources who have leadership skills and who are key persons in the community, were considered to be of great importance. Third, the vegetable crops were harvested by the participants, and utilized as both food crops and cash crops for the residents of the village. It can be one-sidedly regarded that the efforts for the project implementation effect to some extent led to improvement of the residents' cash income and nutrition through their participatory development.

Fourth and finally, as described above, there is a serious problem with the deep well which is a crucial source of agricultural water in the village. In order to pump water from the well with the depth of 45m, the residents usually use a donkey connecting a long rope to the well, and they have the donkey walk until the bucket with water gets to the top of the well from the bottom. This traditional mechanism has been maintained in the village for a long time. However, it has been confirmed that the residents have had opportunities to plan and discuss the repair or replacement of a motor pump with engine power generator even before the project commenced. The motor pump used to be utilized, but from the time it broke down, it has never been repaired. Attempts to repair or replace have not been realized due to prohibitive purchase, repair and maintenance costs for the villagers.

It can be considered that the water supply and acquisition in the village increase the burden of time and labor force on the residents. By implication, this life situation suffers from a lack of water management for residents and can constitute a limiting factor for improvement of their income and sanitation. This situation can also hinder their motivation to the positive activity through their community-based organizations. On the other hand, the latest field research has reviewed that the president of the GIE has installed and started utilizing another motor pump with solar power engine. This latest finding can indicate the possibility of solving the water supply and management problem of the village in the near future.

4. Conclusion

For low-income people in developing and emerging countries, the suitable way to make their lives better will not always be by means of a great development assistance such as revolutionary program and modernization. It will be more important for the people to achieve sanitation and income improvement by means of a sustainable practice by themselves at the level of their life standard and economic resources. It can be examined that the point of view which can lead us to the direction of making progress on the goal is not the idea of breaking away from the vicious circle of poverty, but instead, the idea should be the quality of life even in their low-income life cycle. In order to improve the people's economy, sanitation, and water supply through their community participation, self-help effort is essential for sustainable development.

References

[1] Nurkse R 1953 Problems of Capital Formation in Underdeveloped Countries. Basil Blackwell.
[2] Saito F ed 2002 Participatory Development (in Japanese). Nippon Hyoron Sha.
[3] Sato H ed 2003 Re-examination of Participatory Development (in Japanese). Institute of Developing Economies – Japan External Trade Organization.
[4] Ikemi M 2011 Econ. *J. of Hokkaido Univ.* **40** 15-28.
[5] Chambers R 1983 Rural Development: Putting the Last First. Pearson Education Limited.
[6] Chambers R 1997 Whose Reality Counts?: Putting the First Last. Intermediate Technology Publications.
[7] Hickey S and Mohan G 2004 Participation: From Tyranny to Transformation? Exploring New Approaches to Participation in Development. Zed Books.
[8] Ikemi M 2013 Keizaigaku Kenkyu. Graduate School of Economics and Business Administration, Hokkaido University. **62(3)** 139-152.
[9] Ikemi M 2011 The Society of Economic Sociology Annual Report. **33** 302-313.

IOP Conf. Series: Earth and Environmental Science **60** (2017) 012035 doi:10.1088/1755-1315/60/1/012035

Status of heavy metal in sediment of Saguling Lake, West Java

E Wardhani[1*], D Roosmini[2] and S Notodarmojo[2]
[1]Environmental Engineering Program ITB, Post-Graduate Academy, Bandung, Indonesia
[2]Department of Environmental Engineering, ITB Bandung, Indonesia
*E-mail : ekw_wardhani@yahoo.com

Abstract The quality of water in the Saguling lake has been monitored since 1990 by the state agency for environmental control. However, no data on heavy metal in sediment had been reported. Metal pollutans have receives considerable attention due to their persistence, biogeochemical recycling, and environmental risk. The objective of this study was to assess the level contamination of heavy metals (Cd, Cr, Cu, and Pb) in the surface sediment of the Saguling Lake West Java. Surface sediment samples were collected from 10 location of the Saguling Lake on July and November 2015. The concentration of heavy metals in the surface sediment on July and November 2015 decreased in the order of Cu>Cr>Pb>Cd and Cr>Cu>Pb>Cd respectively. Mean metals concentrations (mg/kg) in July 2015 were in the range of: Cd: 10.69-16.65, Cr: 76.67-138.38, Cu:106.02-229.54, Pb: 23.93-80.17. Mean metals concentrations (mg/kg) in November 2015 were in the range of Cd: 18.64-23.25, Cr: 152.16-197.98, Cu: 63.32-152.53, Pb: 20.31-32.74. Geochemical approaches such as contamination factor, and pollution load index were exploited for the assessment of the contamination and enrichment level of heavy metals in the lake sediment. Contamination factor and pollution load index values indicated that surface sediments around Saguling Lake were polluted with heavy metals. The finding of this study would help in formulating guidelines to control the pollution and suggested for Saguling lake revitalization.

1. Introduction

Lake Saguling was built on The Citarum River. The Citarum Watershed is home to the largest industrial area in West Java Province. Several economic activity are performed along Citarum River, including mining and agriculture. Originally, the Saguling was planned for a single purpose lake to generate electricity. Later, considering the environmental problems of the area, the Saguling was re-planned as a multipurpose lake such as raw water, fisheries, aquaculture, and tourism [1].

Lakes are very important and significant bodies in preserving fresh-water and replenishing underground water, also act as a key role in adjusting local climate and ameliorating environment, consequently, they are considered one of the most versatile ecosystems in the world [2]. Recent years, with the rapid urbanization and intensive development of industry, the pollution of the lake environment with heavy metals has become a worldwide problem because they are indestructible and most of them have toxic effects on organisms [3] Heavy metals enter in lake from natural and anthropogenic sources [4]. Generally, in unaffected environment, most of heavy metals are in very low concentration, and natural geological weathering of rocks and soil, directly exposed to surface waters, is usually the largest natural source [5].

Main anthropogenic sources of heavy metal contamination are mining, disposal of untreated and partially treated effluents contain toxic metals, as well as metal chelates from different industries and indiscriminate use of fertilizer containing heavy metals [6], [7]. Heavy metals discharged in to aquatic system by natural or anthropogenic sources during their transport are distributed between the aqueous phase and sediments. Because of adsorption, hydrolysis and co-precipitation only a small portion of free metal ions stay dissolved in water and a large quantity of them are deposited in these sediment. Although heavy metals are refractory though natural processes in the environment, they can be chemically altered by organisms and converted into organic complexes, some of which may be more hazardous to animal and human life [8]. While metals settled in sediments may be re-suspended and cause secondary contamination to the water environment, because sediments act both as a sink and a source for metals in the aquatic environment [9].

Over the past decade, a number of indices have been developed to assess heavy metal contamination of sediments. These indices, mainly including sediment quality guide-lines (SQGs) , the relative enrichment factors (REF), the pollution load index (PLI) and index of geo-accumulation (I-geo), provide useful information and could be easily communicated to local managers and decision makers [10].

Hence, the main aspects of the present work are to (1) determine the content of heavy metals (Cd, Cr, Cu, and Pb) in Saguling lake surface sediment. (2) Calculate the pollution load index (PLI) in order to assess quality of the sediment.

2. Experimental

2.1. Study area

The Saguling is an artificial lake. The lake was built on the Citarum River, the largest river in West Java Province, in February 1985. The Saguling Lake area is 4,710 Ha the water storage capacity is 730.5 million m^3. The Saguling Lake is an embankment Lake on the headwater of Citarum River in West Java, Indonesia. The primary purpose of the lake is hydroelectric power generation but it also provides for water supply and aquaculture. The surrounding area of Saguling lake is hilly, while the river has many tributaries at this location. This makes the shape of Saguling lake very irregular or dendritic, with many extended bays. The lake area was formerly densely populated by farmer population with extensive agricultural lands. The catchment areas of the lake or the upper Citarum River basin are faced with high population pressure. This is because over 50% of the population consists of farmers with a high annual growth rate (2.34% is the national average). The growth of farmer population has caused the decrease of landholding and this condition forced them to extend their agricultural land by forest clearing and utilize marginal lands. As a consequence, there is a prevailing problem of floods accompanied by accelerated soil erosion in rainy seasons. Due to the high population density in the upper catchment of the lake, extensive agricultural land, soil erosion and the presence of industries, the lake water became polluted and eutrophic. The growth of aquatic weeds has been accelerated, with frequent blooms of *Microcystis* algae. The water is also contaminated by heavy metals, pesticides, etc. An extensive growth of water-hyacinth is maintained by fencing to reduce pollutant contents of water at inlets of the Citarum River [11].

2.2 Sample collection and chemical analysis

Surface sediment samples were collected from ten different sampling in Saguling lake in July and November 2015, sampling location are presented in Table 1. The point chosen because of their classification as having poor water quality accoding to water monitoring report Indonesia Power as a Saguling lake authority. The sampling locations were recorded (latitudinal and longitudinal position) using hand-held Global Positioning System (GPS). Two stations (1A and 1B) were located in the Citarum River, eight stasions (2-9) located in the Saguling lake.

Surface Sediment samples were collected at each station by mixing four to five random samples, whereas composite sediments were collected from the top 10 cm of the surface with a grab sampler. Samples were then placed into polyethylene bags, and refrigerated at 4°C. They were analysed in the laboratory. Each sediment sample was dried at 50°C, the samples were then grounded using pestle and mortar until all particles passed a 200-mesh nylon sieve. This fraction is mainly contained silt and clay; thus, this fraction may adsorb a higher heavy metal concentration. The extraction of heavy metals were made using conventional digestion procedure of sediment samples, which consists of digesting aliquots of 1 g of dry sediment was weighted into Teflon beakers, to which a mixture of concentrated HNO_3 : HCl (3:1) was added and digested in a water bath during 3 h at 120°C. The concentration of Cd, Cr, Cu, and Pb were determined for each sample using an Inductively Coupled Plasma Optical Emission Spectrometry or *ICP-OES*.

Table 1. Sampling stations in Saguling lake.

Station	Location	Coordinate	
		South	East
1A	Citarum River Nanjung section	06°56'29.8"	107°32'10.7"
1B	Citarum River section Trash Boom Batujajar	06°54'58.9"	107°28'35.0"
2	Cihaur Village Cipeundeuy	06°53'13.5"	107°28'32.3"
3	Cimerang	06°53'13.4"	107°27'09.0"
4	Cihaur Estuary Maroko Village	06°54'13.0"	107°25'54.4"
5	Cipatik Estuary	06°56'07.6"	107°27'25.5"
6	Ciminyak Estuary-fishing floating located	06°57'14.6"	107°26'03.8"
7	Cijere Estuary	06°56'14.9"	107°24'50.8"
8	Cijambu Estuary	06°56'00.4"	107°22'22.4"
9	Near Intake Structure	06°54'54.4"	107°22'26.3"

3. Result and discussion

3.1. Heavy metals concentration in Saguling lake

The concentrations of heavy metals (Cd, Cr, Cu, and Pb) in the sediment of the Saguling lake in July and November 2015 are presented in Table 2. The concentration of heavy metals in the surface sediment on July 2015 decreased in the order of Cu>Cr> Pb>Cd. Cu was the most abundant metal in the sediment, with measured concentrations ranging from 106.02 to 229.54 mg/kg follows by Cr (76.67-138-38 mg/kg), Pb (23.93-80.17 mg/kg), and Cd (10.69-16.65 mg/kg). The highest concentration of Cd, Cr, Cu, Pb found in the station 4, 1A, and 1B respectively, with a value of Cd 16.65 mg/kg, Cr 138.38 mg/kg, Cu 229.54 mg/kg, and Pb 80.17 mg/kg. The average value were determined to be as follows (mg/kg): Cd: 13.54±1.98, Cr: 100.83±21.94, Cu: 165.57±42.57, and Pb: 38.32±16.94. When compared with the standard based on ANZECC ISQG-Low for all heavy metals examined, only Pb concentration accordance sediment quality standard. Three heavy metals Cd, Cr, and Cu did not. appropriated sediment quality standard in all sampling locations.

The concentration of heavy metals in the surface sediment on November 2015 decreased in the order of Cr>Cu>Pb>Cd. Cr was the most abundant metal in the sediment, with measured concentrations ranging from 152.16 to 197.98 mg/kg follow by Cu (63.32-152.53 mg/kg), Pb (20.31-32.74 mg/kg), and Cd (18.64-23.25 mg/kg). The highest concentration of Cd, Cr, Cu, Pb found in the station 4, 1A, and 1B respectively, with a value of Cd 23.35 mg/kg, Cr 197.98 mg/kg, Cu 152.53 mg/kg, and Pb 32.74 mg/kg. For all station, the average value were determined to be as follows (mg/kg): Cd: 21.08±4.63, Cr: 174.10±15.86, Cu: 100.22±33.04, and Pb: 26.12±3.78. When compared with the standard based on ANZECC ISQG-Low for all heavy metals examined, only Pb concentration accordance sediment quality standard. Three heavy metals Cd, Cr, and Cu did not. appropriated sediment quality standard in all sampling locations.

Table 2. Range concentration of heavy metals in Saguling lake.

Station	Heavy metal concentration in July 2015 (mg/kg)				Heavy metal concentration in November 2015 (mg/kg)			
	Cd	Cr	Cu	Pb	Cd	Cr	Cu	Pb
1A	15.64	113.75	225.36	80.17	23.25	197.98	63.32	23.79
1B	16.65	125.09	229.54	51.61	22.92	184.39	152.53	25.79
2	13.27	81.00	178.66	42.06	20.92	160.10	135.35	29.11
3	12.86	106.18	175.55	25.69	20.62	169.78	64.96	20.31
4	16.05	138.38	182.88	35.00	22.90	191.02	87.25	32.74
5	11.31	76.67	119.93	36.90	22.00	174.25	82.92	28.92
6	12.88	83.94	106.02	31.65	19.15	157.97	132.80	27.90
7	10.69	82.74	116.70	23.93	20.61	189.56	83.04	25.18
8	13.36	83.74	158.95	30.15	18.64	152.16	73.36	21.18
9	12.70	116.83	162.09	26.10	19.81	163.75	126.71	26.27
Mean	13.54	100.83	165.57	38.32	21.08	174.10	100.22	26.12
Maximum	16.65	138.38	229.54	80.17	23.25	197.98	152.53	32.74
Minimum	10.69	76.67	106.02	23.93	18.64	152.16	63.32	20.31
Standard deviation	1.98	21.94	42.57	16.94	1.63	15.86	33.04	3.78
Standard*	1.50	80.00	65.00	50.00	1.50	80.00	65.00	50.00

*[12]

Total concentration Cd ranged from 10.69-16.65 mg/kg in July and 18.64-23.25 mg/kg in November. Sampling site 1B contained maximum Cd concentration. Higher Cd concentration might be related to industrial activity, atmospheric emission, and deposition of organic and fine grain sediments, leachates from defused Ni-Cd batteries and Cd plated items.The concentration of Cr in sediment ranged from 76.67-138.38 mg/kg in July and 152.16-197.98 mg/kg in November. The sampling point 4 located in Cihaur estuary area contains maximum amount of Cr in July. The sampling point 1A located in Citarum river Nanjung section contain maximum amount of Cr in November. The untreated Cr containing huge leather tanning wastes mix with the lake water might results in theses extreme Cr contamination. The US EPA has classified Pb as a probable human carcinogen. Average Cu content in the sediment was 165.57 and 100.22 mg/kg for July and November respectively. High level of Cu indicates its higher input in these sites, which might be originated from urban and industrial wastes. Lead is a good indicator of traffic related sources or battery recycling plant, and is considered as an indicator of pollution by urban runoff water [13]. A comparison of metal concentration in sediment with reference in our study can indicated that metal accumulation in sediment lake was either by natural and anthropogenic source.

Table 3 presented mean concentration of measured elements in lake sediment samples from different countries along with present study. Comparing with the data from other country, the heavy metals concentration in Saguling lake were found higher. Heavy metals concentration in Saguling Lake must be considered from the government because the lake Saguling functions are related to public health as a source of irrigation water, fisheries, and raw water sources of drinking water for Bandung Basin and surrounding region.

Table 3.
Mean concentration of measured elements in lake sediment samples from different countries along with present study.

No	Name of the lakes	Heavy metals concentrations (mg/kg)			
		Cd	Cr	Cu	Pb
1	East lake, Wuhan, China[1]	-	-	54.8	40.3
2	Songkhla lake, Southern Thailand[2]	0.1-2.4	-	1.8-125	8.2-131
3	Laguna lake Philippines[3]	0.02-0.09	-	9.7-18.7	17-23
4	Manchar lake, Pakistan[4]	4.9-9.7	14.7-26.8	15.6-29.7	14.6-20.9
5	Taihu lake, China[5]	0.94	56.2	36.7	51.8
6	Hussainsagar lake, India[6]	-	40-60	-	40-60
7	Saguling lake, Indonesia[7]				
	July 2015	10.69-16.65	76.67-138.38	106.02-229.54	23.93-80.17
	November 2015	18.64-23.25	152.16-197-98	63.32-152.53	20.31-32.74

[1-6] [14]; [7]present study

3.2. *Assessment of sediment contamination in Saguling lake*

The level of contamination of lake sediment by given toxic substance (metals) is often expressed interms of a contamination factor and is calculated as follows [15].

$$\boldsymbol{CF} = \frac{\boldsymbol{CMetal}}{\boldsymbol{Cbackground}} \tag{1}$$

Where, CF is the contamination factor, C_{metals} is the concentration of pollutants in sediment, $C_{background}$ is the background value for the metals. Background concentration of Cd, Cr, Cu, and Pb was respectively 0.30 mg/kg, 90 mg/kg, 45 mg/kg, and 20 mg/kg [16]. With the help of contamination factor values the intensity of contamination can be inferred from Hakanson, 1980 as follows: low contamination, CF < 1; moderate contamination 1 < CF <3; considerable contamination, 3 < CF < 6; and very high contamination, CF > 6. Pollution load index (PLI) is defined as the root of the product of the CF.

Tomlinson, 1980 has developed a simple method based on PLI to assess the degree of pollution by metals in aquatic sediments, PLI was calculated using the equation 2 where n the number of metals. A value of 0 on the PLI would indicated no pollutants; a value of 1 would suggest the presence of baseline levels of pollutants, while value that exceed 1 would indicated a progressive deterioration of the site and lake [17].

$$\boldsymbol{PLI} = (\boldsymbol{CF_1} \times \boldsymbol{CF_2} \times \boldsymbol{CF_3} \cdots\cdots\cdots \boldsymbol{CF_n})^{\frac{1}{n}} \tag{2}$$

Contamination factor value calculated is summarized in Table 4 Contamination factor value in Saguling lake. The contamination factor value Cd in July ranged between 35.62 to 55.48 with average value 45.13. The contamination factor value Cd in November ranged between 62.14 to 77.49 with average value 70.27. Based on contamination factor indicated that Saguling surface sediment in July and November catagorized very high contamination by Cd. The contamination factor value Cr in July ranged between 0.85 to1.54 with mean value 1.12. The contamination factor value Cr in November ranged between 1.69 to 2.20 with mean value 1.93. Based on contamination factor indicated that Saguling surface sediment in July catagorized low to moderate contamination and November catagorized moderate contamination by Cr. The contamination factor value Cu in July ranged between 2.36 to 5.10 with mean value 3.68. The contamination factor value Cu in November ranged between 1.41 to 3.39 with mean value 2.23. Based on contamination factor indicated that Saguling surface sediment in July and November catagorized moderate to considerable contamination by Cu.

The contamination factor value Pb in July ranged between 1.20 to 4.01 with mean value 1.92. The contamination factor value Pb in November ranged between 1.06 to 1.46 with mean value 1.31. Based on contamination factor indicated that Saguling surface sediment in July catagorized moderate to considerable contamination and November catagorized moderate contamination by Pb. Contamination factor catagory is showed in Table 5 Contamination factor catagory in Saguling lake.

Table 4. Contamination Factor value in Saguling lake.

Station	July 2015				November 2015			
	Cd	Cr	Cu	Pb	Cd	Cr	Cu	Pb
1A	52.13	1.26	5.01	4.01	77.49	2.20	1.41	1.19
1B	55.48	1.39	5.10	2.58	76.41	2.05	3.39	1.29
2	44.22	0.90	3.97	2.10	69.73	1.78	3.01	1.46
3	42.87	1.18	3.90	1.28	68.73	1.89	1.44	1.02
4	53.48	1.54	4.06	1.75	76.33	2.12	1.94	1.64
5	37.68	0.85	2.67	1.84	73.33	1.94	1.84	1.45
6	42.93	0.93	2.36	1.58	63.83	1.76	2.95	1.39
7	35.62	0.92	2.59	1.20	68.69	2.11	1.85	1.26
8	44.52	0.93	3.53	1.51	62.14	1.69	1.63	1.06
9	42.32	1.30	3.60	1.31	66.04	1.82	2.82	1.31
Mean	45.13	1.12	3.68	1.92	70.27	1.93	2.23	1.31

Table 5. Contamination Factor catagory in Saguling lake.

Station	July 2015				November 2015			
	Cd	Cr	Cu	Pb	Cd	Cr	Cu	Pb
1A	Very High	Moderate	Considerable	Considerable	Very High	Moderate	Moderated	Moderate
1B	Very High	Moderate	Considerable	Moderate	Very High	Moderate	Considerable	Moderate
2	Very High	Low	Considerable	Moderate	Very High	Moderate	Considerable	Moderate
3	Very High	Moderate	Considerable	Moderate	Very High	Moderate	Moderate	Moderate
4	Very High	Moderate	Considerable	Moderate	Very High	Moderate	Moderate	Moderate
5	Very High	Low	Moderate	Moderate	Very High	Moderate	Moderate	Moderate
6	Very High	Low	Moderate	Moderate	Very High	Moderate	Moderate	Moderate
7	Very High	Low	Moderate	Moderate	Very High	Moderate	Moderate	Moderate
8	Very High	Low	Considerable	Moderate	Very High	Moderate	Moderate	Moderate
9	Very High	Moderate	Considerable	Moderate	Very High	Moderate	Moderate	Moderate

According to the classification carried out on the basis of CF, bed sediment are very high contaminated with Cd in July and November 2015, low-moderate contaminated with Cr in July 2015 and moderate contaminated with Cr in November 2015. Bed sediment are moderate-considerable contaminated with Cu in July and November 2015. Bed sediment are moderate-considerable contaminated with Pb in July 2015 and moderate contamination in November 2015. Sediment re-suspension can be caused by storm water run off, sediment re-suspension caused by storm water runoff increased its potential for heavy metals release in sediment to water column. During re-suspension of anoxic sediment, the physical and chemical environment surrounding sediment bound metals will change. Bedded sediments is drawn up into the water column which is often fully saturated with dissolved oxygen [18].

PLI represents the number of times by which the metal content in the sediment exceeds the background concentration and give a summative indication of the overall level of heavy metal toxicity in a particular sample. PLI < 1 indicates no sediment pollution, whereas PLI > 1 points to pollution.

The value of PLI obtained in The Saguling sediments in July and November are 4.21 and 4.39 respectively. According to PLI catagorized Saguling lake sediment was polluted with heavy metals. Higher concentration of heavy metals may be caused by external distinct anthropogenic source lie agricultural runoff, industrial waste water dircharge and atmospheric dry deposition [19]. It also provide valuable information to the policy makers on the pollution levels in the study area.

In the Saguling catchment located 556 industries which are 442 textile industries, 8 pulp and paper industries, 11 tanning industries, 16 rubber and plastic industries, 14 chemical industries 40 electroplating and metals industries, and 25 food and beverage industries.The growth of industries population has caused the decrease of landholding. As a consequence, there is a prevailing problem of floods accompanied by accelerated soil erosion in rainy seasons. Due to the high population density in the upper catchment of the lake, extensive agricultural land, soil erosion and the presence of industries, the lake water became polluted and eutrophic. The water is also contaminated by heavy metals, pesticides, etc. The study suggests that the urban wastewater and effluents emerging from the industries must be monitored periodically for maintaining the standards prescribed by the pollution control board for various industries in the region. Further treatment of industrial effluents particularly from textile mills, electroplating, tanning manufacturing and chemical industries before discharging in the upper catchment is prerequisite for controlling the level of pollution in the Saguling lake.

4. Conclusion

The heavy metals were identified in the Saguling lake surface sediment. The total heavy metals contents in the surface sediment in July and November decrease in order of Cu>Cr>Pb>Cd and Cr>Cu>Pb>Cd respectively. Two pollution indices were used for the environmental assessment of Saguling lake surface sediment are contamination factor (CF) and pollution load index (PLI). Based on PLI value Saguling lake surface sediment was polluted by heavy metals. The possible main source of the metals contamination in the sediment are municipal and industrial wastewater discharge, agriculture runoff, and atmospheric input. A variety of useful tools, methods, and indices were used to evaluate heavy metal pollution in the surface sediment of the Saguling lake. However, heavy metals can transfer from the sediments to other ecosystem components such as underground water, crops an animals and affect human health throught the water supply and food chain. Thefore, it is necessary to design an environmental strategy not only manage but also monitor the same with priority.

Acknowledgement

Authors would like to acknowledge State Minister for Research and Technology and Higher Education for financial support for the research.

References

[1] Soemarwoto edited by Barry A Costa-Pierce, Otto 1990. Jakarta, Indonesia: Perusahaan Umum Listrik Negara. p. 21. ISBN 971-10-2250-8.
[2] Forghani G, Moore F, Lee S, Qishlaqi A 2009. *Environ. Earth Sci.* **59 (1)**,173-184.
[3] Hu Y, Qi, S H, Wu C X, Ke Y P, Chen J, Chen W, Gong X Y 2012. *Front. Earth Sci.* **6 (1)**, 39-47.
[4] Reza R, Singh G 2010. *Int. J. Environ. Sci.Tech.***7 (4)**,785-792.
[5] Karbassi A R, Monavari S M, Nabi Bidhendi, G R, Nouri J, Nematpour K 2008. *Environ. Monitor. Assess.* **147 (1-3)**, 107-116.
[6] Nouri J, Mahvi A H, Jahed G R, Babaei AA, 2008. *Environ. Geo.* **55 (6)**, 1337-1343.
[7] Malik N, Biswas A K, Qureshi T A, Borana K, Virha R, 2010. *Environ. Monitoring and Assess.* **160**, 267-276.
[8] Su LY, Liu, J L, Christensen P, 2011. *Ecotox. Ecol.* **20**,1107-1116.
[9] Beutel M W, Leonard T M, Dent S R, Moor B, 2008. *Water Res.* **42**, 1953-1962.
[10] Sundaray S K, Nayak B B, Lin S, Bhatta D, 2011. *J. Hazard. Matter.* **186 (2-3)**, 1837-1846.

[11] Soemarwoto, edited by Barry A. Costa-Pierce, Otto 1990. Jakarta, Indonesia: Perusahaan Umum Listrik Negara. p. 21. ISBN 971-10-2250-8.
[12] ANZECC (Australian and New Zealand Environment and Conservation Council) 2003 ANZECC interim *sediment quality guidelines*.
[13] Yin H, Gao Y, Fan C, 2011. *Environ. Res. Lett.* **6**, 044012 (11 pp).
[14] Varol M, Şen B, 2012. *Catena* **92**,1-10.
[15] Hakanson L, 1980. *Water Res.***14 (8)**, 975-1001.
[16] Turkerian KK and Wedepohl KH 1961. *Bull geol. Soc Am*, **72**, 175-192
[17] Tomilson D C, Wilson J G, Haris C R and Jefferey DW, 1980. *Helgol. Wiss. Meeresunters*, **33:1-4**, 566-575
[18] Gaur V K, Gupta S K, Pandey S D, Gopal K, Misra V, 2005. *Environ Monit Assess.***102**, 419-433.
[19] Fernandes HM. 1997. *Environmental Pollution*, **97 (3)**: 317-325

Determination of total arsenic in fish by hydride-generation atomic absorption spectrometry: method validation, traceability and uncertainty evaluation

W C Nugraha[1*], C Elishian[2] and R Ketrin[2]
[1]Research Center for Clean Technology, Indonesian Institute of Sciences, Jalan Cisitu Sangkuriang Bandung 40135, Indonesia
[2]Research Center for Chemistry, Indonesian Institute of Sciences, Kawasan Puspiptek Serpong, Tangerang Selatan 15314, Indonesia
[*]Email : will003@lipi.go.id

Abstract. Fish containing arsenic compound is one of the important indicators of arsenic contamination in water monitoring. The high level of arsenic in fish is due to absorption through food chain and accumulated in their habitat. Hydride generation (HG) coupled with atomic absorption spectrometric (AAS) detection is one of the most popular techniques employed for arsenic determination in a variety of matrices including fish. This study aimed to develop a method for the determination of total arsenic in fish by HG-AAS. The method for sample preparation from American of Analytical Chemistry (AOAC) Method 999.10-2005 was adopted for acid digestion using microwave digestion system and AOAC Method 986.15 – 2005 for dry ashing. The method was developed and validated using Certified Reference Material DORM 3 Fish Protein for trace metals for ensuring the accuracy and the traceability of the results. The sources of uncertainty of the method were also evaluated. By using the method, it was found that the total arsenic concentration in the fish was 45.6 ± 1.22 mg.Kg^{-1} with a coverage factor of equal to 2 at 95% of confidence level. Evaluation of uncertainty was highly influenced by the calibration curve. This result was also traceable to International Standard System through analysis of Certified Reference Material DORM 3 with 97.5% of recovery. In summary, it showed that method of preparation and HG-AAS technique for total arsenic determination in fish were valid and reliable.

1. Introduction

It is well recognized that arsenic is present in the environmental and in biological systems in various chemical forms, both organic and inorganic forms. The organic forms of arsenic, such as arsenobetaine found in ubiquitous of edible marine animals has no toxic effect. However, the inorganic forms, arsenite (As^{3+}) and arsenate (As^{5+}), are known to be the most toxic. Food and water are significant sources of arsenic intake, accounting 77.5% in children [1]. Levels of arsenic in most food are fairly low, as it is not added due to its toxicity, however the levels of arsenic in fish and seafood may be high because fish can absorb and accumulate arsenic from the water they live in. Inorganic arsenic typically accounts for one to three per cent of the total arsenic found in food [2]; it is found at a higher content in such marine organisms as fish, crab, shrimp, oyster [3].

Hydride generation (HG) coupled with atomic absorption spectrometric (AAS) detection is currently one of the most popular techniques used for the quantification of arsenic in a variety of matrices. HG is a powerful technique of which utilizes chemical with typical properties of the metalloid group of elements such as As, Bi, Sb, Se, Te, Ge and Sn to form volatile hydrides (arsine in the case of As). These hydrides are carried using a stream of inert gas to the atomizer in Flame AAS [4]. The method offers several advantages, including improved detection limits for hydride-forming elements and reduced matrix effects. Although the method is simple, it is not totally free from interferences in the matrices. Several authors have reported that interferences such as transition metals [5-6], mutual hydride-forming elements [7-8], and conditioning of the quartz cell surface can interfere in HGAAS technique [9]. Therefore, it is necessary to validate the method for the determination of total arsenic concentration in fish.

To achieve comparability of results over space and time, it is essential to compare all the individual measurement results to standards [10]. Traceability is accordingly intimately linked to uncertainty. Traceability provides the means of placing all related measurements to reference standards while an uncertainty characterizes the agreement to be expected between laboratories making similar measurements. The quoted uncertainty value combines all uncertainties, either arising from random variation or systematic effects [11-12]. Standard procedure to estimate the measurement uncertainty was adopted from the GUM (Guide to the expression uncertainty in measurement) and EURACHEM/CITAC Guide [10, 12].

In this study, determination of total arsenic in fish by HG-AAS was developed and validated. The sources of uncertainty of the method was also evaluated. Certified Reference Material DORM 3 Fish Protein for trace metals was used for ensuring the accuracy and the traceability of the results.

2. Experimental

2.1 Instruments

All samples were digested using a Milestone MLS 1200 microwave digestion system (Milestone, Bergamo, Italy) and dried ashing in a Vulcan Furnace model A550 (Dentsply International, York, United Stated of America). Arsenic concentrations was measured using a Hitachi Z-5000 Atomic Absorption Spectrometer (Hitachi Instruments Co. Tokyo, Japan) coupled with Hydride Generation system. The operating conditions of AAS are listed in Table 1.

Table 1. HG-AAS operating condition.

System	Operation condition
Atomizer	Standard
Flame type	Air-acetylene
Fuel-flow	1.7 L min^{-1}
Carrier gas flow	0.1 L min^{-1}
Oxidant pressure	160 kPa (1.6 kg cm^{-2})
Ar gas pressure	150 kPa (1.5 kg cm^{-2})
Burner height	Cell position
Sampling time	40 s
Reaction time	30 s
Measurement time	15 s

2.2 Reagents and solutions

SRM 83d standard reference material of arsenic trioxide, As_2O_3 powder 99.9926 ± 0.0030 wt% was purchased from National Institute Standard and Technology (NIST), United State of America. A stock solution of arsenic was prepared by dissolving 0.132 g arsenic trioxide in water containing 0.4 g NaOH, and diluted to 100 g. An intermediate solution was prepared by diluted 1 g stock solution in 100 g of 0.1 M HCl. Accurate dilution 1 g intermediate solutions with 100 g of 0.01 M HNO_3 was carried out to prepare working solutions. Appropriate amount of working solution was diluted in 0.01 M HNO_3 to prepare calibration solution. Magnesium nitrate, $Mg(NO_3)_2$ and sodium borohydride, $NaBH_4$ powder with purity of ≥ 98.0 % was purchased from Sigma-Aldrich Inc. A solution of 7.5% (w/v) $Mg(NO_3)_2$ was prepared by dissolving 7.5 g $Mg(NO_3)_2$ in 100 mL ultrapure water, while sodium borohydride solution was prepared fresh daily by dissolving 2 g $NaBH_4$ in 200 mL 1 M NaOH.

IOP Conf. Series: Earth and Environmental Science **60** (2017) 012036 doi:10.1088/1755-1315/60/1/012036

Potassium Iodide for general purpose reagent was purchased from BDH Chemical Ltd, Poole, England. A pre-reductant solution with concentration of 10% (w/v) KI was prepared from 50 g KI dissolved in 500 mL water. Nitric acid and hydrochloric acid for poisonous metal analysis from Kanto Chemicals, Tokyo, Japan were diluted with ultrapure water to give a 0.01 M, a 0.1 M of nitric acid and a 2M, a 8M of hydrochloric acid solutions. Concentrated hydrogen peroxide and sodium hydroxide were purchased from E-Merck, Germany. All water used for analysis is ultrapure water (18.3 MΩ cm^{-1} resistivity) that was prepared by a Milli-Q Plus 185 from Millipore cooperation, Bedford, United State of America. DORM-3 Fish Protein Certified Reference Material for trace metals from National Research Council, Canada, was used for ensuring the accuracy and traceability of the method.

2.3 Method

2.3.1 Sample preparation

Sample preparation methods were adopted from American of Analytical Chemistry (AOAC) Method 999.10-2005 for acid digestion using microwave digestion system and AOAC Method 986.15 – 2005 for dry ashing [13].

- Solution A

An amount of 0.5 g sample was weighed accurately in a digestion vessel. Sample was added with 5 mL of HNO_3 14 M and 2 mL of concentrated H_2O_2 and then the vessel was closed. The closed vessel was kept for one night before run in a microwave digestion system. Sample was transferred to a 50 mL Nalgene bottle with a screw cup and diluted until 25 g with ultrapure water.

- Solution B

Accurate amount of 2 g solution A was transferred into the porcelain crucible and added with 1 mL 7.5% (w/v) $Mg(NO_3)_2$, and then heated on a hotplate at slow heat to dryness. The crucible was then placed in a furnace at 250-450 $^{\circ}$C with maximum heating rate of 100 $^{\circ}$C/h. After being cooled, the residue was dissolved with 2 mL 8M HCl and diluted with ultrapure water to 20 g.

Accurate amount of 2.5 g solution B was added with 5 mL concentrated HCl and 5 mL 10% (w/v) KI, and then stand for 45 min before diluted 50 g with ultrapure water, and measured by HGAAS.

2.3.2 Standard preparation

Accurate amount of 2.5 g each calibration solution was added with 5 mL concentrated HCl and 5 mL 10% (w/v) KI, and then stand for 45 min before diluted into 50 g with ultrapure water, and measured by HGAAS.

2.3.3 Moisture content

Parallel with sample preparation, the moisture content of samples was examined by drying 1 g of sample at room temperature in desiccators over $CaSO_4$ anhydrous for 10 days to obtain constant weight.

3.Result and discussion

3.1 Method validation for arsenic determination by HGAAS

3.1.1 Optimizing HGAAS parameters

Some parameters of HGAAS such as molarity of hydrochloric acid or sodium tetrahydroborate may vary depending on the type of apparatus, therefore it should be optimized. Other parameters for reducing arsenate (pentavalent arsenic) to arsenite (trivalent arsenic) in the solution also have been optimized. The optimized values were marked with red bold line as shown in Figure 1.

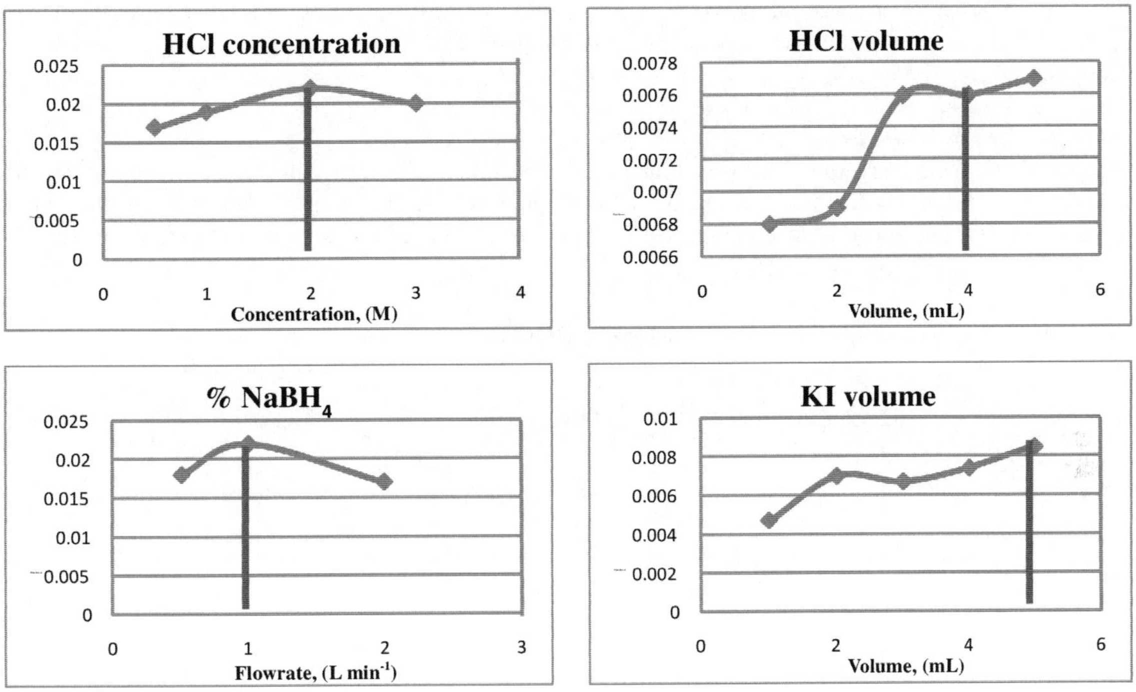

Figure 1. Effect of HGAAS parameters to the measured absorbance. The optimized values are shown by the red lines.

3.1.2 Method validation parameter

The optimized method validation parameters are shown in Table 2. The detection and quantitation limit of the proposed method were examined under the optimal experimental conditions.

Table 2. Optimization of validation parameters method.

No.	Parameter	Examined	Optimized value
1	Linearity (μg.Kg^{-1}) (w/w)	0 – 20	0.5– 5
2	Detection limit (μg.Kg^{-1}) (w/w)	3 sd of 1 μg L^{-1}, 10 replicates	0.17
3	Quantitation limit (μg.Kg^{-1}) (w/w)	10 sd of 1 μg L^{-1}, 10 replicates	0.58
4	Precision (%RSD)	2 μg L^{-1}, 7 replicates	7.5
5	Accuracy (DORM 3*, mg/kg) (w/w)	6.88 ±0.30**	6.71±0.32***

* Fish Protein CRM for trace metals, National Research Council of Canada
** Value from certificate
*** Average and standard deviation from 2 replicates

The detection and quantitation limit were 0.17 and 0.58 μg L^{-1}, respectively, which were calculated as the three times and ten times of the standard deviation of ten replicates of 1 μg.L^{-1} standard. Its precision was calculated as relative standard deviation (%RSD) from sample measurement with 7 replicates.

3.1.3 Effect of sample matrix
Since the method was developed from routine for food sample, the validity of the method was verified to be fit for fish sample by determining the matrix interference. In Figure 2, two calibration curves prepared by external and matrix matched calibration were compared to examine the effect of sample matrix.

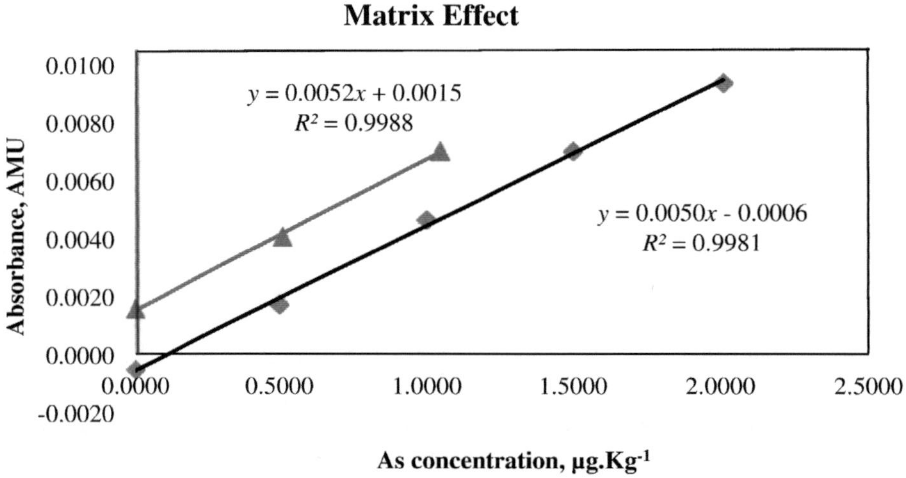

Figure 2. Effect of sample matrix. Red sign is sample spiked with arsenic standard; Blue sign is arsenic standard diluted in HNO_3 0.2 M.

Figure 2 shows that the slope of both calibration curves are similar (0,050 vs 0,052).The similarity indicates that sample matrix does not interfere with the measurement of arsenic and therefore HGAAS method is valid for fish sample.

3.2 Traceability and accuracy
Traceability can be obtained by reference measurement or certified reference materials. In this study, traceability was obtained by analyzing DORM-3 Fish Protein Certified Reference Material. As shown in Table 2, the result obtained by the method was in good agreement with the certified value. This result indicates that the present method was accurate and no bias.

3.2.1. Sources of uncertainty
Sources of uncertainty and the biggest uncertainty component for the determination of total arsenic in fish have been identified using bottom-up approach. This estimation evaluation was based on ISO/IEC Guide 98-3:2008, JCGM 100:2008 and Eurachem/Citac Guide 2012. There are four main steps to estimate the uncertainty components according to Eurachem/Citac Guide: (1) specification of the measurand, (2) identification of uncertainty sources, (3) quantification of uncertainty components and convert to standard uncertainty, and (4) combination of standard uncertainties.

3.2.2. Specification of the measurand

In this step, the estimation begins with making a summary of measurement which is shown in flow chart on fig 1. A dry-weight content of total arsenic (C_{As}) was calculated from the concentration of As in digested solution (C_o), the solvent mass (M), the mass of sample (m) and the dilution factor (DF) which is converted to method recovery (Rec) and moisture content (MC).

$$C_{As} = \frac{C_o \text{ x M x DF}}{\text{m}} \text{ x } \frac{1}{Rec} x \frac{100}{100 - \text{MC}} \text{ x}(Rep) \quad (1)$$

Precision of each uncertainty source was not calculated because it was included in all method precision (Rep).

3.3.3. Identification of uncertainty sources

All relevant sources of measurement uncertainty were identified through cause-effect diagram (Figure 3).

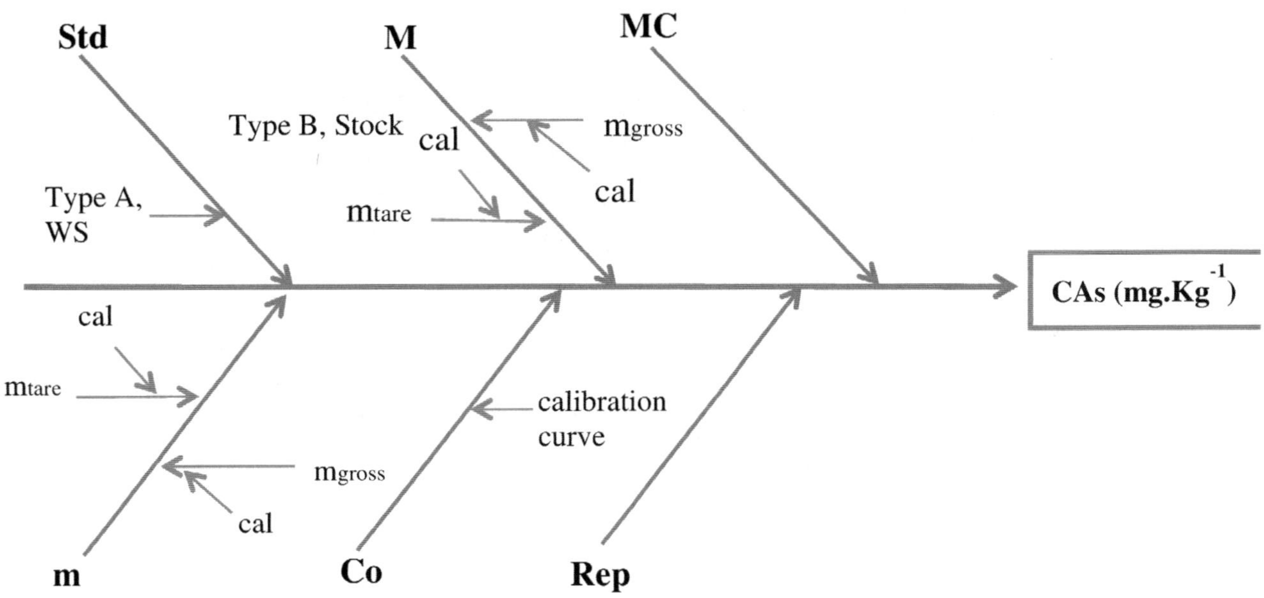

Figure 3. Cause and effect diagram for the determination of total arsenic concentration in fish.

The cause and effect diagram is a very convenient way of listing the uncertainty sources, showing the relation of each component and their influence on the uncertainty results.

3.3.4. Quantification of uncertainty components and convert to standard uncertainty

The quantification of all uncertainty sources are divided into each parameter and then converted into standard uncertainties.

3.3.4.1. Arsenic concentration in digested solution (C_o)

This component was obtained from interpolated concentration in calibration curve equation, y = Ax + B, where y is the degree of absorption from the calibration standard, A is the angular coefficient, x is the concentration of the calibration standard, and B is the point of intersection with the y axis [14]. The calibration curve was made from five standard solutions that prepared manually (w/w), with concentration 0.0, 0.5, 1.0, 2.5, and 5.0 μg.Kg^{-1}, respectively. The calibration standards were prepared from 1000 mg.Kg^{-1} arsenic stock solution and measured three times. The standard uncertainty from Co was calculated using equation (2):

$$\mu C_o = \frac{S}{B_1} x \sqrt{\frac{1}{p} + \frac{1}{n} + \frac{(C_o - \bar{C})^2}{S_{xx}}} \quad (2)$$

with the residual standard deviation S (μg.Kg^{-1}) and S_{xx} ((μg.Kg^{-1})^2) are given by

$$S = \sqrt{\frac{\sum_{j=1}^{n} [A_j - (B_o + B_1\, x\, C_j)]^2}{n-2}} \quad (3)$$

$$S_{xx} = \sum_{j=1}^{n} (C_j - \bar{C})^2 \quad (4)$$

Where

A_j	: measurement of absorbance
B_o	: intercept
B_1	: Slope of calibration curve
p	: number of measurements to determine C_o
n	: number of measurements for the calibration
$\bar{C}$	: mean value of the different calibration standards (n number of measurements)
i	: index for the number of calibration standards
j	: index for the number of measurements to obtain the calibration curve

3.3.4.2. Mass

The standard uncertainty related to the mass of sample (m) and solvent (M) were identified from the uncertainty sources for the tare and gross weighing. Each source takes into account the contribution from the repeatability, the readability (digital resolution) of the balance scale, and the calibration function. There are two components that contributed to the calibration function, identified as the sensitivity and the linearity of the balance. The sensitivity component was neglected because the mass by difference is done on the same balance over a very narrow range [15]. The repeatability contribution from mass was considered as a whole method precision (Rep), therefore the standard uncertainty from this source was estimated using the data from analytical balance calibration certificate, as of 0.0001 g on 95% confidence level that accounted for twice, once for the tare and other for the gross mass.

$$u_{mass} = \sqrt{u_{gross}^2 + u_{tare}^2} \quad (5)$$

3.3.4.3. Dilution factor (DF)

In this method, the digested sample is diluted with water before measured using HG -AAS based on weight per weight calculation. The dilution factor (DF), is given by

$$DF = \frac{w_2}{w_1} x \frac{w_4}{w_3} \quad (6)$$

The weight is determined using the same calibrated analytical balance which has standard uncertainty 0.0001g on 95% confidence level. Therefore, total standard uncertainty μ DF) from this sources is calculated taking into account all those weight uncertainties:

$$\frac{u(DF)}{DF} = \sqrt{\left(\frac{u(w_1)}{w_1}\right)^2 + \left(\frac{u(w_2)}{w_2}\right)^2 + \left(\frac{u(w_3)}{w_3}\right)^2 + \left(\frac{u(w_4)}{w_4}\right)^2} \quad (7)$$

The dilution factor of sample was 200 times which is from 2g (w_1) of solution A was diluted to 20g (w_2, solution B) and then 2.5g (w_3) of solution B was diluted to 50g (w_4).

3.3.4.4. Method recovery (Rec)
The overall recovery, R, is an estimation of the "method recovery", because it has been obtained by representatively varying the factors, such as matrix, concentration and analyte that can be affect to recovery. Therefore, proportional bias can be estimated in terms of the overall recovery [16]. The overall recovery is assessed using DORM-3 Fish Protein Certified Reference Material (CRM), from National Institute of Standard and Technology (NIST). The uncertainty estimation for this parameter was used for the comparison of obtained values with the certificates ones. The uncertainty associated with recovery, μ(R), is evaluated as follows:

$$u(R) = R\sqrt{\left(\frac{s^2_{obs}}{nC^2_{obs}}\right) + \left(\frac{u(C_{CRM})}{C_{CRM}}\right)^2} = 0.028 \quad (8)$$

The method recovery is calculated as the ratio

$$R = \frac{C_{obs}}{C_{CRM}} = 0.943 \quad (9)$$

Not all the recovery parameter have to used in calculation result, to determine whether the parameter is significantly different from one or not a student's t test is used. The t value is compared with the 2-tailed critical value $t_{crits}$for n-1 degrees of freedom at 95% confidence level. If the t value is greater or equal than the t_{crit} then the recovery parameter is significantly different from 1 and has to include in calculation result.[15]
The equation to calculate the t value is given by:

$$t = \frac{|1 - R|}{u(R)} = 2.036 \quad (10)$$

This value is compared with the 2-tailed critical value *t*crit, for *n*–1 degrees of freedom at 95 % confidence (where *n* is the number of results used to estimate *Rec*). If *t* is greater or equal than the critical value *t*crit than *Rec* is significantly different from 1, therefore *Rec* is explicitly included in the calculation of the result. From the experimental, the t value is smaller than $t_{crit;5}$ (2.571), then it can be concluded that the procedure does not have any significant proportional bias. Therefore, the recovery parameter can be neglected in future results obtained with the procedure.

3.3.4.5. Moisture content (MC)

The moisture content of the sample is obtained by drying 1 g of sample at room temperature in desiccators over $CaSO_4$ anhydrous for 10 days. The percentage of moisture content in sample is quantified in the equation (11)

$$\%\,MC = \frac{(M_a - M_b)}{M_a} \times 100\% \quad (11)$$

Sample mass before (M_a) and after (M_b) drying is weighted using the analytical balance which has standard uncertainty 0.0001 from calibration certificate on 95% confident level. This uncertainty sources has contributed to combined uncertainty which calculated as below

$$\mathrm{uMC} = \mathrm{MC}\ \mathrm{x} \sqrt{\frac{(\mathrm{uM_a})^2 + (\mathrm{uM_b})^2}{(\mathrm{M_a} - \mathrm{M_b})^2} + \frac{(\mathrm{uM_a})^2}{\mathrm{M_a}^2}} \qquad (12)$$

The result is shown in Table 3.

3.3.4.6. Method precision (Repeatability)

The results of the method validation experiments show repeatability for the arsenic determination from four kind samples using HGAAS, which is 0.35% as RSD. The entire individual sources precision included in this value. The value can be used directly for the combined standard uncertainty calculation associated with the different precision terms. [16]

3.3.4.7. Standard

The uncertainty component from concentration of calibration standards are also contributed to combined uncertainty. There are two uncertainty sources of this component, uncertainty from standard stock solution (Type B Effect) and uncertainty of working solutions from dilution and random use of analytical balance (Type A Effect). The type B effect is considered the effect of the working standards in the same way and included shifting of calibration curve, such as purity, balance bias, and temperature whereas type A effect is considered the effect of each working solution differently.

3.3.4.8. Combination of standard uncertainties

The relative combined standard uncertainty for the method as whole is evaluated though all individual sources of uncertainty using the following equation

$$\left(\frac{u_{C(As)}}{C_{As}}\right)^2 = \left(\frac{u_{C_{Calstd(type\ A)}}}{C_{Calstd\ (type\ A)}}\right)^2 + \left(\frac{u_{C_{Stock(type\ B)}}}{C_{Stock\ (type\ B)}}\right)^2 + \left(\frac{u_{C_o}}{C_o}\right)^2 + \left(\frac{u_M}{M}\right)^2 + \left(\frac{u_m}{m}\right)^2 + \left(\frac{u_{DF}}{DF}\right)^2 + \left(\frac{u_{MC}}{MC}\right)^2 + \left(\frac{u_{Rep}}{Rep}\right)^2 \quad (13)$$

Detail of each uncertainty sources shown in table 4, where C_{As} represents the dry-weight content of total arsenic and calculated using eq. 1 without recovery factor.

Table 3. Uncertainties sources in total arsenic in fish determination.

Uncertainty components	Value (Xi)	Standard uncertainty (u Xi)	Unit	RSU (u Xi/Xi)
Calibration standard type A	0.0010	0.000000159	$mg.Kg^{-1}$	0.000159
Calibration standard type B	1865	0.00150	$mg.Kg^{-1}$	0.000000804
Calibration curve (C_o)	0.00390	0.0000503	$mg.Kg^{-1}$	0.0129
Mass of solvent (M)	25.9	0.0000707	g	0.00000273
Mass of sample (m)	0.517	0.0000707	g	0.000137
Dilution factor (DF)	200.0	0.00909	-	0.0000454
Moisture content in sample (MC)	14.4	0.000100	%	0.00000699
Repeatability of method (Rep)	-	-	-	0.00353
C_{As}	45.6		$mg.Kg^{-1}$	
$u_c(As)$	0.611		$mg.Kg^{-1}$	
U	1.22		$mg.Kg^{-1}$	

Figure 4 shows the majority of uncertainty contribution generates from calibration curve and method precision, in contrast with uncertainties from calibration stock type B and solvent mass that have not significant contribution to overall standard uncertainty.

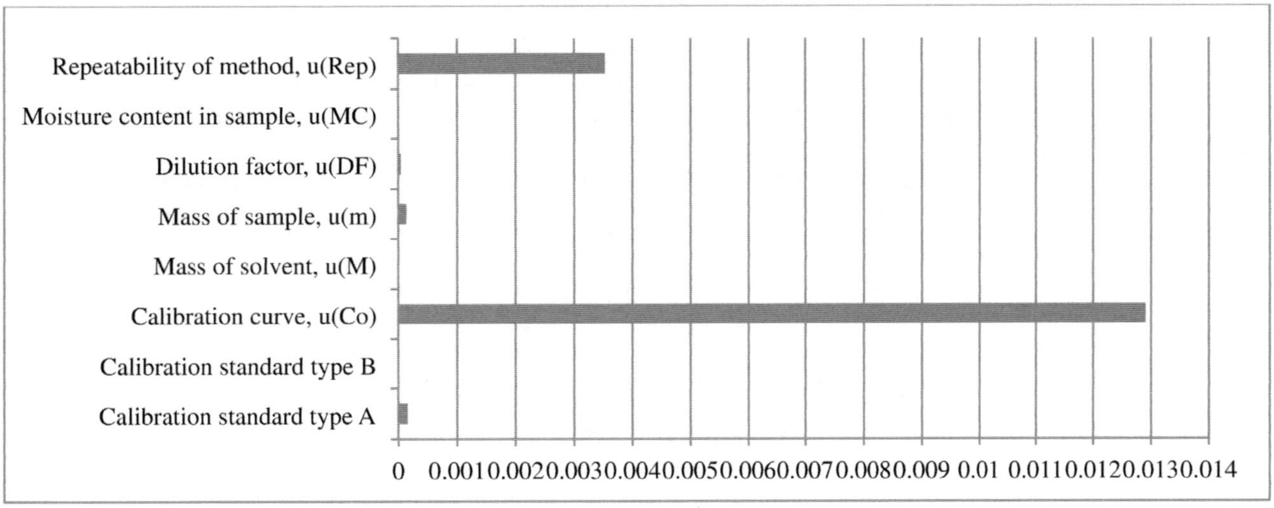

Figure 4. Uncertainty budget of measurement of arsenic in fish.

The expanded uncertainties, U_{As}, is obtained by multiplying the combined standard uncertainty by the coverage factor of k=2 which gives a confidence level of approximately 95%.

$$U_{As} = k \times u_{C(As)} = 2 \times 0.611 = 1.22 \text{ mg.Kg}^{-1}$$

Therefore, the measured total arsenic concentration in fish was 45.6 ± 1.22 mg.Kg^{-1} (w/w).

4. Conclusion

The developed method for determination of total arsenic in fish product was valid and traceable to International Standard. The result shows that the total arsenic concentration found in fish was 45.6 ± 1.22 mg.Kg^{-1}, with a coverage factor of *k* equal to 2 with 95% confidence level and %recovery obtained was 97.5% which is in good agreement with analysis of Certified Reference Material DORM 3.

Acknowledgement

This work was supported by Daftar Isian Pelaksanaan Anggaran (DIPA) for Metrology in Chemistry research activity in Research Center for Chemistry – Indonesia Institute of Sciences (LIPI).

References

[1] NSW technical report *Inorganic arsenic in seaweed and certain fish* October 2010
[2] Food Standard Agency (FSA) *Arsenic in fish and shellfish* 2005. Retrieved March 25, 2013 from http://www.food.gov.uk/multimedia/pdfs/fsis8205.pdf
[3] J. Yoshinaga, Y. Shibata, T. Horiguchi and M. Morita 1997 *Accred Qual Assur.* **2** 154–156.
[4] K.A. Hudson-Edwards , S.L. Houghton and A. Osborn 2004 *Trends in Anal. Chem.* **23** 10–11.
[5] T. Wickstrom, W. Lund, and R. Bye 1996 *Analyst.* **121 (2)** 201–204.
[6] U. Ay and E. Henden 2000 *Spectrochim Acta Part B.* **55** 951–958.
[7] M. Walcerz, E. Bulska and A. Hulanicki. 1993 *J Anal Chem.* **346** 622–626.
[8] N. Erdem and E. Henden 2004 *Anal Chim Acta.* **505** 59–65.
[9] A. Ramesh and K. A P. Riyazuddin 2007 *Accred. Qual. Assur.* **12** 455–458.
[10] EURACHEM/CITAC Guide : Traceability in Chemical Measurement, 2003
[11] International vocabulary of basic and general terms in metrology. ISO, Geneva (1993)
[12] JCGM 100:2008 Evaluation of measurement data — Guide to the expression uncertainty in measurement.
[13] American Of Analytical Chemistry (AOAC) Method 999.10-2005 and AOAC Method 986.15

– 2005.
[14] A. K. Avila, T. O. Araujo, R. M. H. Borges, P. R. G. Couto, M. F. G. Rosias, O. G. F. Rocha, D. S. Vaitsman and V. Dias 2005 *Accred. Qual. Assoc.* **10**: 277-282.
[15] EURACHEM/CITAC Guide CG 4: Quantifying Uncertainty in Analytical Measurement. (2012).
[16] A. Maroto, R. Boque, J. Riu and F. Xavier Rius 2001 *Analytica Chimica Acta.* **440** 171-184.

IOP Conf. Series: Earth and Environmental Science **60** (2017) 012037 doi:10.1088/1755-1315/60/1/012037

Acute toxicity (LC_{50}) of ammonia to carp fish (*Cyprinus carpio*) at different pH Levels

Ardeniswan[1*], F Dara[1] and F Sukmawati[2]
[1]Research Unit for Clean Technology, Indonesian Institute of Sciences (LIPI) Jalan Cisitu 21/154 D, Bandung 40135, Indonesia
[2]Chemistry Department, Universitas Jenderal Ahmad Yani, Jalan Terusan Jendral Sudirman, Cimahi, Indonesia
[*]Email : ardeniswan_r@yahoo.co.id

Abstract. Mass death of carp fish in reservoirs and lakes in Indonesia is due to the up-welling phenomenon of ammonia generates from the decomposition of fish feed remaining in the bottom of the lakes/reservoirs by microbes. The formation of ammonia gas is very dependent on pH. Most of the ammonia gas is formed form at the high pH value. Ammonia concentration can be determined with indophenol blue method using UV-Vis spectrophotometer. Carp fish (*Cyprinus carpio* of 5-6 gr was exposed to the three concentration of ammonia. Acute toxicity (LC_{50}) of ammonia (NH_3) was tested on similar sizes (5-6 g) of carp fish was maintained at three different pH levels within range of 7-9 for 96-h. Results showed that the concentration of ammonia increased at higher pH. The 96-h LC_{50}'s for exposure to ammonia were 60% (NH_3 1.85 ppm at pH 8), 100% (NH_3 2.16 ppm at pH 9), and insignificant result at NH_3 1.68 ppm with pH 7.

1. Introduction

Ammonia occurs in natural water in unionized (NH_3) and ionized (NH_4^+) forms and can be a serious toxicant to fishes and other aquatic species. It is a product of biological metabolism and is the principal nitrogenous excretory product of freshwater teleost [1]. It also enters water systems from accumulated fish food which is not consumed by fish. The massive dead fish at Maninjau, Saguling, and Cirata lake in 2008, could be caused by fish feeding activity [2].

The toxicity of ammonia to different fish species has been extensively investigated [3,4]. The toxicity of NH_3 has been ascribed to the fact that this unionized form of ammonia can readily diffuse across gill membranes due to lipid solubility and lack of charge, whereas the ionized form occurs as a larger hydrated form with charged entities which cannot readily pass through the hydrophobic micropores in the gill membrane [5].

The carp fish *(Cyprinus carpio),* is on a world-wide basis, one of the most extensively cultivated fish species. Based on statistical data from Dinas Kelautan dan Perikanan in 2011, the carp fish production reaches 120 thousand tons per year, but the number of dead fish also high, reaching 30 thousand tons per year in West Java [6]. In view of suppress the dead fish number and the lack of information on ammonia toxicity to crap fish, the present investigation aimed to carried out determination of the acute toxicity of ammonia to carp fish at different pH levels. Cage culture is a system of cultivation in a container of floating net cage (FNC) placed in lakes, reservoirs, lagoons, straits and bays [7].

A mass death of carp fish happened in Maninjau, Toba, Cirata, and Saguling in December 2008-early 2009 [8]. This was due to the ammonia resulted from the decomposition of accumulated feed waste in the bottom part of the lake or reservoirs [9]. When temperature of lake surface is lower than the bottom part, up-welling phenomenon happened frequently in which ammonia gas arises from bottom part which is toxic for fish. LC_{50} of ammonia based on NH_4Cl residual in the aquarium has been studied to carp fish has been evaluated [10]. Current study aimed to determine of ammonia formed at various pH by using ammonia generator which was then absorbed by dilute sulfuric acid solution. The resulted gas absorbed at various pH was used to simulate LC_{50} to carp fish.

2. Experimental

2.1. Material

Carp fish was obtained from fisheries at Ciparay (West Java). Prior to test, the fish was maintained and acclimatization in tanks for about 96 hr. The average weight of the fish for testing was of 5-6 gr. Ammonium chloride, sodium hydroxide, sulfuric acid, phenol, trisodium phosphate, hydrogen chloride, methanol, sodium nitroprusside, sodium hypochloride were pro analysis grade (E.Merck).

2.2. Optimization ammonia formation at Different pH levels

A few volume of 1N NaOH solution was added into ammonia generator containing of 10 mL NH_4Cl solution to produce 2 ppm of ammonia gases. The ammonia gas was then absorbed by 100 mL of 1N H_2SO_4 in the Impinger bottle into ammonium sulfate using a sampling pump with a flow rate of 0.5 L/min.

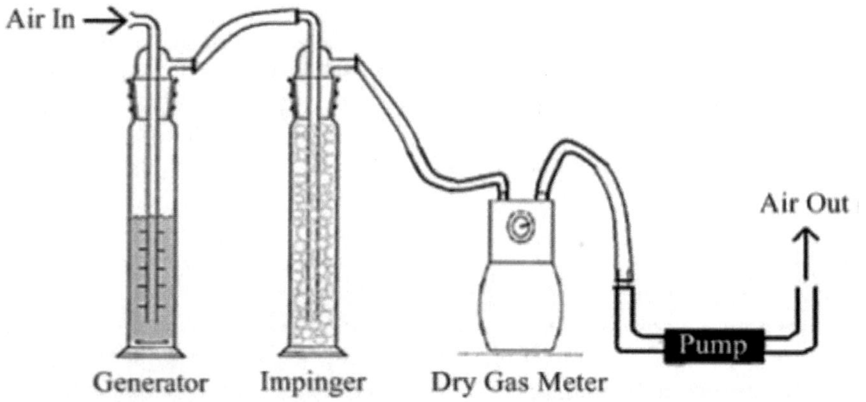

Figure 1. Scheme of ammonia generator using Impinger.

Determination of ammonium using indophenol method was carried out with spectrophotometric[11]. A calibration curve was made by preparing various ammonia chloride solution of 0.05 to 0.3 mg NH_3-N/L. Its absorbance was measured at 630 nm. Optimization was carried out using the same method with calibration using 90 mL solution at pH 7-9 with the addition of 10 ml NH_4Cl 25 ppm.

2.3. Acute Toxicity (LC50) Test

Before the start of each test, carp fish were placed in the aquarium in water of pH 7. Prepare 4 aquarium and adjust the pH of the water (10 L) in each aquarium with a variety of pH 7-10. Add the amount of volume (ml) of NH_4Cl solution (scale-up from Preliminary test) in each aquarium. Put the 10 fish in each 4 aquarium. Observe the fishing mortality every 2 hr for 96 hr. Measuring pH, temperature, and DO every 2 hours for 96 hr. Specify LC50

3. Result and discussion

A calibration curve was made by preparing various ammonia chloride solution of 0.05 to 0.3 mg NH_3-N/L. Its absorbance was measured at 630 nm. Results of calibration process shown as a curve of ammonia concentration (Figure 1), with correlation coefficient is 0.9991, indicating its reliability for quantitative determination.

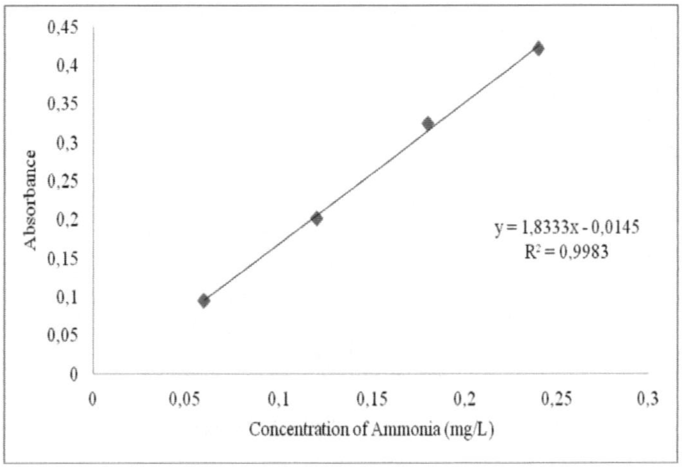

Figure 2. Calibration curve for ammonia gas formation.

Figure 3 shows that the ammonia gas formation was obtained of 1.09 ppm in 10 min and reached the maximum level of 2.16 ppm in 30 min. Its production was formed constanly afer 30 min.

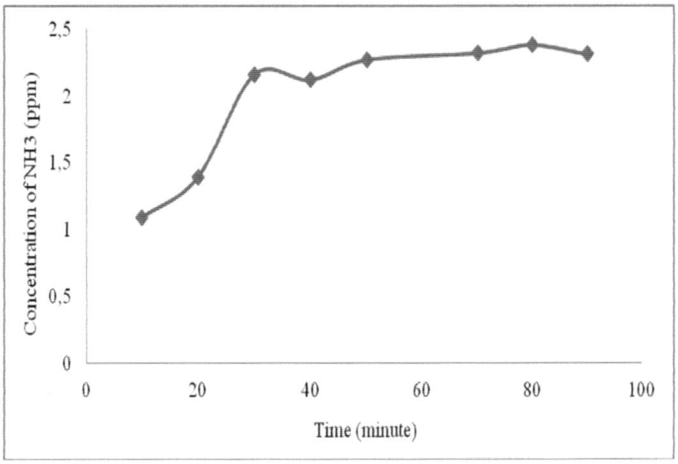

Figure 3. Time requirement of ammonia generator for NH_3 formation.

Ammonia formation at difference pH levels in shown in Figure 4. The ammonia concentration from the results of experiment using Impinger bottle showed an increase in ammonia concentration of 1,68 ppm at pH 7 becomes 2.35 ppm at pH10. The results of this experiments are used to determine LC50 in carp fish with and without the addition of ammonium chloride at difference pH levels.

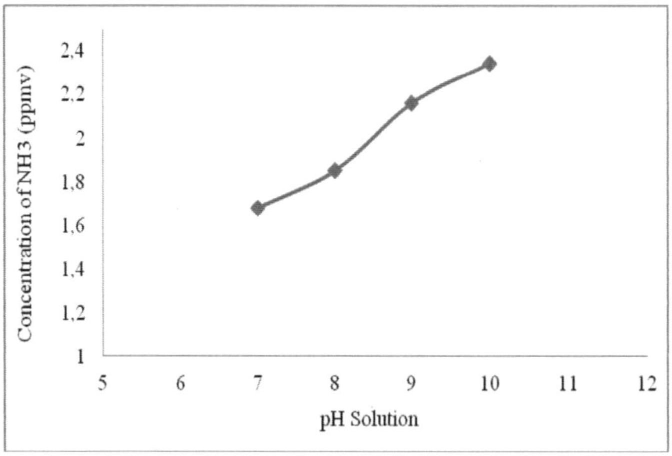

Figure 4. Formation of NH_3 gas at different pH levels in the Impinger bottle.

Table 1 shows that % mortality of carp fish reached 100% at pH 10 without ammonium chloride addition. At this point, all of fishes were no longer alive. This due to the excessive mucus with red spots on their gills and scales chipped. These characteristic revealed that all fish sufferred from ammonia poisoning [12]. Moreover, LC_{50} value is shown in Table 1.

Table 1. pH Effect (LC_{50}) and ammonium gas on carp fish mortality.

No.	pH Solution in Aquarium	LC50-Mortality of Carp Fish (%)	
		Without Adding of NH_4Cl 25 ppm	With Adding of NH_4Cl 25 ppm
1	7	0	0 (NH_3 1.68 ppm)
2	8	0	60 (NH_3 1.85 ppm)
3	9	0	100 (NH_3 2.16 ppm)
4	10	100	-

Notes : - Water temperatur in aquarium 27-28 °C;
- Dissolved oxygen (DO) 3-5 ppm

The acute toxicity of NH_3 in 96 hr exposure times increased within higher pH Levels, which is indicated by the lethal fish in aquaria. The results showed that no mortality at pH 7 after 96 hr exposure but increased to 60 and 100% at pH 8 and pH 9, respectively.

4. Conclusion

It was concluded that the carp (*C. carpio*) are more sensitive to water containing ammonium chloride at a pH of 9 and 8 as compared to water pH 7. They show changes in behavior caused by bleeding in the gills and its scales chipped.

Acknowledgment
Authors are grateful Indonesian Institute of Sciences through Mandiri Project for financial support and to Mr. Rusvirman Muchtar, Chemistry Department, University of General Ahmad Yani and Mrs.Wartini,Waste Treatment- Research Unit for Clean Technology for their assistance.

IOP Conf. Series: Earth and Environmental Science **60** (2017) 012037 doi:10.1088/1755-1315/60/1/012037

References

[1] Hasan M R and Macintosh D J 1986 *J. Aquaculture.* **54**(1-2) 97-107.
[2] Erlania, Rusmaedi, Anjay BP, Joni H 2010 *Impact of feed activity management on raising tilapia fish in keramba cage to the quality of water of maninjau lake* (in Bahasa Indonesia). Pusat Riset Perikanan Budidaya, Jakarta
[3] Sampaio L A, Wasielesky W, Miranda-Filho K C 2002 *Bull. of Env. Contamination & Toxicol.* **68**(5) 668-674.
[4] Chew S F, Hong L N, Wilson J M, Randall D. J, Ip Y K 2003 *Phy. and Biochem. Zool.* **76**(2) 204-214.
[5] Fromm P O, and Gillette J R 1968 *Compar. Biochem. and Phys.* **26**(3) 887-896.
[6] Dinas Kelautan dan Perikanan 2011 *Statistik Kelautan dan Perikanan Jakarta.* Dinas Kelautan dan Perikanan
[7] Effendi H 2003 *Telaah Kualitas Air Bagi Pengelolaan Sumber Daya dan Lingkungan Perairan". Kanisius.* Yogyakarta
[8] Kusdiarti 2011 *Tesis.* Institut Pertanian Bogor
[9] Kuncoro E B 2008 *Aquascape.* Kanisius Yogyakarta
[10] Zonneveld N and Rasyid Fadholi 1991 *Agricultural.* Vol. 99(1)83-94.
[11] Lodge J P Jr 1989 *Methods of Air Sampling and Analysis.* 3rd Edition. Intersociety Committee. Lewis Publishers Inc. Michigan.
[12] Maloedyn 2002 *Mengatasi Penyakit dan Hama Pada Ikan.* Agromedia Pustaka Tangerang.